GENETICS, EVOLUTION AND BIOLOGICAL CONTROL

Genetics, Evolution and Biological Control

Edited by

L.E. Ehler
University of California, Davis, USA

R. Sforza
USDA, Montpellier, France

and

T. Mateille
IRD, Montpellier, France

CABI Publishing

CABI Publishing is a division of CAB International

CABI Publishing
CAB International
Wallingford
Oxon OX10 8DE
UK

Tel: +44 (0)1491 832111
Fax: +44 (0)1491 833508
E-mail: cabi@cabi.org
Web site: www.cabi-publishing.org

CABI Publishing
875 Massachusetts Avenue
7th Floor
Cambridge, MA 02139
USA

Tel: +1 617 395 4056
Fax: +1 617 354 6875
E-mail: cabi-nao@cabi.org

A catalogue record for this book is available from the British Library, London, UK.

Library of Congress Cataloging-in-Publication Data

Genetics, evolution, and biological control/edited by L. E. Ehler, R. Sforza, and T. Mateille.
p. cm
Papers from an International Organization for Biological Control symposium held in Montpellier, France, 2002.
Includes bibliographical references and index.
ISBN 0-85199-735-X (alk. paper)
1. Pests--Biological control--Congresses. 2. Biological pest control agents--Congresses. I. Ehler, Lester E. II. Sforza, R. (Rene) III. Mateille, T. (Thierry) IV. International Organization for Biological Control. V. Title.
SB975.G.46 2004
632′.96--dc21

2003009939

ISBN 0 85199 735 X

Typeset by MRM Graphics Ltd, Winslow, Bucks
Printed and bound in the UK by Cromwell Press, Trowbridge

Contents

Contributors

J.K. Brown, *Department of Plant Sciences, University of Arizona, Tucson, AZ 85721, USA. E-mail: jbrown@ag.arizona.edu*

J.J. Burdon, *CSIRO-Plant Industry and CRC for Australian Weed Management, GPO Box 1600, Canberra, ACT 2601, Australia. E-mail: Jeremy.Burdon@csiro.au*

K.J. Evans, *Cooperative Research Centre for Australian Weed Management, Tasmanian Institute of Agricultural Research, New Town Research Laboratories, 13 St Johns Avenue, New Town, Tasmania 7008, Australia. E-mail: Kathy.Evans@dpiwe.tas.gov.au*

D.R. Gomez, *Department of Applied and Molecular Ecology, University of Adelaide, Waite Campus, PMB 1, Glen Osmond, South Australia 5064, Australia. New address: School of Agriculture and Wine, Plant and Pest Sciences, University of Adelaide, Waite Campus, PMB 1, Glen Osmond, South Australia 5064, Australia. E-mail: don.gomez@adelaide.edu.au*

J. Heraty, *Department of Entomology, University of California, Riverside, CA 92521, USA. E-mail: john.heraty@ucr.edu*

A. Hilbeck, *Geobotanical Institute, Swiss Federal Institute of Technology (ETH), Zürichbergstrasse 38, CH-8044 Zürich, Switzerland. E-mail: hilbeck@geobot.umnw.ethz.ch*

A.R. Kraaijeveld, *NERC Centre for Population Biology and Department of Biological Sciences, Imperial College at Silwood Park, Ascot, Berkshire SL5 7PY, UK. E-mail: a.kraayeveld@imperial.ac.uk*

J.E. Losey, *Department of Entomology, Cornell University, Ithaca, NY 14853, USA. E-mail: jel27@cornell.edu*

H. Müller-Schärer, *Université de Fribourg, Département de Biologie, Unité Ecologie et Evolution, chemin du Musée 10, CH-1700 Fribourg, Switzerland. E-mail: heinz.mueller@unifr.ch*

J.J. Obrycki, *Department of Entomology, Iowa State University, Ames, IA 50011, USA. New address: Department of Entomology, University of Kentucky, S-225 Agricultural Science Center – North Lexington, KY 40546, USA. E-mail: johnobrycki@uky.edu*

G.M Poppy, *Biodiversity and Ecology Division, School of Biological Sciences, University of Southampton, Bassett Crescent East, Southampton SO16 7PX, UK. E-mail: gmp@soton.ac.uk*

W. Powell, *Plant and Invertebrate Ecology Division, Rothamsted Research, Harpenden, Herts AL5 2JQ, UK. E-mail: Wilf.powell@bbsrc.ac.uk*

G.K. Roderick, *201 Wellman Hall MC 3112, Environmental Science, Policy and Management, Division of Insect Biology, University of California, Berkeley, CA 94720-3112, USA. E-mail: roderick@nature.berkeley.edu*

J.R. Ruberson, *Department of Entomology, University of Georgia, Tifton, GA 31793, USA. E-mail: ruberson@tifton.cpes.peachnet.edu*

T. Steinger, *Université de Fribourg, Département de Biologie, Unité Ecologie et Evolution, chemin du Musée 10, CH-1700 Fribourg, Switzerland. E-mail: thomas.steigner@unifr.ch*

R. Stouthamer, *Department of Entomology, University of California, Riverside, CA 92521, USA. E-mail: richard.stouthamer@ucr.edu*

P.H. Thrall, *CSIRO-Plant Industry and CRC for Australian Weed Management, GPO Box 1600, Canberra, ACT 2601, Australia. E-mail: Peter.Thrall@csiro.au*

E. Wajnberg, *INRA, 37 Boul. du Cap, 06600 Antibes, France. E-mail: wajnberg@antibes.inra.fr*

Preface

Fifteen years ago, discussions of genetics and evolution in biological control typically dealt with issues such as genetic drift, inbreeding and natural selection, and how these processes could alter genetic variation for attributes that affect the success of agents in the field. The last decade of the 20th century brought major change, with revolutionary advances in molecular biology that opened up vast new areas for basic and applied research. This has provided new issues for each of the major approaches in applied biological control, such as: (i) the use of molecular genetics to trace the origin of target pests in classical biological control; (ii) the potential of mass-reared, transgenic agents in augmentative biological control; and (iii) the compatibility of transgenic crops and natural enemies in conservational biological control.

Following successful IOBC International Symposia at Montpellier devoted to technology transfer (1996) and indirect effects (1999) in biological control, the third International Symposium was convened in October 2002 and addressed recent developments in genetics and evolutionary biology, and their relevance to biological control. This symposium was co-organized by the Complexe International de Lutte Biologique-Agropolis (CILBA). The aims of the symposium were to acquaint biological-control workers with the latest advances in genomics and to explore ways that these advances could be put to practical use in biological control.

This book is the result of the 12 keynote addresses from the symposium. These presentations were organized around the following themes: genetic structure of pest and natural enemy populations (Burdon and Thrall, Wajnberg), molecular diagnostic tools in biological control (Heraty, Evans and Gomez), tracing the origin of pests and natural enemies (Roderick, Brown), predicting evolutionary change in pests and natural enemies (Müller-Schärer and Steinger,

Kraaijeveld), compatibility of transgenic crops and natural enemies (Obrycki *et al.*, Hilbeck), and genetic manipulation of natural enemies (Poppy and Powell, Stouthamer).

We wish to thank the following members of the Organizing Committee for their efforts in making the symposium a success: T. LeBourgeois, A. Gassmann, M.-C. Bon, A. Kirk, P.C. Quimby, K.A. Hoelmer, M.-S. Garcin, E. Wajnberg, J.K. Scott, M. Navajas, C. Silvy and M. Montes de Oca. We also acknowledge the following colleagues, who provided external reviews of the chapter manuscripts: D. Bourguet, D. Coutinot, K.J. Evans, R.A. Hufbauer, A. Kirk, D.G. Luster, M. Navajas, A.H. Purcell, J.A. Rosenheim, R.T. Roush, L. Smith, E. Wajnberg and P.S. Ward.

L.E. Ehler
R. Sforza
T. Mateille

1

Genetic Structure of Natural Plant and Pathogen Populations

J.J. Burdon and P.H. Thrall

CSIRO-Plant Industry and CRC for Australian Weed Management, GPO Box 1600, Canberra, ACT 2601, Australia

Introduction

The enemy release hypothesis – that species become invasive as a consequence of escape from their natural enemies and a resultant enhanced competitive ability (Keane and Crawley, 2002) – and its converse – that invasive species may be brought under control through re-establishment of host–parasite associations – is the basic cornerstone of biological control. However, despite the widely recognized importance of biological control as a tool for managing and controlling invasive species, sourcing biological control agents for a particular application is still generally done in a rather *ad hoc* fashion. Overall, there has been relatively little effort made to explore systematically plant–natural enemy associations to understand the factors that determine the success or failure of biological control programmes. For example, a wide range of insect herbivores and fungal pathogens have been introduced into invasive populations of *Mimosa pigra* in Australia, with limited success in most cases (Steinbauer *et al.*, 2000; Paynter and Hennecke 2001). What, then, are the factors that need to be taken into account when sourcing pathogens for biocontrol? In particular, what features of the invasive plant, e.g. life history, resistance to natural enemies, spatial distribution, may act to limit the post-release effectiveness of fungal biological control agents?

As part of developing a more general conceptual framework that would have predictive value for biocontrol management (Shea *et al.*, 2000), it is important to know for a host with a given set of life-history features, what types of pathogens are likely to be most effective with respect to long-term reductions in host population size. This involves identifying important host (e.g. longevity, dispersal mode) and pathogen (e.g. transmission mode, stage-specificity of attack) life-history features that can distinguish qualitative classes of weeds and biocontrol agents, and that are likely to significantly impact on initial rates of agent spread and ability to persist. However, the success of biocontrol programmes depends not only on what happens within single populations, but also on regional (metapopulation) persistence. In such situations, how the agent is deployed (e.g. the number and distribution of initial release sites) may influence overall dynamics and rates of spread, and therefore impacts on the host. In turn, these may determine the rate at which resistance evolves in the target host. Do optimal release strategies (e.g. few sites, with large amounts of inoculum, through to multiple sites, with little inoculum) vary according to the life-history features of target hosts and biocontrol agents? Given particular life histories and dispersal scales, how do different deployment strategies influence long-term persistence and overall impact? How does the rate at which host resistance evolves depend on the spatial scales at which hosts and biocontrol agents interact, and on the severity of fitness effects? In essence, is the optimal biocontrol agent always the one with the greatest immediate impact on host fitness?

We highlight these interlocking questions not only to illustrate the complexities and challenges facing biologists studying biocontrol-related issues, but also to draw attention to the potential for comparative analyses using data from previous biocontrol releases to better understand interactions between hosts and pathogens. Moreover, despite the enormous gaps in our understanding, a range of information is becoming available from basic coevolutionary studies of natural plant–pathogen interactions. In particular, this is showing how temporal and spatial variability in host resistance and pathogen virulence interact to influence disease dynamics and persistence. The primary aim of this chapter is to describe briefly some of these results, and indicate how incorporating such information may increase our ability to deploy effectively fungal pathogens as biocontrol agents.

Natural Variation in Host–Pathogen Associations

Disease in natural plant populations is characterized by unpredictability in its occurrence and severity. Diseased individuals may be infected because they lack any resistance whatsoever, or because of the presence of pathotypes of the pathogen that are capable of overcoming any resistance that is present. Similarly, healthy individuals may be free of disease through the possession of resistance gene(s) effective against the pathotypes present, or simply through chance escape. In plants, disease resistance may be controlled by the action of single

genes/alleles with a major phenotypic effect (qualitative resistance); through the action of many genes, each of small effect (quantitative resistance); or through a combination of both qualitative and quantitative resistance. In pathogens confronted by significant levels of qualitative resistance, virulence (the ability to attack individual host genotypes) is similarly controlled by single genes – these are typified by the biotrophic interactions of rusts and mildews, for which the gene-for-gene system is the classic model (Flor, 1955). On the other hand, for pathogens confronted by quantitative resistance, all isolates tend to be virulent but vary greatly in their aggressiveness (ability to grow and sporulate on the host). These pathogens are typically necrotrophic – killing tissue as they spread through the host.

Disease resistance in plant populations in their native range

Detailed analyses of several plant–pathogen systems have shown that variation for disease resistance is widespread in natural plant populations, although the genetic basis of the resistance present (quantitative, qualitative) influences its phenotypic expression. Thus when De Nooij and van Damme (1988) examined the resistance of populations of *Plantago lanceolata* to the necrotrophic fungus *Phomopsis subordinaria*, they found that all individuals contracted the disease, but the severity of symptoms varied considerably. Lesions on the most susceptible individual were more than 25% larger than those on the most resistant plant. However, the resistance responses of other plants fell between these two extremes, with the population as a whole showing a semi-continuous distribution of responses ranging from resistance to susceptibility. This continuum of responses was detected using a single isolate of the pathogen, strongly suggesting that the resistance involved was quantitative in nature.

In qualitatively based systems, where single resistance genes have large phenotypic effects, patterns of resistance may be very simple or equally complex. Thus challenging populations of *Amphicarpaea bracteata* with a single isolate of the pathogen *Synchytrium decipiens* resulted in the classification of all host individuals into one of two classes, with individuals showing either high resistance (no lesions) or high susceptibility (Parker, 1985). In *Glycine argyrea*, extensive testing with nine distinct pathotypes of *Phakopsora pachyrhizi* again demonstrated the occurrence of two host classes (resistant/susceptible), and subsequent genetic studies found that this resistance was conferred by a single resistance gene, which was present in either a homozygous or a heterozygous state (Jarosz and Burdon, 1990). However, in many populations protected by major resistance genes, the pattern of resistance and susceptibility is far more complex, shifting substantially as the population is challenged by a series of different pathotypes, each with a different virulence combination. Thus in *Linum marginale*, individual populations may be totally susceptible to some isolates of the rust pathogen *Melampsora lini*, be completely resistant to others, and show varying frequencies of resistance/ susceptibility to yet other pathotypes (Jarosz and Burdon, 1991; Burdon *et al.*, 1999; Thrall *et al.*, 2001) (Fig. 1.1). Detailed phenotypic and genetic analyses of these resistances have shown the presence of many different alleles, giving either complete or partial resistance. In most cases, individual lines carry single resistance alleles (Burdon, 1994). In the *Senecio vulgaris–Erysiphe fischeri* association, interactions can be further complicated by a layering of major and minor genes on top of each other (Harry and Clarke, 1986, 1987).

The nature of resistance found in plant populations varies considerably

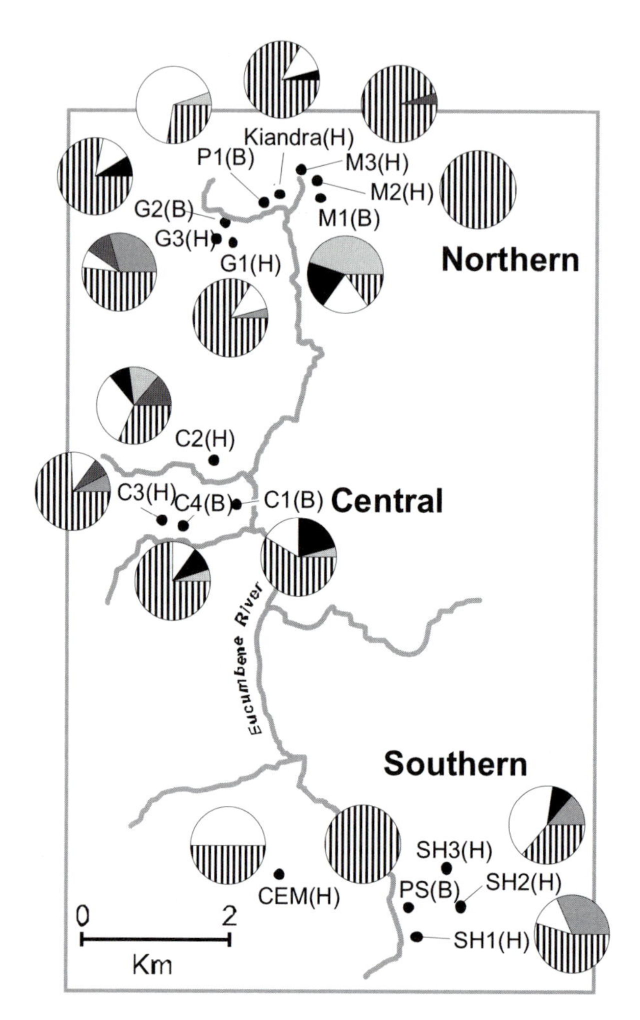

Fig. 1.1. Variation in the resistance structure of multiple populations of a *Linum marginale–Melampsora lini* metapopulation growing in the Kiandra region of New South Wales, Australia. The shading of the pie diagrams represents the relative frequency, in each population, of the six most common resistance phenotypes found across the metapopulation as a whole.

among associations, with life-history features of both partners and environmental factors playing a significant role in its genetic basis. In turn, this is likely to influence the speed and extent to which populations respond to pathogen-imposed selective pressures. Thus marked differences in the epidemiology of *Melampsora occidentalis* on *Populus trichocarpa* on the eastern and western sides of the Canadian Rocky Mountains, and of *M. lini* in montane and plains populations of *L. marginale* in Australia, are intriguingly correlated with differences in the genetic basis of resistance and the partitioning of resistance within populations of these two systems, respectively (Hsiang and Chastagner, 1993; Burdon *et al*., 1996, 1999).

Variation in space: host–pathogen interactions as metapopulation associations

That variation for disease resistance occurs in natural plant populations is not surprising. Of far greater interest is how resistance is partitioned within and between populations, and how different host populations relate to one another in terms of the pathogen pressure to which they are subject and the resistances that are present. At regional and continental geographic scales, forestry provenance trials provide a plethora of examples of differences in resistance structure shown by widely separated populations when they are exposed to the same pathogen population in a common environment. In these cases, host populations are frequently exposed to pathogen populations that they would never encounter naturally, and differences in resistance structure are probably the rule rather than the exception.

At a much more restricted spatial scale, host–pathogen associations reflect a metapopulation structure, with each host population evolving semi-independently of its neighbours. Certainly, gene flow occurs, but closely adjacent populations frequently suffer different levels of disease (Ericson *et al*.,1999) and show different resistance and virulence structures in host and pathogen populations, respectively. In the *Linum–Melampsora* model system, Thrall *et al*. (2001) found considerable variation in the resistance of 16 populations in three local groupings spread out over 12 km (Fig. 1.2). Although within local population groups, distinct differences were found in the occurrence and frequency of specific resistance phenotypes (Fig. 1.1), overall there was evidence for an isolation-by-distance effect, such that plant populations in close proximity were more similar in their level of resistance to specific pathotypes than were more distantly placed populations. Similarly, a wide range of pathotypes occurred across the metapopulation; many pathotypes were found in some populations in all areas but overall there was evidence for significant local adaptation of the virulence structure of pathogen populations to the resistance structure of sympatric host populations. More susceptible host populations supported 'simpler' pathogen populations with lower average virulence than more resistant and higher diversity host populations where more complex pathotypes predominated (Thrall *et al*., 2002).

The evolutionary and selective/adaptive relationships that develop between host populations and their pathogens have significant implications for the selection of pathogen pathotypes for use in biological control. While it is not necessary to identify the precise host population from which a weed infestation originated to ensure pathogen isolates possess appropriate virulence combinations, as the source of the biological control agent gets further from that of the origin of the weed, there is an increasing probability of a mismatch of pathogen to host. A good example of this is provided by broad differences in the virulence of isolates of *M. lini* from eastern and western Australia. Glasshouse inoculation

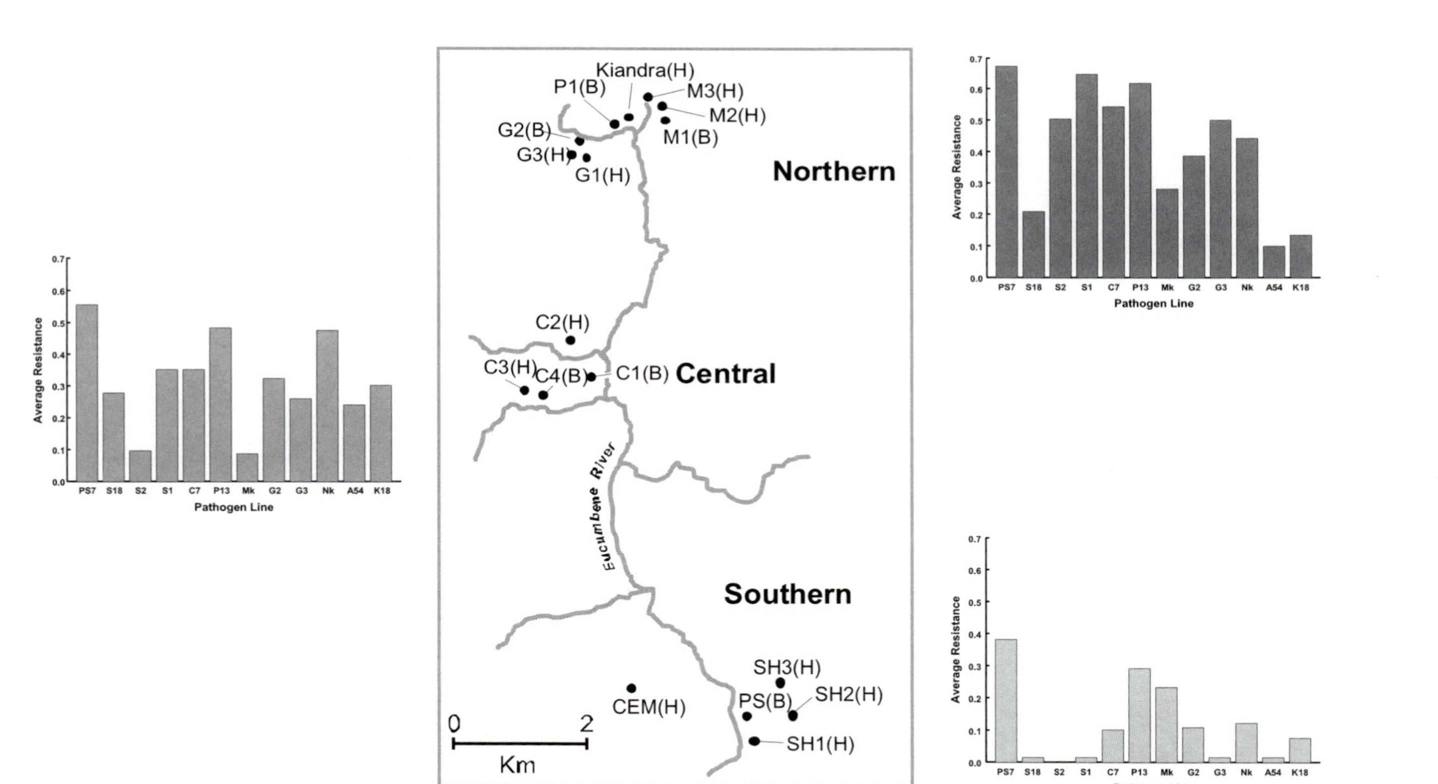

Fig. 1.2. Regional differences in the resistance response of multiple populations of *Linum marginale* to 12 pathotypes of *Melampsora lini* occurring within the metapopulation. The average resistance represents the mean resistance response of each population to each pathotype.

trials showed that isolates from both sides of the continent were equally able to attack and induce disease on lines of *L. marginale* carrying no resistant genes. However, pathotypes collected in Western Australia were virulent on a significantly lower proportion of eastern Australian host lines than Western Australian ones (17.6% versus 75%). This pattern was also true in the reverse direction (Burdon *et al.*, 2002).

Clearly, knowledge of geographic origin is central to choosing appropriate pathogens/pathotypes as biocontrol candidates. In addition, the consequences of spatial structure for disease dynamics and coevolutionary interactions at more local scales suggests that these may also be important considerations. For example, Carlsson-Granér and Thrall (2002) used a spatial simulation approach to show that equilibrium patterns of disease incidence and prevalence across a metapopulation were highly dependent on host population structure. When there was little or no isolation among local host populations, disease incidence (presence/absence in local demes) was high, but prevalence (percentage of infected individuals) was very low. The opposite pattern was evident when host populations were isolated. The underlying reason for this was that host resistance evolved rapidly when local populations were highly connected (thus limiting disease levels but not pathogen movement) but very slowly (or not at all) in the more isolated case. While this model did not include pathogen variation, the implication is that we might generally expect to find more virulent and/or aggressive pathogens in regions where host populations are more densely clustered, as opposed too those where, on average, host populations are more isolated.

Host and Pathogen Variation in New Environments

All invasive species go through a similar series of steps as they invade new environments: (i) migration into the new habitat; (ii) initial colonization and establishment; and (iii) subsequent widespread dispersal (Sakai *et al.*, 2001). Through all these processes there is considerable potential for genetic changes to occur through founder effects, drift and selection – genetic changes that may well impact on the vulnerability of invasive species to the re-establishment of links with host-specific pathogens and herbivores. It is important to remember that these issues also apply to the pathogen biocontrol agent, although the founding process is under the direct control of the practitioner to a much greater extent, including where the agent is sourced from, where it is introduced (e.g. environment, geographic area), how it is introduced (e.g. in one vs. several populations), and how much genetic diversity is represented in the released population of the agent.

What is the resistance structure of invasive host populations?

Colonization events usually involve population bottlenecks, and genetic drift during colonization may further reduce genetic variation in newly established populations. Such reductions in variation have been documented for a range of neutral allozyme and molecular markers in a broad diversity of weedy species (e.g. Amsellem *et al.*, 2000). Given this, it is not unreasonable to assume a similar loss of genetic variation in respect to resistance, although this may have a greater impact where resistance is coded by a few genes with major phenotypic effect than when resistance is more broadly based on the action of many genes of minor effect. However, as with all other forms of variation, when multiple colonization events have occurred, especially from a diverse range of sites of origin or over an extended period of time (e.g. Schierenbeck *et al.*, 1995; Pappert *et al.*, 2000), the possibility exists for significant levels of resistance to have been imported.

Empirical data that shed light on this question are very limited. Biological control screening procedures that collect seed from a limited number of invasive plant populations and use these for screening pathogen virulence and aggressiveness prior to importation certainly suggest that little or no resistance is present in at least some invasive species. However, even fewer data are available on the resistance patterns of specific biological control targets in their home range (or the range of virulence and/or aggressiveness in pathogens where major genes predominate). Furthermore, a major problem that confronts virtually all such comparisons is that we very rarely know from precisely where the original weedy immigrants originated.

Bearing these important caveats in mind, there are a few examples of weed–pathogen combinations for which we have information about the resistance structure of the weed in both its exotic home and at least part of its native range. One of the best documented of these is the interaction occurring between the composite apomict *Chondrilla juncea* and its host-specific rust pathogen *Puccinia chondrillina*. In Australia and the USA, 3 and 3–4 apomictic races of *C. juncea* occur, respectively (Burdon *et al.*, 1981; Emge *et al.*, 1981), each of which shows a unique pattern of response to a range of different pathogen isolates. In contrast, towards the eastern end of its native range, in Eurasia, studies have shown that most populations contained a number of different apomictic races (Chaboudez and Burdon, 1995), with some populations showing considerable complexity, with at least eight different resistance phenotypes being present (Espiau *et al.*, 1998). Other examples of significant levels of resistance occurring in invasive populations far beyond their native range are found in the resistance of *Avena* species to the rust fungus *Puccinia coronata* (resistance in exotic populations, Burdon *et al.*, 1983; in native Israeli populations, Dinoor, 1977); in *Bromus tectorum* to *Ustilago bullata* in the USA (Meyer *et al.*, 2001); and in *Hordeum leporinum* in Australia to *Puccinia striiformis* and *Rhynchosporium secalis* (Jarosz and Burdon, 1996; Wellings and Burdon, unpublished data). From a handful of such examples it is not possible to gain a true picture of the extent to which invasive species retain detectable levels of resistance. However, it is clear that a complete spectrum of possibilities, from no resistance through to the presence of multiple, distinct resistance phenotypes, is likely to be encountered.

Finding surrogate measures to easily identify invasive species or groups of species that might carry a greater range of resistance into a new environment, or once there be able to respond rapidly to the re-imposition of pathogen-induced selection pressures, has proved difficult. In a review of all biological control (using both fungi and insects), Burdon and Marshall (1981) found evidence supporting Levin's hypothesis (1975) for a link between the mating system of the target species and control success. Invasive species likely to have more uniform populations (apomictics, inbreeding species) were apparently easier to control than outcrossing ones. Subsequent re-analysis of this proposition has thrown some doubt on this result (Chaboudez and Sheppard, 1996). However, clear evidence exists for a positive link between increased diversity within crops (mixtures, multilines) and reduced disease severity (Wolfe, 1985), and similar relationships have been found in more natural associations (Schmid, 1994; Thrall and Burdon, 2000). Given this, it seems likely that the disagreement over the merits of using mating system as a surrogate simply reflects its poor ability to predict diversity within broad categories such as inbreeders, rather than that the genetic diversity of weedy stands is unimportant.

How important is continuing natural selection for the maintenance of resistance in introduced populations?

Seen from the point of view of the biological control practitioner, how quickly will any resistance originally introduced with an invasive species decline? Formulating even a theoretical answer to this question is beyond our current capabilities, as the eventual outcome of this scenario is one that is affected by a wide range of factors, including: (i) the genetic basis of the resistance involved (quantitative, qualitative); (ii) the intensity of the genetic bottleneck involved in the original introduction(s); (iii) the rate of expansion of the species; (iv) year-to-year dynamics and amplitude of populations in their new environment; (v) its spatial distribution in the new environment (one continuous population vs. multiple, small, isolated populations); and, finally, (vi) whether there is a fitness cost associated with the resistance mechanisms in question. Which factors predominate will vary among invasion events; however, theoretical models support the idea that the extent to which invasive species are dispersed across the landscape, and the degree of among-population isolation, will strongly influence the rate at which resistance reduces to levels indistinguishable from zero. For example, metapopulation models based on the *Silene–Ustilago* plant–pathogen interaction showed that, even in the absence of disease and assuming a 10% fitness cost associated with resistance, the resistant allele could take orders of magnitude longer to be lost from the system than would be predicted by a classical genetic formulation based on a single population with no spatial structure (Thrall and Antonovics, 1995) (Fig. 1.3).

Many of the factors listed above are clearly ones that are species and circumstance specific. Thus different species frequently suffer founder effects of different intensity when introduced into the same environment, while the same

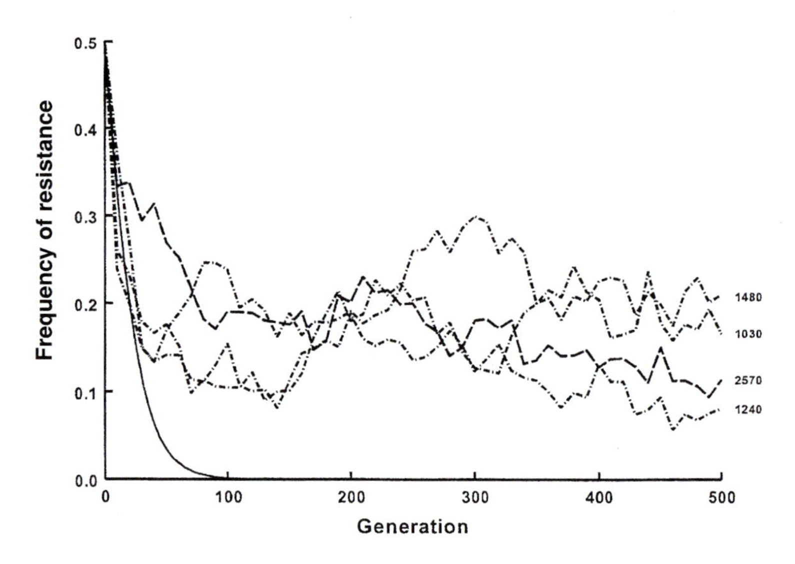

Fig. 1.3. Changes in the mean frequency of resistance in populations of host over time in the absence of disease and the existence of a 10% fitness penalty associated with resistance. The solid line is the predicted time for complete loss of resistance in a single-population, deterministic model, while the dotted lines represent four random runs of a simulation model (the numbers to the right-hand side of the graph are the number of generations to fixation for each simulation run). (Redrawn from Thrall and Antonovics, 1995.)

species may experience quite different population bottlenecks when introduced into different environments. Similarly, the rate of expansion of an exotic species, its final distribution in its new environment and its demographic cycles are all likely to be significantly affected by environmental circumstances. On the other hand, whether fitness costs are associated with resistance is a more generic question, to which many studies have attempted to provide an answer. Bergelson and Purrington (1996) summarized the results of 88 studies, gleaned from the herbivore–plant, pathogen–plant and herbicide literature, to explore the existence of trade-offs between resistance and fitness traits. Half of these showed some indication of lower fitness associated with resistance, but as they and others have pointed out (e.g. Brown, 2002), many studies are confounded by linkage and other complexities introduced by comparisons of inbred lines where resistance genes may be in disequilibrium with other loci, or where wide crosses are involved. In a study designed to overcome these problems, Mitchell-Olds and Bradley (1996) established a series of artificial selection experiments and examined changes in the resistance/susceptibility of *Brassica rapa* to the pathogens *Leptosphaeria maculans* and *Peronospora parasitica* over three cycles of selection. At the end of that time they then compared dry mass production of resistant and susceptible lines in the absence of the pathogens. For *L. maculans* they were unable to detect any difference between resistant and susceptible lines; however, a significant genetic fitness cost was detected with respect to resistance (6% lower growth) to *P. parasitica*. More studies of this type are needed to assess accurately the extent to which resistance is truly associated with a fitness cost, whether this varies according to the genetic basis of resistance, and the impact of the host and pathogen involved.

How rapidly will resistance respond to the re-application of pathogen selective pressures?

Lack of detailed long-term assessments of individual populations means that very few data are available with which to answer this question directly. Despite this, there are a number of examples from natural host–pathogen systems that show that pathogen pressure can induce marked and rapid changes in existing resistance structures. Thus, in a population of *Linum marginale* containing multiple different resistance phenotypes, a major epidemic of the pathogen rust *Melampsora lini* caused significant levels of differential adult plant mortality (~70%), which resulted in a general increase in the resistance diversity of the population as the frequency of the most common resistance phenotypes declined and rarer ones increased (Burdon and Thompson, 1995). In a related study, Thrall and Burdon (2000) showed that, in epidemic years, disease development was poorer in *L. marginale* populations containing a broad range of resistance phenotypes (and those showing higher average resistance) than in more uniform ones.

In situations involving invasive/weedy species in new environments, there are a small number of relevant examples. In eastern Australia the introduced weeds *Avena fatua* and *Avena barbata* show significant levels of resistance to the rust pathogen *Puccinia coronata* (from Eurasia). Because of their life-history characters and association with agriculture, it is highly likely that the current gene pool of both of these species in Australia is derived from multiple introductions spaced over many years. As a consequence, the existence of significant levels of resistance in both these species is not surprising. However, the distribution of resistance shows distinct spatial features, with levels of resistance being higher in areas generally more suitable for pathogen growth than in harsh ones (Burdon *et al.*, 1983). While this example suggests that selection has favoured the accumulation of resistance in some areas, it is not possible to determine the rate at which such changes have occurred. That information is available from another example, involving the use of the rust pathogen *Puccinia chondrillina* to control the weedy composite *Chondrilla juncea* in southern Australia. *C. juncea* is an apomictic weed, three genotypes of which were accidentally introduced into Australia. In 1968 one of these genotypes was widespread across south-eastern Australia, while the other two were restricted to a small area of central New South Wales. At that time a single pathotype of *P. chondrillina*, capable of attacking the most widespread genotype, was released (Cullen *et al.*, 1973), rapidly became established, and during subsequent years had a very significant impact on its host genotype.

As the vigour and size of populations of the susceptible form of *C. juncea* declined, both of the remaining resistant genotypes gained a major competitive advantage (Burdon *et al.*, 1984) with one, in particular, expanding rapidly into the partially vacated niche (Burdon *et al.*, 1981), so that, less than 15 years later, little evidence was left of the temporary population reduction.

Improving Biological Control Strategies

Despite its spectacular nature, control that gives a short-term gain but no longer-term benefit is less valuable than control that is slower to reach its peak but then persists. However, given that it would be a long and difficult process to determine before a release event the full range of resistance that may be present in invasive populations, can we make any predictions about the most appropriate way in which to release biological control agents to maximize both the short- and the longer-term benefits of control? This question can be further subdivided to address the initial process of selecting pathogens for the biological control of specific hosts – are there, for example, particular combinations of control agent and target weed life-history characters that will enhance the probability of a successful control outcome (e.g. Smith and Holt, 1997)? And, secondly, in addressing the release phase, is a rapid 'knock-down' of population size better than a slower one? That is, what are the selective consequences of rapid versus slow declines in population numbers, particularly with respect to the evolution or re-emergence of resistance?

How do host and pathogen life-history features and the environment interact to determine the genetic basis of resistance?

Are there patterns in the distribution of resistance with respect to significant life-history aspects of the host (e.g. annual–perennial; herb–woody, etc.) that can be used to guide control strategies? A broad survey of animal host–pathogen interactions showed that diseases can be differentiated into distinct classes on the basis of their mode of transmission – a key point for selection with respect to their persistence (Lockhart *et al.*, 1996). Thus sexually transmitted animal diseases differ from other infectious diseases for a wide range of characteristics, having smaller host ranges, longer infectious periods, and being less likely to increase host mortality. In plant–pathogen associations, a number of small studies have suggested that particular features of plants – for example, propensity to form rhizomes or stolons (Wennström, 1999) or host longevity (Thrall *et al.*, 1993) – might be associated with particular patterns of disease persistence and dynamics. Unfortunately, however, this possibility has not been addressed in a systematic way, nor has the possibility that particular combinations of environmental and host–pathogen life-history features, by affecting pathogen persistence and severity, may favour the evolution of different resistance mechanisms (Burdon *et al.*, 1996). If the latter possibility is the case, and environmental/life-history features that result in stochastic epidemiology favour the evolution of qualitative resistance in the host, while features leading to more predictable coexistence favour other resistance mechanisms (Burdon *et al.*, 1996), then it should be possible to select combinations of different types of pathogen to maximize control efficiency.

An example of the way in which this might work would be through the simultaneous release of both a biotrophic and a necrotrophic pathogen. In this scenario, assuming the isolate(s) of the biotroph released were selected for their wide virulence, most individuals in the weedy target population would be susceptible, and major epidemics would lead to rapid reductions in population size. As population sizes fall, and populations become more fragmented, the frequency of epidemics and their severity will decline. At this point a 'rebound' in weed numbers, resulting from the expansion of previously rare resistant phenotypes (cf. the situation in *Chondrilla juncea* detailed above), would be much reduced by the impact of the necrotrophic pathogen, which, by virtue of its general ability to survive prolonged periods on dead plant tissue, would maintain strong disruptive selection pressure on the target species.

What are the selective consequences of rapid versus slower declines in population size?

Biological control frequently focuses on the immediate ecological process of reducing population sizes to levels at which they have sub-economic effects or, in the case of invaders of natural environments, where they are deemed to no longer pose a threat to individual native species, communities or aesthetic features of the environment. However, while this focus is understandable, it de-emphasizes potentially important evolutionary interactions that may affect both the initial chances of control success (finding and utilizing suitable pathogen isolates) and the longer-term outcomes of the interaction.

Although many ecologists tend to ignore the possibility of evolutionary change, as Thompson (1998) has pointed out, such change can occur over time scales that are deceptively short. Indeed, very many examples exist among invasive plants of non-adaptive geographic patterning of selectively neutral allozyme markers. Furthermore, even though there have been considerably fewer studies of natural host–pathogen interactions, examples exist of both maladaptive and adaptive evolution in plant disease defences (*Amphicarpaea–Synchytrium*, Parker, 1991; *Linum–Melampsora*, Burdon and Thompson, 1995) and overall changes in the outcome of the interaction (rabbit–myxoma virus, Kerr and Best, 1998; Best and Kerr, 2000; Saint *et al.*, 2001). Given the possibility of evolutionary response by the target organism, is there any advantage to be gained by influencing the rate of decline of the target population? In other words, will a highly virulent and aggressive pathogen that generates repeated severe epidemics place more or less selective pressure on the target organism than a pathogen whose effect is more incremental, or will there be no long-term difference?

If a rapid 'knock-down' approach has the effect of reducing populations to such a low ebb that stochastic environmental variation results in frequent local extinction, then this would be a potentially attractive option. Where resistance is qualitative, a rapid knock-down effect may well be potentially possible, so long as the pathogenicity spectrum of the release agents covers all the resistance genes in the target area. By rapidly reducing population sizes, such an approach would reduce the possibility of new resistance combinations being generated through outcrossing within weed populations, or of new specificities evolving through intragenic recombination. However, if the control agents released do not carry all necessary combinations of virulence to match all genotypes in the invasive population, such a rapid reduction in fitness of susceptible genotypes would give a massive selective advantage to any resistant types currently present at low frequency, and could well result in the 'bounce-back' effect seen in *Chondrilla juncea* in Australia (Burdon *et al.*, 1983)

What is likely to be the outcome of such interactions when resistance is quantitatively based and where all isolates of a pathogen are capable of causing disease but the rate at which disease develops is measured by its aggressiveness? Would the selective pressures exerted by pathogen on host and, subsequently, host on pathogen differ, and result in different coevolutionary trajectories, if we deployed a highly aggressive pathogen rather than one with a more intermediate level of aggressiveness? There are no empirical examples that address this issue in plant–pathogen associations. However, in the interaction that occurred between rabbits and the introduced myxoma virus in Australia, over time the morbidity associated with infection has declined as selection in the virus favoured less pathogenic strains that had higher transmission rates, and selection in the rabbits has favoured intermediate levels of resistance.

Conclusions

Biological control is deceptively simple; however, lying behind the simple concept of finding and releasing natural enemies to re-establish predatory or parasitic relationships is a raft of complex interactions. Moreover, these interactions between plants and their fungal pathogens are characterized by a wide range of potential outcomes. While environmental conditions set the broad parameters for these interactions, life-history features of both the biocontrol agent and the target species have a major impact on the ecological and evolutionary trajectory of individual host–pathogen associations. By studying the interplay of life-history characters in multiple populations scattered through spatially heterogeneous environments, we will develop a better understanding of plant–pathogen associations and hence gain insight into the strategies most likely to increase the rate of their successful re-establishment in new environments.

Acknowledgement

PHT acknowledges the support given by a Queen Elizabeth II Fellowship.

References

Amsellem, L., Noyer, J.L., Le Bourgeois, T. and Hossaert-McKey, M. (2000) Comparison of genetic diversity of the invasive weed *Rubus alceifolius* Poir. (Rosaceae) in its native range and in areas of introduction, using amplified fragment length polymorphism. *Molecular Ecology* 9, 443–455.

Bergelson, J. and Purrington, C.B. (1996) Surveying patterns in the cost of resistance in plants. *American Naturalist* 148, 536–558.

Best, S.M. and Kerr, P.J. (2000) Coevolution of host and virus: the pathogenesis of virulent and attenuated strains of myxoma virus in resistant and susceptible European rabbits. *Virology* 267, 36–48.

Brown, J.K.M. (2002) Yield penalties of disease resistance in crops. *Current Opinion in Plant Biology* 5, 1–6.

Burdon, J.J. (1994) The distribution and origin of genes for race specific resistance to *Melampsora lini* in *Linum marginale*. *Evolution* 48, 1564–1575.

Burdon, J.J. and Marshall, D.R. (1981) Biological control and the reproductive mode of weeds. *Journal of Applied Ecology* 18, 649–658.

Burdon, J.J. and Thompson, J.N. (1995) Changed patterns of resistance in a population of *Linum marginale* attacked by the rust pathogen *Melampsora lini*. *Journal of Ecology* 83, 199–206.

Burdon, J.J., Groves, R.H. and Cullen, J.M. (1981) The impact of biological control on the distribution and abundance of *Chondrilla juncea* in south-eastern Australia. *Journal of Applied Ecology* 18, 957–966.

Burdon, J.J., Oates, J.D. and Marshall, D.R. (1983) Interactions between *Avena* and *Puccinia* species. I. The wild hosts: *Avena barbata* Pott ex Link, *A. fatua* L. and *A. ludoviciana* Durieu. *Journal of Applied Ecology* 20, 571–585.

Burdon, J.J., Groves, R.H., Kaye, P.E. and Speer, S.S. (1984) Competition in mixtures of susceptible and resistant genotypes of *Chondrilla juncea* differentially infected with rust. *Oecologia* 64, 199–203.

Burdon, J.J., Wennström, A., Elmqvist, T. and Kirby, G.C. (1996) The role of race specific resistance in natural plant populations. *Oikos* 76, 411–416.

Burdon, J.J., Thrall, P.H. and Brown, A.H.D. (1999) Resistance and virulence structure in two *Linum marginale–Melampsora lini* host–pathogen metapopulations with different mating systems. *Evolution* 53, 704–716.

Burdon J.J., Thrall, P.H. and Lawrence, G.J. (2002) Coevolutionary patterns in the *Linum marginale–Melampsora lini* association at a continental scale. *Canadian Journal of Botany* 80, 288–296.

Carlsson-Granér, U. and Thrall, P.H. (2002) The spatial distribution of populations, disease dynamics and evolution of resistance. *Oikos* 97, 97–110.

Chaboudez, P. and Burdon, J.J. (1995) Frequency-dependent selection in a wild plant–pathogen system. *Oecologia* 102, 490–493.

Chaboudez, P. and Sheppard, A.W. (1996) Are particular weeds more amenable to biological control? A reanalysis of mode of reproduction and life history. In: Delfosse,

E.S. and Scott, R.R. (eds) *Biological Control of Weeds: Proceedings of the VIII International Symposium on Biological Control of Weeds.* CSIRO, Melbourne, pp. 95–102.

Cullen, J.M., Kable, P.F. and Catt, M. (1973) Epidemic spread of a rust imported for biological control. *Nature* 244, 462–464.

De Nooij, M.P. and van Damme, J.M.M. (1988) Variation in host susceptibility among and within populations of *Plantago lanceolata* L. infected by the fungus *Phomopsis subordinaria* (Desm.) Trav. *Oecologia* 75, 535–538.

Dinoor, A. (1977) Oat crown rust resistance in Israel. *Annals of the New York Academy of Science* 287, 357–366.

Emge, R.G., Melching, J.S. and Kingsolver, C.H. (1981) Epidemiology of *Puccinia chondrillina,* a rust pathogen for the biological control of rush skeleton weed in the United States. *Phytopathology* 71, 839–843.

Ericson, L., Burdon, J.J. and Müller, W.J. (1999) Spatial and temporal dynamics of epidemics of the rust fungus *Uromyces valerianae* on populations of its host, *Valeriana salina. Journal of Ecology* 87, 649–658.

Espiau, C., Riviere, D., Burdon, J.J., Gartner, S., Daclinat, B., Hasan, S. and Chaboudez, P. (1998) Host–pathogen diversity in a wild system: *Chondrilla juncea–Puccinia chondrillina. Oecologia* 113, 133–139.

Flor, H.H. (1955) Host–parasite interaction in flax rust – its genetics and other implications. *Phytopathology* 45, 680–685.

Harry, I.B. and Clarke, D.D. (1986) Race-specific resistance in groundsel (*Senecio vulgaris*) to the powdery mildew *Erysiphe fischeri. New Phytologist* 103, 167–175.

Harry, I.B. and Clarke, D.D. (1987) The genetics of race-specific resistance in groundsel (*Senecio vulgaris*) to the powdery mildew *Erysiphe fischeri. New Phytologist* 107, 715–723.

Hsiang, T. and Chastagner, G.A. (1993) Variation in *Melampsora occidentalis* rust on poplars in the Pacific northwest. *Canadian Journal of Plant Pathology* 15, 175–181.

Jarosz, A.M. and Burdon, J.J. (1990) Predominance of a single major gene for resistance to *Phakopsora pachyrhizi* in a population of *Glycine argyrea. Heredity* 64, 347–355.

Jarosz, A.M. and Burdon, J.J. (1991) Host–pathogen interactions in natural populations of *Linum marginale* and *Melampsora lini*: II. Local and regional variation in patterns of resistance and racial structure. *Evolution* 45, 1618–1627.

Jarosz, A.M. and Burdon, J.J. (1996) Resistance to barley scald (*Rhynchosporium secalis*) in wild barley grass (*Hordeum glaucum* and *H. leporinum*) populations in south-eastern Australia. *Australian Journal of Agricultural Research* 47, 413–425.

Keane, R.M. and Crawley, M.J. (2002) Exotic plant invasions and the enemy release hypothesis. *Trends in Ecology and Evolution* 17, 164–170.

Kerr, P.J. and Best, S.M. (1998) Myxoma virus in rabbits. *Revue Scientifique et Technique de L' Office International des Epizooties* 17, 256–268.

Levin, D.A. (1975) Pest pressure and recombination systems in plants. *American Naturalist* 109, 437–451.

Lockhart, A., Thrall, P.H. and Antonovics, J. (1996) Sexually-transmitted diseases in animals: ecological and evolutionary implications. *Biological Reviews of the Cambridge Philosophical Society* 71, 415–471.

Meyer, S.E., Nelson, D.L. and Clement, S. (2001) Evidence for resistance polymorphism in the *Bromus tectorum–Ustilago bullata* pathosystem: implications for biocontrol. *Canadian Journal of Plant Pathology* 23, 19–27.

Mitchell-Olds, T. and Bradley, D. (1996) Genetics of *Brassica rapa.* 3. Costs of disease resistance to three fungal pathogens. *Evolution* 50, 1859–1865.

Pappert, R.A., Hamrick, J.L. and Donovan, L.A. (2000) Genetic variation in *Pueraria lobata*

(Fabaceae), an introduced clonal, invasive plant of the southeastern United States. *American Journal of Botany* 87, 1240–1245.

Parker, M.A. (1985) Local population differentiation for compatibility in an annual legume and its host-specific fungal pathogen. *Evolution* 39, 713–723.

Parker, M.A. (1991) Nonadaptive evolution of disease resistance in an annual legume. *Evolution* 45, 1209–1217.

Paynter, Q. and Hennecke, B. (2001) Competition between two biological control agents, *Neurostrota gunniella* and *Phloeospora mimosae-pigrae*, and their impact on the invasive tropical shrub *Mimosa pigra*. *Biocontrol Science and Technology* 11, 575–582.

Saint, K.M., French, N. and Kerr, P. (2001) Genetic variation in Australian isolates of myxoma virus: an evolutionary and epidemiological study. *Archives of Virology* 146, 1105–1123.

Sakai, A.K., Allendorf, F.W., Holt, J.S., Lodge, D.M., Molofsky, J., With, K.A., Baughman, S., Cabin, R.J., Cohen, J.E., Ellstrand, N.C., McCauley, D.E., O'Neil, P., Parker, I.M., Thompson, J.N. and Weller, S.G. (2001) The population biology of invasive species. *Annual Review of Ecology and Systematics* 32, 305–332.

Schierenbeck, K.A., Hamrick, J.L. and Mack, R.N. (1995) Comparison of allozyme variability in a native and an introduced species of *Lonicera*. *Heredity* 75, 1–9.

Schmid, B. (1994) Effects of genetic diversity in experimental stands of *Solidago altissima* – evidence for the potential role of pathogens as selective agents in plant populations. *Journal of Ecology* 82, 165–175.

Shea, K., Thrall, P.H. and Burdon, J.J. (2000) An integrated approach to management in epidemiology and pest control. *Ecology Letters* 3, 150–158.

Smith, M.C. and Holt, J. (1997) Analytical models of weed biocontrol with sterilizing fungi: the consequences of differences in weed and pathogen life-histories. *Plant Pathology* 46, 306–319.

Steinbauer, M.J., Edwards, P.B., Hoskins, M., Schatz, T. and Forno, I.W. (2000) Seasonal abundance of insect biocontrol agents of *Mimosa pigra* in the Northern Territory. *Australian Journal of Entomology* 39, 328–335.

Thompson, J.N. (1998) Rapid evolution as an ecological process. *Trends in Ecology and Evolution* 13, 329–332.

Thrall, P.H. and Antonovics, J. (1995) Theoretical and empirical studies of metapopulations: population and genetic dynamics of the *Silene–Ustilago* system. *Canadian Journal of Botany* 73, S1249–S1258.

Thrall, P.H. and Burdon, J.J. (2000) Effect of resistance variation in a natural plant host–pathogen metapopulation on disease dynamics. *Plant Pathology* 49, 767–773.

Thrall, P.H., Biere, A. and Antonovics, J. (1993) Plant life-history and disease susceptibility – the occurrence of *Ustilago violacea* on different species within the Caryophyllaceae. *Journal of Ecology* 81, 489–498.

Thrall, P.H., Burdon, J.J. and Young, A.G. (2001) Variation in resistance and virulence among demes of a single host–pathogen metapopulation. *Journal of Ecology* 89, 736–748.

Thrall, P.H., Burdon, J.J. and Bever, J.D. (2002) Local adaptation in the *Linum marginale–Melampsora lini* host–pathogen interaction. *Evolution* 56, 1340–1351.

Wennström, A. (1999) The effect of systemic rusts and smuts on clonal plants in natural systems. *Plant Ecology* 141, 93–97.

Wolfe, M.S. (1985) The current status and prospects of multiline cultivars and variety mixtures for disease resistance. *Annual Review of Phytopathology* 23, 251–273.

2

Measuring Genetic Variation in Natural Enemies Used for Biological Control: Why and How?

E. Wajnberg

INRA, 37 Boul. du Cap, 06600 Antibes, France

Introduction

Biological control programmes of pests with the use of natural enemies have been employed for more than a century. However, on average, only between 10 and 35% of the introduced natural enemies established successfully (Force, 1967; Huffaker and Messenger, 1976; Hall and Ehler, 1979), and only a fraction of these led to economic control (Mackauer, 1972; Hall *et al.*, 1980). There is thus considerable room for improvement and, as pointed out by Roush (1990a) and Hopper *et al.* (1993), the importance of genetic aspects as a means of improving the success of this crop protection strategy has always attracted a great deal of attention (Mally, 1916; Simmonds, 1963; Mackauer, 1976; Messenger *et al.*, 1976; Hoy, 1985, 1992; Roush, 1990b). The central issue in this context is based on the analysis of genetic variation in the key biological attributes of the biocontrol agents (Roush, 1990a). An important effort of research has been devoted repeatedly to the quantification of the *inter*specific variability of some of these attributes, and results are supposed to provide some help in selecting the right species for controlling an identified pest in a given environment (Messenger and van den Bosch, 1971). Other works have aimed at describing the *intra* specific genetic variability in the biological traits studied. In this case, most authors tried to compare different populations of the same species originating from different geographical locations (Simmonds, 1963; Messenger and van den Bosch, 1971; Diehl and Bush, 1984; Caltagirone, 1985). The aim of comparing different populations of the same species is usually to identify the strain that is most adapted to the environment where it will be released, and thus to improve the success of the biological control programme (Messenger and van den Bosch, 1971). Finally, some work has been done to quantify the *intra*specific, *intra*population genetic variabil-

 Genetics, Evolution and Biological Control (eds L.E. Ehler, R. Sforza and T. Mateille)

ity in the biological attributes of the natural enemies, but only little effort has been made at this level of genetic variation (Lewis *et al.*, 1990; Hopper *et al.*, 1993).

The present chapter will discuss the intrapopulation level of genetic variation in natural enemies, and more specifically insect parasitoids. Genetic comparison between populations will not be considered here, nor will studies that assessed genetic variation by analysing laboratory populations generated by the amalgamation of geographically different stains (e.g. Simmonds, 1947; Parker and Orzack, 1985; Antolin, 1992a,b). Not all biological characteristics are polygenic (i.e. influenced by many loci, each of small effect), but such an assumption is probably true for most traits important for the efficiency of natural enemies in biological control programmes (Roush, 1990a). Thus, only intrapopulation genetic variability in continuously variable, polygenic traits will be considered here. Genetic variation in molecular markers (enzymatic or nucleic) will thus not be discussed, nor variability in pesticide resistance that is repeatedly considered to be due to allelic variation at just a few major genes (Roush and McKenzie, 1987; Hoy, 1990a,b).

Among all potential natural enemies that can be used for pest control, insect parasitoids are the most important. They can be found in nearly all ecological systems, and there are probably almost 2 million species on Earth (Godfray, 1994). Most of them reproduce through arrhenotoky, a special kind of parthenogenesis in which mated females can produce either unfertilized eggs, giving rise to haploid males, or fertilized eggs, leading to diploid females. This peculiar haplo-diploid reproductive system often precludes application of the usual procedures for the estimation of quantitative genetic variation (Carton *et al.*, 1989; Sequeira and Mackauer, 1992). However, although a low level of genetic variability is expected within parasitoid populations (Crozier, 1977; Hoy, 1990a; Legner, 1993), some authors have argued that, as in any insect species, some significant genetic variation should be found (Ayala, 1982; Bartlett, 1985), especially when behavioural traits are being quantified (Barinaga, 1994; Pompanon *et al.*, 1999).

After discussing why it is usually considered to be of prime importance to quantify the intrapopulation genetic variability in the biological attributes of insect parasitoids, the different experimental and statistical methods that can be used to quantify such a level of genetic variation in quantitative biological traits will be presented. Then, an exhaustive analysis of the results available in the literature will be discussed. In a final part, the choice of the biological attributes that need to be studied for better pest-control efficacy by the parasitoids will be considered.

Why Measure Intrapopulation Genetic Variation in Natural Enemies?

Despite the fact that genetic variability in mass-produced and released natural enemy populations remains poorly described, an accurate estimation of such genetic variation has always been considered to be important, for several complementary reasons. As pointed out by Remington (1968), a population introduced in a new environment will have to cope with a community composed largely of organisms not present in its origin location. It will therefore interact with environmental constraints for which individuals' genotypes will not have been previously selected. Thus, in order to maximize its ability to establish, it is generally admitted that the released population of natural enemies should have the maximum possible genetic variability (Simmonds, 1963; Force, 1967; Mackauer, 1972; Lewis *et al.*, 1990; Wajnberg, 1991). In order to check this, such genetic variation has to be quantified accurately.

Measuring the level of genetic variation in biological attributes of natural enemies should also lead to a better estimate of their ability to evolve in response to the environmental characteristics of the system used to rear and produce them. Since colonies maintained and produced in the laboratory usually experience constant environments that are obviously different from those encountered in the field, selection could occur for laboratory-adapted genotypes. The response to selection will be more rapid when genetic variation is high. In turn, such selection might rapidly reduce both the genetic variance in the biological traits involved (Bulmer, 1976) and the pest-control efficacy during field release (Mackauer, 1972, 1976; Bartlett, 1984; Joslyn, 1984; Roush, 1990b; Hopper *et al.*, 1993).

During the production phase of insects in the laboratory, some individuals can contribute randomly to the next generation more than others, and this sampling effect can lead to a random loss of some alleles in each generation. This process is called genetic drift, and can lead to a substantial reduction in the genetic variability within the reared population (Joslyn, 1984; Roush, 1990b; Hopper *et al.*, 1993). However, it should be noted that genetic drift is important only in small populations, and theoretical developments indicate that a loss in genetic variability due to such a process is not always a matter of concern (Nei *et al.*, 1975; Hopper *et al.*, 1993). As pointed out by Wajnberg (1991), even if this loss is sometimes of minor importance, it cannot be prevented. At best, it can be reduced as much as possible. For this, an accurate quantification of the genetic variability in the reared population can also provide some fruitful information. More generally, measuring the genetic variation in important attributes of the natural enemies should allow us to better define the rearing conditions used to produce them before field release (Boller, 1979).

As a general rule, an accurate quantification of genetic variability should lead to better estimates of the survival potential of the founder population during the rearing and production process (Mackauer, 1972), and in the field after being released (Hopper *et al.*, 1993). An accurate estimation of the intrapopulation genetic variation should thus help in better defining how new natural enemies' biotypes should be found and collected, and also to optimize methods and timing used to release them.

Perhaps the most discussed reason why it is important to measure the genetic variability in biological traits within populations of natural enemies is that the existence of significant genetic variation can allow commencement of a breeding selection programme aimed at improving the efficacy of the released animals to control the target pests (Hoy, 1976; Wajnberg, 1991). Such an improving selection procedure is equivalent to that used for domestic animals and plants (Wilkes, 1947). In the case of natural enemies, the traits that are the subject of improvement should be related to insectary production and/or to field effectiveness (Hoy, 1976). The usefulness of the method has been discussed by several authors (for a review, see Mackauer, 1972, 1976), but some arguments against selection breeding of natural enemies have also been raised on several occasions. As a general rule, a breeding selection programme of a strain of natural enemies is usually considered a difficult task (Simmonds, 1963), and White *et al.* (1970) thus recommended the use of selection for improving pest-control efficacy only when there is no chance of finding a better-adapted species. Moreover, the laboratory procedure employed to select an improved strain will probably reduce genetic variation, which goes against the need to release a population showing the maximum possible genetic variability (Wilson, 1965). Also, the idea of successfully selecting a natural enemy for improving its pest-control efficacy supposes an accurate understanding of the different biological attributes that determine effectiveness, a task which is repeatedly considered as extremely difficult (Wilson, 1965; Hoy, 1976; Mackauer, 1976). Finally, successfully selecting a population for an accurately identified attribute could lead to the possible occurrence of correlated, pleiotropic variations in other biological traits (Simmonds, 1963). As a result, the character selected might indeed be improved, but some undetected accumulation of other characteristics, which might be disadvantageous in the field, could appear.

In the past few decades an important body of data has accumulated on the ecological characteristics of insect parasitoids. For this, optimal foraging theory was used, allowing us to identify the optimal biological characteristics these insects should adopt in order to maximize their reproductive potential (Stephens and Krebs, 1986; Godfray, 1994). Such a theoretical approach provided information of prime importance for defining what the optimal features of a potential biocontrol agent should be (Waage, 1983, 1990). However, talking about optimal biological traits supposes that these traits have been settled progressively by natural selection, and thus that there is/was, in the natural enemy population, genetic variability upon which natural selection could act. Quantifying such intrapopulation genetic variability would thus also provide important information for confirming the relevance of the theory and its ability to produce results useful for improving the efficacy of biological control programmes.

Methods for Measuring Intrapopulation Genetic Variation

Variation among individuals, for any kind of quantitative trait, is a common feature of all biological studies (Bartlett, 1985; Lewis *et al.*, 1990). Moreover, in most cases, measuring such variation is considered to require experimental protocols based on many replicates to collect solid estimates of means (Roitberg, 1990). Such phenotypic variability is known to be the result of interactions between the genotype, or genetic make-up, of each organism and the environment that it lives in (Collins, 1984). The genetic source of variability is passed from individuals to their progeny, which is not the case for the environmental (i.e. non-genetic) source of variation. The aim of the methods to be presented briefly here is both to separate these two sources of variability, genetic and environment, and to see whether the genetic part is of statistically significant importance in contributing to the phenotypic variation observed (Falconer, 1989).

The variability among all individuals in a population can be quantified with the so-called variance, and the notation V_P is used to describe the variance of the phenotypes. Using statistical models, such phenotypic variance can be divided into V_E, the variance due to environmental effects, and V_G, the variance of the genotypes. In turn, V_G can be further subdivided into the variance due to additive genetic effects (V_A), the variance due to dominance effects (V_D), and the variance due to the interaction between the loci involved in determining the trait under consideration (= epistasis) (V_I) (see, for example, Falconer, 1989, for a detailed presentation). So, the basic model used in this case is:

$$V_P = V_E + V_A + V_D + V_I \qquad (2.1)$$

Of the three types of genetic variance, V_A is considered to be the most important, since it defines the breeding value for the trait on the organisms. In other words, selection – artificial or natural – will essentially act on V_A, and the methods that have been defined to estimate the genetic variability in quantitative biological traits are all built to estimate it, in a direct or indirect way.

In theory at least, perhaps the simplest way to estimate V_A for a given trait would be to measure two populations in several different environments. The first population would consist of mixed individuals of different genotypes, the other would consist of individuals all having the same genotype. Results obtained for the first population will enable us to estimate the total phenotypic variance, V_P. Those obtained for the other would give an estimate of the environmental variance, V_E, only. The difference between these two variances would give an estimate of the additive genetic component of the variance, V_A (Collins, 1984). However, in most cases, using such a simple method is not realistic and other more feasible, accurate methods are available. They will be presented briefly below. These methods are usually used to estimate the genetic variation in ecological (e.g. behavioural) traits in insect populations. However, they can also be used to quantify any kind of quantitative genetic variability, at the molecular, cytological, physiological or morphological levels (Parsons, 1980). They can be used to estimate the genetic variation, and thus to assess the genetic structure of natural and mass-reared populations, even those of species where details of the genome are not well known (Parsons, 1980).

Parent–offspring regression

If the variation in a quantitative trait is, at least in part, genetically determined, then offspring should resemble their parents. Based on this basic statement, Sir F. Galton, more than a century ago, proposed a study of the transmissibility of a trait over two successive generations. For this, the trait is quantified in a set of parents (mothers, fathers or both), and also in their progenies. Then, the slope of the regression line of the offspring's values on that of their parents is computed and its statistical significance tested. If both parents are quantified, their average value is used. Only one offspring can be measured for each parent, but, if several offspring are quantified, then their average value is also used to estimate the slope of the regression line. Finally, in this latter case, if the number of offspring is not constant, the regression can be weighted by the actual number of offspring measured for each parent (Falconer, 1989).

In most cases, only parasitoid females are useful for controlling a target pest in a biological control programme, and, thus, the trait studied can sometimes be measured in females only (e.g. fecundity, host attack rate, etc.). In this case, the parent–offspring regression analysis becomes a mother–daughter regression analysis.

Sib analysis

Another method consists of mating a number of males (sires) with a number of females (dams). Each sire has to be mated to more than one dam, but each dam is mated with one sire only. The trait under study is then measured in the offspring produced by the mated females. Finally, an analysis of variance (ANOVA) is used to quantify the variation among sires, among dams (within sires), and within the progeny of each dam. In turn, the estimated parameters of the ANOVA can be used to test the significance of the genetic variation among dams and among sires for the quantified traits.

Despite the fact that one advantage of this method is in excluding possible maternal effects through the comparison of different sires (Falconer, 1989; Hopper *et al.*, 1993), it has almost never been used to quantify the genetic variability in biological attributes of natural enemies.

Family analysis

A related, and more commonly used, method is the isofemale strains method, also called isofemale lines or family analysis (Parsons, 1980; Hoffman and Parsons, 1988). In this method, an array of families (or lines, or strains) is founded, each family from a single mated female, and the trait under study is quantified in several offspring produced by each female in the F_1 generation. Finally, a one-way ANOVA is used to test for a significant difference among average values of the different families compared, which will indicate a significant genetic variability in the quantified trait. The main difference between this method and the previous one is that females used to found each family are not mated with clearly identified males. On the contrary, they are supposed to be mated randomly.

A compromise should be found between the number of families compared and the number of individuals measured in each family. Of course, small or moderate-size datasets could lead to imprecision in estimating the genetic variation (Shaw, 1987; Falconer, 1989) and/or to poor power of the statistical test used to detect it. The number of replicates used usually depends on the time needed to measure each individual and how difficult it is to found and rear the different families compared.

Despite the fact that it could lead to laborious experimental protocols, the family analysis is considered to be more simple to use than the other methods available. However, the variation between families can be caused by a mixture of additive, dominance and epistatic components (Falconer, 1989). Hence, despite giving a crude estimate of the genetic variation in the trait studied, this method does not provide direct information regarding the ability of the population to respond to natural or artificial selection, and other more accurate methods should be used.

Finally, it has to be noted that the different families compared are just a sample of all possible families in the entire population studied. Therefore, the 'family' effect in the ANOVA used to test the difference among the families' averages values should be considered a random effect and should be treated accordingly. Occasionally, this might raise some problems, especially when generalized linear models are used for handling traits that are not distributed according to a Gaussian distribution (McCullagh and Nelder, 1991).

Breeding selection

Despite the fact that the modification of the average value of a trait through several generations of breeding selection does not accurately represent a way to quantify its genetic variability in a population, it proves that the observed phenotypic variation is at least partly under genetic control (Roush, 1990a). Such a procedure has thus been used by several authors to demonstrate the existence of significant genetic variation in several biological attributes of natural enemies. Briefly, two categories of methods can be used here. In mass selection, the individuals used to found the next generation are chosen according to their own phenotype. In family selection, the individuals are chosen according to the average value of the family from which they come (Collins, 1984; Falconer, 1989; Wajnberg, 1991).

All the methods presented above can be used to estimate the heritability of the trait studied, defined in either its narrow or its broad sense. In its narrow sense, this is the ratio of additive genetic variance to the total phenotypic variance. In the broad sense, it is defined as the ratio of the total genetic variance (i.e. additive, dominance and epistatic) to the total phenotypic variance (Hoffmann and Parsons, 1988; Falconer, 1989). Also, these different methods are sometimes combined together into a single experimental set-up. For example, a mother–daughter regression analysis can be performed over two successive generations, with several offspring measured for each mother. At the F_1 generation, the daughters represent different isofemale lines, which are compared by means of a family analysis (e.g. Chassain and Boulétreau, 1991; Bruins *et al.*, 1994). Of course, since the main aim of these methods is to estimate the genetic and environmental components of the phenotypic variation observed, they should all be conducted under conditions where environmental causes of variation are reduced to a minimum. Measurements have to be made in precisely controlled environmental chambers, under constant temperature, humidity, measured at the same age, and so forth. If possible, the method used to quantify the traits should be as simple, fast and cheap as possible in order to perform a large number of replicates in a short time interval. These points, or some of them, might be difficult to solve, and sometimes even be limiting factors.

Finally, some attributes of natural enemies cannot be quantified by a single value. For example, several quantitative parameters are sometimes needed to quantify a single behaviour of a parasitoid female. In this case, all the methods described above can be generalized using multivariate statistical methods that are built to take into account possible correlation among the different traits measured. For example, a multidimensional regression analysis (i.e. a canonical regression analysis) can be used to perform a multivariate parent–offspring regression (e.g. Wajnberg, 1993), or a factorial discriminant analysis can be used to compare the mean-vector describing all isofemale lines in a family analysis.

Intrapopulation Genetic Variation in Insect Parasitoids

A detailed survey was performed over the main scientific publication databases to find all references describing intrapopulation genetic variation in quantitative attributes of insect parasitoids. Only 39 references were found, covering 23 different species names (see Table 2.1). These 39 references were published over a period of almost 60 years (from Wilkes (1942) to Gu and Dorn (2000)), which represents a very low publication rate (about 0.67 publications per year). However, two-thirds of the references appeared during the past decade (1990–2000), suggesting increasing interest in this sort of scientific work.

The process of host exploitation by insect parasitoids is usually described using a series of steps that draw them progressively closer to their hosts and enable the immature stages to develop successfully in them (Vinson, 1975, 1976). A parasitoid female first has to discover a habitat where potential hosts are living. Then, she should discover a host, and should recognize and attack it. The laid progeny should be able to overcome or evade the host's

Table 2.1. List of all species in which intrapopulation genetic variation in quantitative traits was studied.

Family, species	References
Braconidae	
Aphidius ervi	Sequeira and Mackauer (1992), Henter (1995), Gilchrist (1996)
Asobara tabida	Mollema (1991)
Cotesia glomerata	Gu and Dorn (2000)
Cotesia melanoscela	Weseloh (1986), Chenot and Raffa (1998)
Microplitis croceipes	Prévost and Lewis (1990)
Ichneumonidae	
Aenoplex carpocapsae	Simmonds (1947)
Horogenes molestae	Allen (1954)
Microplectron fuscipennis	Wilkes (1942, 1947)
Eucoilidae	
Leptopilina boulardi	Carton *et al.* (1989), Perez-Maluf *et al.* (1998)
Pteromalidae	
Muscidifurax raptor	Geden *et al.* (1992)
Nasonia vitripennis	Orzack (1990), Orzack and Parker (1990), Orzack *et al.* (1991), Orzack and Gladstone (1994)
Trichogrammatidae	
Trichogramma brassicae	Chassain and Boulétreau (1991), Fleury *et al.* (1993), Wajnberg (1993, 1994), Bruins *et al.* (1994), Pompanon *et al.* (1994, 1999), Wajnberg and Colazza (1998)
as *Trichogramma maidis*	Chassain and Boulétreau (1987), Wajnberg (1989), Wajnberg *et al.* (1989)
Trichogramma cacoeciae	Chassain and Boulétreau (1991), Pompanon *et al.* (1994)
Trichogramma carverae	Bennett and Hoffmann (1998)
Trichogramma dendrolimi	Limburg and Pak (1991), Schmidt (1991)
Trichogramma evanescens	Limburg and Pak (1991), Schmidt (1991)
Trichogramma minutum	Urquijo (1950), Liu and Smith (2000)
Trichogramma pretiosum	Ashley *et al.* (1974)
Trichogramma semifumatum	Ashley *et al.* (1974)
Trichogramma voegelei	Mimouni (1991)
Scelionidae	
Telenomus busseolae	Wajnberg *et al.* (1999)
Tachinidae	
Lixophaga diatraeae	Pintureau *et al.* (1995)

internal defence mechanisms and, finally, the parasitoid must find the host nutritionally suitable for complete development, resulting in adult emergence. Figure 2.1 gives the distribution of the different biological traits in which an intrapopulation genetic variability has been studied along such a sequential process of host exploitation by parasitoids. There is a clear lack of studies on biological traits involved in host habitat location by the foraging parasitoid female, and on those involved in evading the host immune system by the laid progeny. This is probably due to the difficulties in measuring these traits in the laboratory and in performing all the replicates needed to identify significant genetic variation.

About 60% of these studies used the family analysis method (see Fig. 2.1). In this case, on average, 19.85 ± 3.70 families were compared, with an average of 12.15 ± 3.73 individuals measured in each family. The parent–offspring

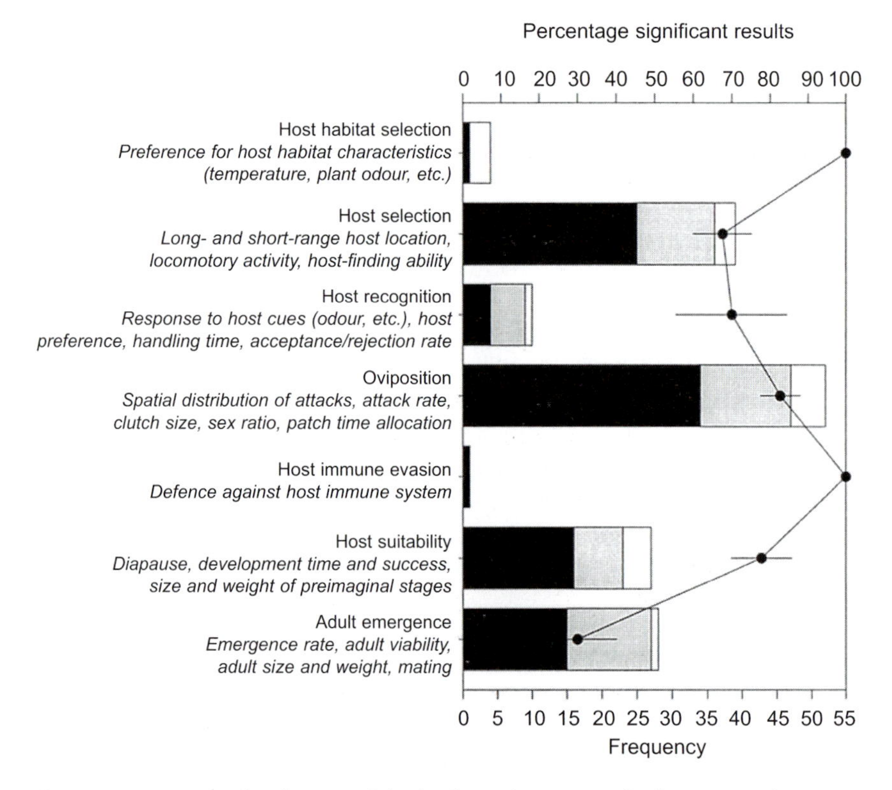

Fig. 2.1. Frequency distribution of the biological traits in which intrapopulation genetic variation has been studied in insect parasitoids. The seven classes of traits correspond to a series of steps usually used to describe the host exploitation process by parasitoids (from top to bottom). Black, grey and white rectangles correspond to the use of family analyses, parent–offspring regressions or breeding selection methods, respectively. The curve represents percentages (± SE) of these studies showing significant genetic variability.

regression method was used in 29.81% of the studies, with an average of 65.31 ± 13.22 parent/offspring couples measured. The remaining studies (i.e. 10.56%) used the breeding selection method. In this case, on average, 9.67 ± 1.50 generations of selection were followed.

About 70% of these studies showed the existence of significant genetic variation. However, as can be seen in Fig. 2.1, the percentage of significant results varies along the different steps of host exploitation by parasitoids, and this variation appears to be statistically significant (Fisher's exact test, $P < 0.005$). Some biological traits seem to present higher genetic variation than others. Traits that are closely related to fitness are often supposed to have been purged of genetic variance by strong directional selection, and hence should present lower genetic variability (Gustafsson, 1986; Mousseau and Roff, 1987; Roff and Mousseau, 1987; for an alternative point of view, see Price and Schluter, 1991; Houle, 1992). This should be the case for the life-history traits of parasitoids, such as body size and development time, and female fecundity. For such life-history traits, 61.76% of the studies showed the existence of significant genetic variation, while this percentage rose to 77.78% for the other biological attributes studied. The difference between these two percentages is statistically significant (one-sided Fisher's exact test, $P = 0.025$). Thus, the hypothesis that traits related to fitness should present lower genetic variation seems to be verified in insect parasitoids. Most of these traits belong to the 'adult emergence' (last) category shown in Fig. 2.1, and this might explain the corresponding decrease in the percentage of significant results obtained in this case.

What Characters Should be Studied?

As we have seen, methods that can be used to quantify the genetic variation in biological attributes of natural enemies are available and have been in use for several years now. However, identifying the biological features in which intrapopulation genetic variation should be studied still remains an open question. Ideally, the traits should be related to pest-control efficacy, and are thus important either during the laboratory production phase or after field release. This is related to the question of the quality of biocontrol agents (Bigler, 1989). As such, it has always been, and still is, an intensively debated issue, and lists of potential characters are found repeatedly in the literature (e.g. Flanders, 1947; Hoy, 1976, 1990b; Mackauer, 1976; Roush, 1979; Waage and Hassell, 1982; Hopper *et al.*, 1993). Briefly, based either on empirical intuition or on theoretical considerations, the main proposed categories of criteria are climatic adaptation, habitat preference, synchrony with hosts, host-searching capacity, specificity, dispersal ability, attack rate, female fecundity and sex ratio.

As a general rule, it has been admitted repeatedly that it is difficult to identify those biological attributes that distinguish efficient biological control agents from unsuccessful ones (DeBach, 1958; Hoy, 1976, 1990b; Roush, 1990a). Another difficulty is the interpretation of laboratory results and their link to field performance (Bigler, 1989). Some authors even assert that the important traits just cannot be identified satisfactorily (for a discussion, see Roush, 1979). Of course, the important attributes of a potential biological control agent should clearly depend on the nature of the pest to be controlled, on the ecological features of the crop to be protected, and also on the type of release (inoculative, inundative, etc.) that is to be used (Hoy, 1976; Roush, 1979). Thus, general statements would probably lead to misinterpretations (Bigler, 1989). Despite this, some authors argue that, in most cases, the important traits should be those implicated in the overall ability and propensity of the released insects to colonize new habitats (Force, 1967).

Identifying the important traits is usually considered to be based on a pre-release evaluation of natural enemies, a procedure that is often perceived more as an art than a science (van Lenteren, 1980). A possible, time-consuming process would be to find different strains or variants for every potential trait (or to create them through breeding selection experiments) and to study the relationship between their average phenotypic value, estimated in the laboratory, and their efficacy in controlling the targeted pest after being released in the field. This process has been used successfully by Bigler *et al.* (1988), who found a significant positive relationship between the walking velocity of the females of different *Trichogramma maidis* (= *T. brassicae*) strains, estimated in the laboratory, and their parasitism rate against the eggs of the European corn borer, *Ostrinia nubilalis*, in the field. This indicates that the walking speed of the parasitoid females should be an interesting trait, and its genetic variation should be quantified in this wasp species (Pompanon *et al.*, 1994).

Finding the most important trait probably needs a costly, labour-intensive process. The use of mathematical models might provide substantial help in determining these traits before the experimental work (Bigler, 1989). To our knowledge, Wajnberg and Colazza (1998) provided the only example relating a modelling approach demonstrating the importance of a biological trait (i.e. the prospected surface per time unit of foraging parasitoid females) to an analysis of its intrapopulation genetic variation. More generally, theoretical approaches using optimality models are now used intensively in order to quantify the importance of foraging decisions in the reproductive ability of insect parasitoids. The traits that appear to be important are probably related to the pest-control efficacy of the parasitoids when they are used in biological control programmes (Waage, 1983, 1990). Such a theoretical approach allows us to identify the traits in which genetic variability should be studied. Among others, the traits that are revealed to be important are host attack rate, spatial aggregation of the attacks, selection of host patches, time allocation in host patches, etc. Wajnberg *et al.* (1999) provide the first attempt to demonstrate significant genetic variation in a trait (patch time allocation) that has been shown by theoretical approaches to be of importance in the reproductive ability of insect parasitoids.

Finally, there is no doubt that identifying the main biological attributes of natural enemies remains the most important issue for improving the efficacy of biological control programmes. It is also the most difficult question to solve and will certainly necessitate, in the future, the solution of numerous challenges, probably through the use of theoretical developments.

Conclusion

Only a limited number of studies of intrapopulation genetic variation in insect parasitoids have been published so far. So, as pointed out by Hopper *et al.* (1993), much work still remains to be done. A lot of information regarding the level of genetic variation in natural or mass-reared populations of natural enemies is still missing if we want to understand the ecology and evolutionary potential of beneficial insects used in biological control programmes. More accurately, the exhaustive bibliographic survey presented in this chapter indicates that some biological traits are studied less than others. For example, we have seen that the behavioural traits involved in the ability of an emerging female parasitoid to find a habitat where potential hosts are located still remain poorly studied. Actually, the research that is needed concerns the 'ecological genetics' of natural enemies, in the sense that the genetic variation should be studied in biological attributes that can be implicated in the ecology of their interactions with their hosts.

During the last decade of the 20th century, the development of new technologies to study numerous variable loci simultaneously led to the possibility of analysing, at the genome level, the genetic determinism of phenotypic variation. This so-called 'population genomics' approach (Black *et al.*, 2001) includes the creation of genetic maps enabling us to identify the major genes involved in the variation observed (i.e. quantitative trait loci, QTL). There is no doubt that such technology will allow us to quantify more rapidly and more accurately the genetic variation in populations of natural enemies. In turn, the results obtained will also provide the means to perform marker-assisted selection of their pest-control efficacy. These results will also certainly lead us to have a better insight into their evolutionary potentialities. The aim of this chapter is to stimulate research in this direction.

Acknowledgements

I thank F. Fleury, T. Guillemaud, R.H. Messing and R.T. Roush for critical reading of the manuscript, and C. Curty and A. Dufay for help with the bibliographic survey.

References

Allen, H.W. (1954) Propagation of *Horogenes molestae*, an Asiatic parasite of the Oriental fruit moth, on the potato tuberworm. *Journal of Economic Entomology* 47, 278–281.

Antolin, M.F. (1992a) Sex ratio variation in a parasitic wasp. I. Reaction norms. *Evolution* 46, 1496–1510.

Antolin, M.F. (1992b) Sex ratio variation in a parasitic wasp. II. Diallel cross. *Evolution* 46, 1511–1524.

Ashley, T.R., Gonzalez, D. and Leigh, T.F. (1974) Selection and hybridization of *Trichogramma*. *Environmental Entomology* 3, 43–48.

Ayala, F. (1982) Evolutionary significance of genetic variation in insects. In: Stock, M.W. and Bartlett, A.C. (eds) *The Evolutionary Significance of Insect Polymorphism*. University of Idaho Press, Moscow, pp. 1–18.

Barinaga, M. (1994) Genes and behavior. From fruit flies, rats, mice: evidence of genetic influence. *Science* 264, 1690–1693.

Bartlett, A.C. (1984) Genetic changes during insect domestication. In: King, E.G. and Leppla, N.C. (eds) *Advances and Challenges in Insect Rearing*. USDA/ARS, New Orleans, pp. 2–8.

Bartlett, A.C. (1985) Guidelines for genetic diversity in laboratory colony establishment and maintenance. In: Singh, P. and Moore, R.F. (eds) *Handbook of Insect Rearing*. Elsevier, New York, pp. 7–17.

Bennett, D.M. and Hoffmann, A.A. (1998) Effects of size and fluctuating asymmetry on field fitness of the parasitoid *Trichogramma carverae* (Hymenoptera: Trichogrammatidae). *Journal of Animal Ecology* 67, 580–591.

Bigler, F. (1989) Quality assessment and control in entomophagous insects used for biological control. *Journal of Applied Entomology* 108, 390–400.

Bigler, F., Bieri, M., Fritschy, A. and Seidel, K. (1988) Variation in locomotion between laboratory strains of *Trichogramma maidis* and its impact on parasitism of eggs of *Ostrinia nubilalis* in the field. *Entomologia Experimentalis et Applicata* 49, 283–290.

Black, W.C., Baer, C.F., Antolin, M.F. and DuTeau, N.M. (2001) Population genomics: genome-wide sampling of insect populations. *Annual Review of Entomology* 46, 441–469.

Boller, E.F. (1979) Ecological genetics and quality control. In: Hoy, M.A. and McKelvey, J.J. Jr (eds) *Genetics in Relation to Insect Management*. A Rockefeller Foundation Conference, Bellagio, Italy, pp. 153–160.

Bruins, E.B.A.W., Wajnberg, E. and Pak, G.A. (1994) Genetic variability in the reactive distance in *Trichogramma brassicae* after automatic tracking of the walking path. *Entomologia Experimentalis et Applicata* 72, 297–303.

Bulmer, M.G. (1976) The effect of selection on genetic variability: a simulation study. *Genetic Research* 28, 101–117.

Caltagirone, L.E. (1985) Identifying and discrimination among biotypes of parasites and predators. In: Hoy, M.A. and Herzog, D.C. (eds) *Biological Control in Agricultural IPM Systems*. Academic Press, Orlando, Florida, pp. 189–200.

Carton,Y., Capy, P. and Nappi, A.J. (1989) Genetic variability of host–parasite relationship traits: utilization of isofemale lines in a *Drosophila simulans* parasitic wasp. *Génétique, Sélection, Evolution* 21, 437–446.

Chassain, C. and Boulétreau, M. (1987) Genetic variability in the egg-laying behaviour of *Trichogramma maidis*. *Entomophaga* 32, 149–157.

Chassain, C. and Boulétreau, M. (1991) Genetic variability in quantitative traits of host exploitation in *Trichogramma* (Hymenoptera: Trichogrammatidae). *Genetica* 83, 195–202.

Chenot, A.B. and Raffa, K.F. (1998) Heritability estimates of development time and size characters in the Gypsy Moth (Lepidoptera: Lymantriidae) parasitoid *Cotesia melanoscela* (Hymenoptera: Braconidae). *Biological Control* 27, 415–418.

Collins, A.M. (1984) Artificial selection of desired characteristics in insects. In: King, E.G.

and Leppla, N.C. (eds) *Advances and Challenges in Insect Rearing.* USDA/ARS, New Orleans, pp. 9–19.

Crozier, R.H. (1977) Evolutionary genetics of the Hymenoptera. *Annual Review of Entomology* 22, 263–288.

DeBach, P. (1958) Selective breeding to improve adaptations of parasitic insects. In: Becker, E.C. (ed.) *Proceedings of the 10th International Congress of Entomology, Montreal*, pp. 759–768.

Diehl, S.R. and Bush, G.L. (1984) An evolutionary and applied perspective of insect biotypes. *Annual Review of Entomology* 29, 471–504.

Falconer, D.S. (1989) *Introduction to Quantitative Genetics*, 3rd edn. Longman, New York.

Flanders, S.E. (1947) Elements of host discovery exemplified by parasitic Hymenoptera. *Ecology* 28, 299–309.

Fleury, F., Chassain, C., Fouillet, P. and Boulétreau, M. (1993) La dispersion spatiale de la pontes des trichogrammes (Hymenoptera: Trichogrammatidae): bases génétiques et épigénétiques de la variabilité. *Bulletin de la Société Zoologique de France* 118, 149–157.

Force, D.C. (1967) Genetics in the colonization of natural enemies for biological control. *Annals of the Entomological Society of America* 60, 722–729.

Geden, C.J., Smith, L., Long, S.J. and Rutz, D.A. (1992) Rapid deterioration of searching behavior, host destruction, and fecundity of the parasitoid *Muscidifurax raptor* (Hymenoptera: Pteromalidae) in culture. *Annals of the Entomological Society of America* 85, 179–187.

Gilchrist, G.W. (1996) A quantitative genetic analysis of thermal sensitivity in the locomotor performance curve of *Aphidius ervi. Evolution* 50, 1560–1572.

Godfray, H.C.J. (1994) *Parasitoids. Behavioral and Evolutionary Ecology.* Princeton University Press, Princeton, New Jersey.

Gu, H. and Dorn, S. (2000) Genetic variation in behavioral response to herbivore-infested plants in the parasitic wasp, *Cotesia glomerata* (L.) (Hymenoptera: Braconidae). *Journal of Insect Behavior* 13, 141–156.

Gustafsson, L. (1986) Lifetime reproductive success and heritability: empirical support for Fisher's fundamental theorem. *American Naturalist* 128, 761–764.

Hall, R.W. and Ehler, L.E. (1979) Rate of establishment of natural enemies in classical biological control. *Bulletin of the Entomological Society of America* 25, 280–282.

Hall, R.W., Ehler, L.E. and Bisabri-Ershadi, B. (1980) Rate of success in classical biological control of arthropods. *Bulletin of the Entomological Society of America* 26, 111–114.

Henter, H.J. (1995) The potential for coevolution in a host–parasitoid system. II. Genetic variation within a population of wasps in the ability to parasitize an aphid host. *Evolution* 49, 439–445.

Hoffmann, A.A. and Parsons, P.A. (1988) The analysis of quantitative variation in natural populations with isofemale strains. *Génétique, Sélection, Evolution* 20, 87–98.

Hopper, K.R., Roush, R.T. and Powell, W. (1993) Management of genetics of biological-control introductions. *Annual Review of Entomology* 38, 27–51.

Houle, D. (1992) Comparing evolvability and variability of quantitative traits. *Genetics* 130, 195–204.

Hoy, M.A. (1976) Genetic improvement of insects: fact or fantasy. *Environmental Entomology* 5, 833–839.

Hoy, M.A. (1985) Improving establishment of arthropod natural enemies. In: Hoy, M.A. and Herzog, D.C. (eds) *Biological Control in Agricultural IPM Systems.* Academic Press, New York, pp. 151–166.

Hoy, M.A. (1990a) Genetic improvement of arthropod natural enemies: becoming a conventional tactic? In: Baker, R.R. and Dunn, P.E. (eds) *New Directions in Biological Control: Alternatives for Suppressing Agricultural Pests and Diseases.* Liss Inc., New York, pp. 405–417.

Hoy, M.A. (1990b) Genetic improvement of parasites and predators. *FFTC-ASPAC International Seminar on the Use of Parasitoids and Predators to Control Agricultural Pests.* National Agricultural Research Centre (NARC), Tsukuba, Japan, pp. 233–242.

Hoy, M.A. (1992) Biological control of arthropods: genetic engineering and environmental risk. *Biological Control* 2, 166–170.

Huffaker, C.B. and Messenger, P.S. (eds) (1976) *Theory and Practice of Biological Control.* Academic Press, New York.

Joslyn, D.J. (1984) Maintenance of genetic variability in reared insects. In: King, E.G. and Leppla, N.C. (eds) *Advances and Challenges in Insect Rearing.* USDA/ARS, New Orleans, pp. 20–29.

Legner, E.F. (1993) Theory for quantitative inheritance of behavior in a protelean parasitoid, *Muscidifurax raptorellus* (Hymenoptera: Pteromalidae). *European Journal of Entomology* 90, 11–21.

Lewis, W.J., Vet, L.E.M., Tumlinson, J.H., van Lenteren J.C. and Papaj, D.R. (1990) Variations in parasitoid foraging behavior: essential element of a sound biological control theory. *Environmental Entomology* 19, 1183–1193.

Limburg, H. and Pak, G.A. (1991) Genetic variation in the walking behaviour of the egg parasite *Trichogramma.* In: Bigler, F. (ed.) *Proceedings of the 5th Workshop on Quality Control of Mass-reared Arthropods, Wageningen,* pp. 47–55.

Liu, F.H. and Smith, S.M. (2000) Measurement and selection of parasitoid quality for mass-reared *Trichogramma minutum* Riley used in inundative release. *Biocontrol Science and Technology* 10, 3–13.

Mackauer, M. (1972) Genetic aspects of insect production. *Entomophaga* 17, 27–48.

Mackauer, M. (1976) Genetic problems in the production of biological control agents. *Annual Review of Entomology* 21, 369–385.

Mally, C.W. (1916) On the selection and breeding of desirable strains of beneficial insects. *South African Journal of Science* 13, 191.

McCullagh, P. and Nelder, J.A. (1991) *Generalized Linear Models,* 2nd edn. Chapman & Hall, London.

Messenger, P.S. and van den Bosch, R. (1971) The adaptability of introduced biological control agents. In: Huffaker, C.B. (ed.) *Biological Control.* Plenum Press, New York, pp. 68–92.

Messenger, P.S., Wilson, F. and Whitten, M.J. (1976) Variation, fitness and adaptability of natural enemies. In: Huffaker, C.B. and Messenger, P.S. (eds) *Theory and Practice of Biological Control.* Academic Press, New York, pp. 209–231.

Mimouni, F. (1991) Genetic variations in host infestation efficiency in two *Trichogramma* species from Morocco. *Redia* 74, 393–400.

Mollema, C. (1991) Heritability estimates of host selection behavior by the *Drosophila* parasitoid *Asobara tabida. Netherlands Journal of Zoology* 41, 174–183.

Mousseau, T.A. and Roff, D.A. (1987) Natural selection and the heritability of fitness components. *Heredity* 59, 181–197.

Nei, M., Maruyama, T. and Chakraborty, R. (1975) The bottleneck effect and genetic variability in populations. *Evolution* 29, 1–10.

Orzack, S.H. (1990) The comparative biology of second sex ratio evolution within a natural population of a parasitic wasp, *Nasonia vitripennis. Genetics* 124, 385–396.

Orzack, S.H. and Gladstone, J. (1994) Quantitative genetics of sex ratio traits in the parasitic wasp, *Nasonia vitripennis*. *Genetics* 137, 211–220.

Orzack, S.H. and Parker, E.D. Jr (1990) Genetic variation for sex ratio traits within a natural population of a parasitic wasp, *Nasonia vitripennis*. *Genetics* 124, 373–384.

Orzack, S.H., Parker, E.D. Jr and Gladstone, J. (1991) The comparative biology of genetic variation for conditional sex ratio behavior in a parasitic wasp, *Nasonia vitripennis*. *Genetics* 127, 583–599.

Parker, E.D. Jr and Orzack, S.H. (1985) Genetic variation for the sex ratio in *Nasonia vitripennis*. *Genetics* 110, 93–105.

Parsons, P.A. (1980) Isofemale strains and evolutionary strategies in natural populations. In: Hecht, M., Steere, W. and Wallace, B. (eds) *Evolutionary Biology*, vol. 13. Plenum Press, New York, pp. 175–217.

Perez-Maluf, R., Kaiser, L., Wajnberg, E., Carton, Y. and Pham-Delégue, M-H. (1998) Genetic variability of conditioned probing to a fruit odor in *Leptopilina boulardi* (Hymenoptera: Eucoilidae), a *Drosophila* parasitoid. *Behavior Genetics* 28, 67–73.

Pintureau, B., Grenier, S., Paris, A. and Ogier, C. (1995) Genetic variability of some biological and morphological characters in *Lixophaga diatraeae* (Diptera: Tachinidae). *Biological Control* 5, 231–236.

Pompanon, F., Fouillet, P. and Boulétreau, M. (1994) Locomotion behaviour in females of two *Trichogramma* species. Description and genetic variability. *Norwegian Journal of Agricultural Sciences* 16, 185–190.

Pompanon, F., Fouillet, P. and Boulétreau, M. (1999) Physiological and genetic factors as sources of variation in locomotion and activity rhythm in a parasitoid wasp (*Trichogramma brassicae*). *Physiological Entomology* 24, 346–357.

Prévost, G. and Lewis, W.J. (1990) Heritability differences in the response of the braconid waps *Microplitis croceipes* to volatile allelochemicals. *Journal of Insect Behavior* 3, 277–287.

Price, T. and Schluter, D. (1991) On the low heritability of life-history traits. *Evolution* 45, 853–861.

Remington, C.L. (1968) The population genetics of insect introduction. *Annual Review of Entomology* 13, 415–426.

Roff, D.A. and Mousseau, T.A. (1987) Quantitative genetics and fitness: lessons from *Drosophila*. *Heredity* 58, 103–118.

Roitberg, B.D. (1990) Variation in behaviour of individual parasitic insects: Bane or Boon? In: Mackauer, M., Ehler, L.E. and Roland, E. (eds) *Critical Issues in Biological Control*. Intercept Ltd, Andover, UK, pp. 25–39.

Roush, R.T. (1979) Genetic improvement of parasites. In: Hoy, M.A. and MacKelvey, J.J. Jr (eds) *Genetics in Relation to Insect Management*. The Rockfeller Foundation, New York, pp. 97–105.

Roush, R.T. (1990a) Genetic variation in natural enemies: critical issues for colonization in biological control. In: Mackauer, M., Ehler, L.E. and Roland, J. (eds) *Critical Issues in Biological Control*. Intercept Ltd, Andover, UK, pp. 263–288.

Roush, R.T. (1990b) Genetic considerations in the propagation of entomophagous species. In: Baker, R.R. and Dunn, P.E. (eds) *New Directions in Biological Control: Alternatives for Suppressing Agricultural Pests and Diseases*. Liss Inc., New York, pp. 373–387.

Roush, R.T. and McKenzie, J. (1987) Ecological genetics of insecticide and acaricide resistance. *Annual Review of Entomology* 32, 361–380.

Schmidt, J.M. (1991) The inheritance of clutch size regulation in *Trichogramma* species

(Hymenoptera: Chalcidoidea: Trichogrammatidae). In: Bigler, F. (ed.) *Proceedings of the 5th Workshop on Quality Control of Mass-reared Arthropods, Wageningen*, pp. 26–37.

Sequeira, R. and Mackauer, M. (1992) Quantitative genetics of body size and development time in the parasitoid wasp *Aphidius ervi* (Hymenoptera: Aphidiidae). *Canadian Journal of Zoology* 70, 1102–1108.

Shaw, R.G. (1987) Maximum-likelihood approaches applied to quantitative genetics of natural populations. *Evolution* 41, 812–826.

Simmonds, F.J. (1947) Improvement of the sex ratio of a parasite by selection. *The Canadian Entomologist* 79, 41–44.

Simmonds, F.J. (1963) Genetics and biological control. *The Canadian Entomologist* 95, 561–567.

Stephens, D.W. and Krebs, J.R. (1986) *Foraging Theory*. Princeton University Press, Princeton, New Jersey.

Urquijo, P. (1950) Aplicacion de le genetica al aumento de la eficacia des *Trichogramma minutum* en la lucha biologica. *Boletin de Patologia Vegetal y Entomologia Agricola (Madrid)* 18, 1–12.

van Lenteren, J.C. (1980) Evaluation of control capabilities of natural enemies: does art have to become science? *Netherlands Journal of Zoology* 39, 369–381.

Vinson, S.B. (1975) Biochemical coevolution between parasitoids and their hosts. In: Price, P.W. (ed.) *Evolutionary Strategies of Parasitic Insects and Mites*. Plenum Press, New York, pp. 14–48.

Vinson, S.B. (1976) Host selection by insect parasitoids. *Annual Review of Entomology* 21, 109–133.

Waage, J.K. (1983) Aggregation in field parasitoid populations: foraging time allocation by a population of *Diadegma* (Hymenoptera: Ichneumonidae). *Ecological Entomology* 8, 447–453.

Waage, J.K. (1990) Ecological theory and the selection of biological control agents. In: Mackauer, M., Ehler, L.E. and Roland, E. (eds) *Critical Issues in Biological Control*. Intercept Ltd, Andover, UK, pp. 135–157.

Waage, J.K. and Hassell, M.P. (1982) Parasitoids as biological control agents – a fundamental approach. *Parasitology* 84, 241–268.

Wajnberg, E. (1989) Analysis of variations of handling time in *Trichogramma maidis*. *Entomophaga* 34, 397–407.

Wajnberg, E. (1991) Quality control of mass-reared arthropods: a genetical and statistical approach. In: Bigler, F. (ed.) *Proceedings of the 5th Workshop on Quality Control of Mass-reared Arthropods, Wageningen*, pp. 15–25.

Wajnberg, E. (1993) Genetic variation in sex allocation in a parasitic wasp. Variation in sex pattern within sequences of oviposition. *Entomologia Experimentalis et Applicata* 69, 221–229.

Wajnberg, E. (1994) Intra-population genetic variation in *Trichogramma*. In: Wajnberg, E. and Hassan, S.A. (eds) *Biological Control with Egg Parasitoids*. CAB International, Wallingford, UK, pp. 245–271.

Wajnberg, E. and Colazza, S. (1998) Genetic variability in the area searched by a parasitic wasp. Analysis from automatic video tracking of the walking path. *Journal of Insect Physiology* 44, 437–444.

Wajnberg, E., Pizzol, J. and Babault, M. (1989) Genetic variation in progeny allocation in *Trichogramma maidis*. *Entomologia Experimentalis et Applicata* 53, 177–187.

Wajnberg, E., Rosi, M.C. and Colazza, S. (1999) Genetic variation in patch-time allocation in a parasitic wasp. *Journal of Animal Ecology* 68, 121–133.

Weseloh, R.M. (1986) Artificial selection for host suitability and development length of the gypsy moth (Lepidoptera: Lymantriidae) parasite, *Cotesia melanoscela* (Hymenoptera: Braconidae). *Journal of Economic Entomology* 79, 1212–1216.

White, E.B., DeBach, P. and Garber, M.J. (1970) Artificial selection for genetic adaptation to temperature extremes in *Aphytis lingnanensis* Compere (Hymenoptera: Aphelinidae). *Hilgardia* 40, 161–192.

Wilkes, A. (1942) The influence of selection on the preferendum of a Chalcid (*Microplectron fuscipennis* Zett.) and its significance in the biological control of an insect pest. *Proceedings of the Royal Society of London Series B* 130, 400–415.

Wilkes, A. (1947) The effects of selective breeding on the laboratory propagation of insect parasites. *Proceedings of the Royal Society of London, Series B* 134, 227–245.

Wilson, F. (1965) Biological control and the genetics of colonizing species. In: Baker, H.G. and Stebbins, G.L. (eds) *Genetics of Colonizing Species*. Academic Press, New York, pp. 307–329.

3

Molecular Systematics, Chalcidoidea and Biological Control

J. Heraty

Department of Entomology, University of California, Riverside, CA 92521, USA

Introduction

Chalcidoidea include some of the most important groups used for the biological control of pest insect populations (Noyes, 1978; Greathead, 1986; LaSalle and Gauld, 1992). Taxa currently placed in Aphelinidae are primarily parasitoids of aphids, scales and whiteflies (Viggiani, 1984). Encyrtidae are parasitic mostly on non-heteropteran Euhemiptera (Noyes and Hayat, 1994). Eulophidae attack a wide variety of insects, primarily the larval stages of Coleoptera, Lepidoptera and Hymenoptera (Goulet and Huber, 1993). Trichogrammatidae are egg parasitoids of primarily Hemiptera, but importantly Lepidoptera. While Trichogrammatidae are important for augmentative control measures (Smith, 1996), Aphelinidae, Encyrtidae and Eulophidae are used primarily for the classical biological control of pests such as cassava mealybug, olive scale, citrus blackfly and purple scale (DeBach, 1971).

Chalcidoidea are recognized to contain approximately 21,000 described species, distributed in 19 families and 89 subfamilies (Noyes, 1990; Gibson *et al.*, 1999). Estimates of the number of species range between 60,000 and 400,000 (Noyes, 1978, 1990, 2000; Gordh, 1979). Ecologically and economically, Chalcidoidea are one of the most important groups for control of insect populations (Noyes, 1978; LaSalle, 1993). The importance of Chalcidoidea in agricultural systems is unchallenged. They have one of the highest success rates in biological control programmes, in terms of both establishment and control of pest populations. Additionally, because of their high degree of host specificity, Chalcidoidea present the least number of problems from introduced species attacking non-target organisms (Noyes, 1978; Greathead, 1986; LaSalle and Gauld, 1992; LaSalle, 1993; Noyes and Hayat, 1994). Perhaps surprisingly, for

 Genetics, Evolution and Biological Control (eds L.E. Ehler, R. Sforza and T. Mateille)

such an important group of insects, the taxonomy and classification of the superfamily is still unresolved, frequently revised, and largely lacking a consensus in understanding of monophyly at higher taxonomic levels (Gibson *et al.*, 1999). In part, this taxonomic confusion stems from an overwhelming number of undescribed species that remain to be collected, curated and compared with existing material, and often the condition of the existing material (LaSalle, 1993). Many cryptic species are known that can be recognized only by their degree of reproductive isolation. For biological control, providing correct names is an essential part of any successful programme. Taxonomic identification is usually a cursory assessment based solely upon morphological distinctness, which is often later corroborated by information on degree of reproductive isolation or other behavioural data. Differentiation of species has also benefited from various molecular methods of analysis, ranging from allozyme profiles to the more recent use of molecular markers such as random amplified fragment polymorphism (RAPD) analysis and restriction fragment length polymorphism (RFLP) analyses (Landry *et al.*, 1993; Vanlerberghe-Masutti, 1994; Antolin *et al.*, 1996; Silva *et al.*, 1999; Unruh and Woolley, 1999; Zhu and Greenstone, 1999; Zhu *et al.*, 2000). Furthermore, because of their small size, convergence of morphological traits and the frequent reduction or loss of features, it can be difficult to assess the relationships of species, or even the relationships between genera, with any degree of confidence. Also, the general lack of available hypotheses of relationships for most taxa prevent application of the predictive power of phylogenetic systematics to practical and theoretical aspects of biological control.

Over the past decade there has been a tremendous change in the use and application of molecular methods both for the recognition of species and for understanding the relationships of Hymenoptera. Of specific interest to biological control are the recent molecular applications within Ichneumonoidea and Chalcidoidea (Noyes and Hayat, 1994); two of the most frequently used groups in classical or augmentative biological control programmes. With the exception of the subfamily Aphidiinae, most of the work in Ichneumonoidea has been focused on the understanding of higher taxonomic relationships (Belshaw and Quicke, 1997, 2002; Gimeno *et al.*, 1997; Belshaw *et al.*, 1998, 2001; Dowton and Austin, 1998; Whitfield and Cameron, 1998; Mardulyn and Whitfield, 1999; Kambhampati *et al.*, 2000; Sanchis *et al.*, 2000; Dowton *et al.*, 2002). Although Chalcidoidea have been included in studies of relationships at the ordinal level (Derr *et al.*, 1992a,b; Dowton and Austin, 1994, 1995, 1998, 2001) and at the superfamily level (Campbell *et al.*, 2000), within this superfamily, there is a much greater emphasis on problems associated with the recognition of species and resolving the relationships between closely related groups of species and genera. The resolution of the higher-level relationships of these groups is important for understanding major evolutionary events within each group, although it is probably of less direct relevance for biological control programmes, which are focused on populations, species, or at most, closely related groups of species. Unruh and Woolley (1999) provided a comprehensive review of molecular methods and their application to biological control. However, even since 1999, there has been a general shift to the application of DNA sequencing over other marker-based techniques (i.e. allozymes, RAPD, RFLP). This new emphasis on sequencing over marker-based techniques has allowed for a greater focus on understanding relationships over recognition, and for Chalcidoidea, the importance of understanding these relationships within the context of biological control needs to be emphasized.

'Molecular methods' refers broadly to techniques used for the recognition of groups of individuals, whether they be populations, species or higher taxonomic groups, and ultimately the understanding of relationships between these different units (Unruh and Woolley, 1999). The term 'molecular systematics' is generally reserved for sequencing technology and the comparison of nucleotide strings of known genetic regions (Fig. 3.1). For the recognition of species, unique strings diagnostic for a group are important. For postulating relationships, the possession of derived features shared with a common ancestor (synapomorphies) are most important. In either case, a single mutational event resulting in a nucleotide change in the common ancestor of a group, which subsequently becomes fixed within a lineage, can be applicable to either recognition or relationships. Subsequent changes at other sites can help to reinforce our concepts, whereas multiple changes at the same sites (homoplasies) can obfuscate our ideas, especially regarding relationships, as these changes become more common. Alone or in combination with morphology or behavioural information, sequence data can be used to develop better phylogenies, classifications and identification keys, which are fundamental to all biological control programmes. Also, the interpretation of environmental or behavioural change on a given phylogeny can improve our knowledge of the rate and means of acquiring novel host associations or other adaptations or features that might improve our evaluation of new control agents.

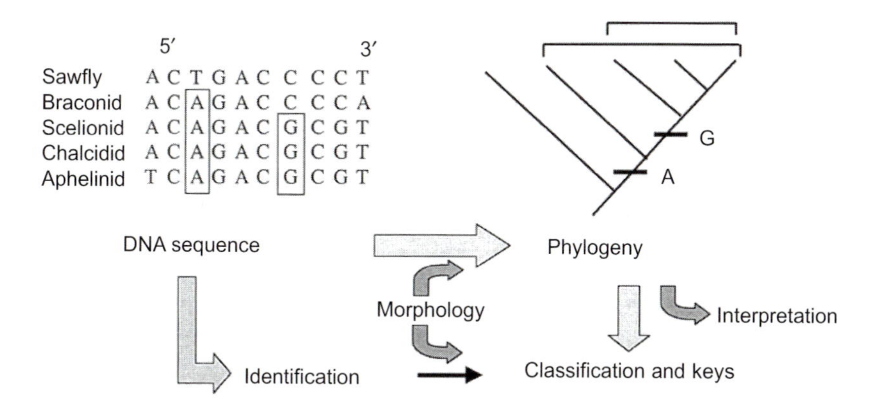

Fig. 3.1. General procedure for comparison of DNA sequences leading to the final development of phylogenies, classifications and identification keys, which hinge on the correct assessment of shared homologous features or synapomorphies (boxes).

Genes of interest

The potential array of genetic regions that have been used in molecular studies of insects was reviewed by Simon *et al.* (1994) and more specifically within the Hymenoptera by Cameron *et al.* (1992). Of the vast array of possibilities, only a few regions have been used for Chalcidoidea or, for that matter, even insects in general (Caterino *et al.*, 2000). The most commonly used regions involve nuclear or mitochondrial DNA sequences that transcribe for ribosomal sequences or mitochondrial coding genes. Each gene region, or in some cases portions of a region, evolves at different rates, depending on the degree of tolerance for mutations before the function of the region is adversely affected. Important considerations for choosing a particular gene are the amount of within-site variation (slow or fast evolving) and changes in length of the region by insertion and deletion events, which ultimately affect our ability to align sequences with each other. The latter case really only affects our ability to analyse the data phylogenetically. Differences in sequence length using specific primers may aid in our ability to discriminate taxa, but only between closely related species where the length differences are well characterized.

Nuclear genes

The 28S rDNA transcript region that codes for ribosomal RNA is most commonly sequenced for studies ranging from species recognition to subfamilial relationships with divergence times ranging from 60 to 200 million years ago (MYA) (Cameron *et al.*, 1992). The 28S region is comprised of a series of highly conserved regions and 11 expansion regions. Each expansion region consists of a series of stem (conserved) and loop (variable) regions, which are useful for assessing different levels of taxonomic divergence. Typically, the D2 expansion region (600–700 bp) has been used in most analyses, with the more conserved and much shorter D1 and D3 expansion regions (each < 350 bp) used in fewer studies, and usually providing supporting evidence for relationships inferred by the D2 region. The 18S rDNA region is highly conserved in insects and generally is used for study of ordinal-level relationships with estimated divergence time of 65–250 MYA (Caterino *et al.*, 2000; Wiegmann *et al.*, 2000). However, the 18S-E23 expansion region can be used for support of family and subfamily group relationships in Eucharitidae and Perilampidae (Heraty unpublished; cf. Figs 3.2, 3.4). Ribosomal genes, both nuclear and mitochondrial, have both a secondary and a tertiary structure that affect the rate of change along the gene, with a series of highly conserved stem regions and more rapidly evolving loop regions. Two internal transcribed spacer regions (ITS1 and ITS2) are both 400–600 bp regions located between 18S and 28S and separated by the 5.8S rDNA region. These are transcribed, but non-coding, and therefore typically tolerate a greater rate of mutation than the surrounding regions, with ITS1 expressing a higher rate of change than ITS2. Both regions are applicable for observing change between closely related species; however, differences in length and rapid changes within the region mean that they could be difficult to align beyond closely related species within a genus. All of the above gene regions usually occur in numerous identical copies within each cell (pleurologous genes) and thus are relatively easy to extract and amplify from single individuals of even the smallest Chalcidoidea. However, even pleurologous genes can have more than one copy in the same individual, which can complicate the search for homologous changes (i.e. ITS2 in *Trichogramma*, R. Stouthamer, California, 2002, personal communication). The problem of dealing with this gene duplication (paralogy) and the evolution of gene families (gene trees) that may or may not correspond to the correct phylogenetic or species tree is well understood (Goodman *et al.*, 1979; Hillis, 1994; Nelson, 1994). The challenge is to find single-copy (orthologous) nuclear-coding genes of similar utility, or at least to be able to recognize the different copies of a paralogous region. Coding genes are much easier to align between different taxa, and orthologous copies better estimate the phylogeny of the taxa being compared (Cameron *et al.*, 1992). Within Hymenoptera, elongation factor 1 (EF-1α, copies F1 and F2; cf. Danforth and Ji, 1998; Danforth, 1999; Rokas *et al.*, 2002), long-wavelength opsin (LW *Rh*; Mardulyn and Cameron, 1999; Rokas *et al.*, 2002) and phosphoenolpyruvate carboxykinase (PEPCK) and DOPA decarboxylase (DDC) (C. Desjardins, Maryland, 2002, personal communication) are potential candidates for exploration.

Mitochondrial genes

Within Chalcidoidea, the 16S rDNA transcript region (~500 bp) has been used only for studies interested in the placement of Chalcidoidea within Hymenoptera (Derr *et al.*, 1992a,b; Dowton and Austin, 1994, 1995, 2001; Dowton *et al.*, 1998). The 12S rDNA transcript region (~350 bp) was used for analyses at the generic level in Agaonidae (Herre *et al.*, 1996; Machado *et al.*, 1996). Three protein-coding regions have been used, cytochrome *b* (Cyt *b*), and cytochrome oxidase I and II (COI, COII) in Chalcidoidea. Cyt *b* sequences (~800 bp) were used to compare species relationships within Agaonidae (Kerdelhue *et al.*, 1999; Lopez-Vaamonde *et al.*, 2001). The COI and COII gene regions (~1400 and 650 bp, respectively) are used for analysis of various levels of diversification within Chalcidoidea, although usually only about half of each region is used in most analyses. COI is considered to be more conserved and has been used in the analysis of ordinal relationships in Hymenoptera (Dowton and Austin, 2001), although it can be variable enough to be useful at the population level (Scheffer and Grissell, 2003).

There are a few fundamental properties that govern the use and application of each gene region. These properties vary at different taxonomic levels, and importantly at different rates in different taxonomic groups. For understanding relationships at most taxonomic levels, single-copy nuclear genes (excluding introns) are probably the best choice. In the Apocrita, nuclear-coding genes have roughly an equal base composition (equal frequency of the four bases) and their alignment, with only rare insertion or deletion events, is trivial and accomplished with most automated alignment packages. For protein-coding mitochondrial genes, Apocrita have a distinct A–T compositional bias, with greater than 70%

of nucleotides either adenine or thymine (Dowton and Austin, 1995, 1997). Mitochondrial ribosomal genes have the same base compositional bias, but for either genome these are more difficult to align. Ribosomal genes have a secondary coding structure that can make them relatively easy to align within the conserved stem or loop regions, but difficult or nearly impossible to align without bias in the variable loop regions, which may have both numerous substitutions and multiple insertion and deletion events. These regions can be hypervariable, with unique long insertions that make them difficult to align, especially between divergent taxa, and these regions are either subjectively aligned by eye or the ambiguous alignment regions are excluded from the analysis (Cameron *et al.*, 1992; Unruh and Woolley, 1999).

Genetic Divergence

Do different genes or gene regions change at the same rate? No, each region maintains mutations at a rate that is dependent on the effect of changes to the function of the region. Non-coding regions (ITS, rDNA loop regions, introns) change at the highest rate. Stem regions of rDNA tolerate the fewest changes. Unruh and Woolley (1999) proposed three questions to be asked for selecting a molecular marker: (i) is the rate at which mutations are fixed appropriate for the question being asked; (ii) can homologous features be compared; and (iii) are the analytical methods appropriate for the rates of evolution? The first two questions can relate directly to the alignment of sequences between individuals. The third question relates to methods used to determine relationships based on these aligned sequences and how to accommodate properties of a particular region such as base composition and the potential for multiple substitutions at a single site (saturation). The three questions are essential for addressing phylogenetic relationships, but perhaps less important when using sequence data for purposes of identification, which involve minor differences between closely related species. While most of the following discussion is focused on the analysis and evaluation of relationships, many of the properties of the data are also relevant for molecule-based diagnostic taxonomy. For thorough discussions of the applications and appropriateness of various methods of analysis, see Swofford *et al.* (1996). For discussions pertaining more specifically to Hymenoptera, see Cameron *et al.* (1992) and Unruh and Woolley (1999).

The 'zone of optimal divergence' was coined by Brower and DeSalle (1994) for the ideal region in which the sequences have diverged enough to provide phylogenetic signal but not enough for homoplasy (noise) to be a problem. The authors refer to two fundamental problems with increasingly divergent sequences: variability of sequence length and multiple substitutions. Length changes affect our ability to correctly align homologous sites. Protein-coding sequences are usually invariant in length and can be aligned without gaps over very divergent taxa. Regions that are not translated (ITS, introns), or regions with minimal impact on transcribed sequences (loop regions of rDNA), are often highly variable and tolerate numerous insertions or deletions of nucleotides. Insertions or deletions of nucleotides can make sequences extremely difficult to align, to the point where they are often excluded from phylogenetic analyses.

Only four bases are possible for DNA sequences, therefore multiple substitutions at individual sites can be a problem. Assuming that random substitutions at novel sites are common after the initial divergence of taxa, then all changes can be considered as homologous. However, enough substitutions must be accumulated and fixed within the taxon of interest to be detected in the regions of interest, and to provide phylogenetic signal. If not enough changes are fixed between lineages, then random non-homologous changes may easily confuse any phylogenetic signal. In this 'zone of optimal divergence', the number of expected and observed base changes is approximately equal (Fig. 3.2). With increasing divergence time and increasing numbers of substitutions, nucleotides may change more than once, resulting in a decrease in the number of observed versus expected changes (Fig. 3.2, points fall below line of equality). The zone of optimal divergence on Fig. 3.2 refers to a balance of changes in which there is sufficient divergence to establish relationships among closely related taxa, and yet not too much of a saturation of changes that cause relationships of more divergent taxa to be obscured. As the observed and expected changes begin to deviate more dramatically, then the number of unobserved changes increases, finally to the point where the sequence cannot be used to analyse a particular set of relationships.

Owing to functional constraints, not all gene regions accept mutations at the same rate (Fig. 3.2). Within the same taxonomic comparisons, ribosomal genes may change very quickly (28S-D2), at an intermediate rate (28S-D3), or very slowly (18S). Because of codon bias, protein-coding genes diverge at different rates for the three codon positions, with the third base the most free to change and the second position most conserved (cf. *Encarsia* COI, Fig. 3.2). In this latter example, the divergence in the first base positions of COI between species of *Encarsia* is intermediate to that of 28S-D2 and D3. Although divergence rates appear to be stable within taxonomic groups (cf. D2 and D3 for the eucharitid/perilampid and *Encarsia* comparisons, Fig. 3.2), they may differ between groups. The divergence of 28S-D2 of Eulophidae is comparable in divergence to the more conserved 28S-D3 region of Eucharitidae + Perilampidae, and divergence between species of *Encarsia* for 28S-D2 and D3 is comparable to, or faster than, the family-level divergence in other Chalcidoidea (Table 3.1, Fig. 3.2).

The diversification of Chalcidoidea is coincident with the late Cretaceous explosion of angiosperms, with representation by Mymaridae, Trichogrammatidae and Tetracampidae in Canadian amber (Yoshimoto, 1975) and what appears to be a species of Torymidae from late Cretaceous (Turonian) compression fossils in Botswana (D. Brothers, South Africa, 2002, personal communication). Given a similar potential age of origin, amazing differences are found in the amount of sequence divergence between taxonomic groups in Chalcidoidea. Divergence can range from 14.8% for Signiphoridae to 36.2% for Pteromalidae (Table 3.1). It should be noted, however, that comparisons of diver-

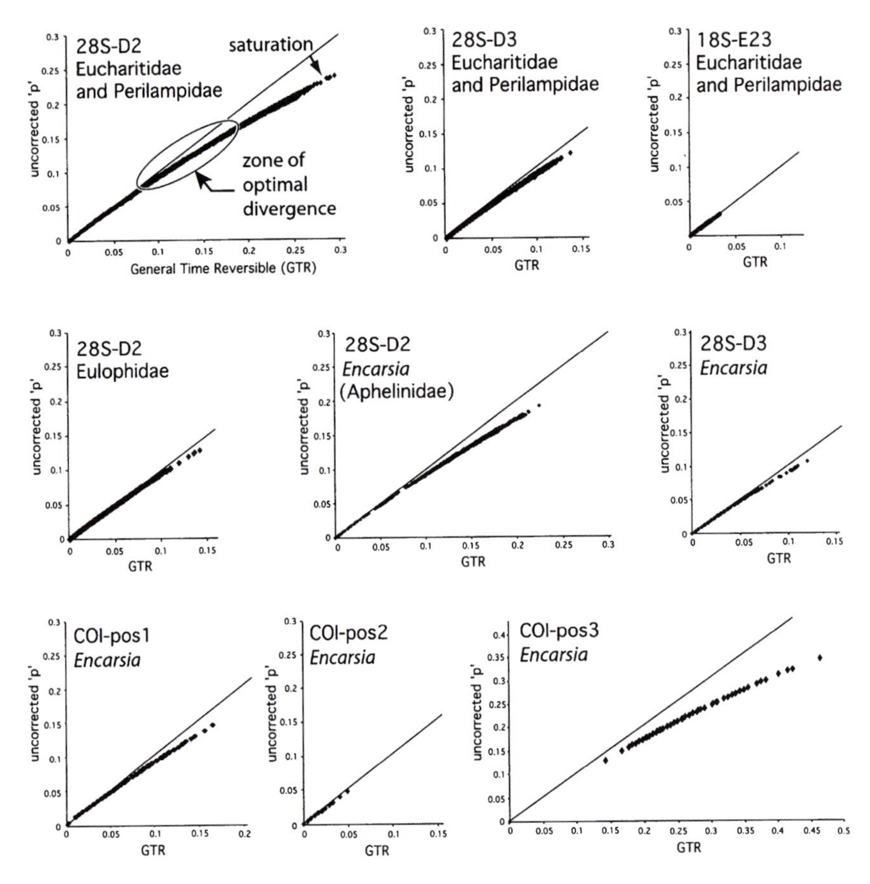

Fig. 3.2. DNA saturation curves for divergence of nucleotides between species for various gene regions and different taxonomic groups within Chalcidoidea. Pairwise uncorrected and general time reversible models of character state change were estimated in PAUP 4.0*b9. Changes within COI are separated into the three codon positions. In the region of optimal divergence, observed and estimated changes are roughly equal and phylogenetic relationships are estimated relatively easily using various phylogenetic methods. As regions become saturated, procedures that estimate multiple substitutions per site may better resolve relationships between taxa.

gence in Aphelinidae (28.8%) and Pteromalidae are a bit unfair, since these are probably not monophyletic families, thus indicating a problem of taxonomy and not divergence. However, Chalcididae and Encyrtidae are demonstrably monophyletic groups that have divergence rates of 25.9–28.6%. Trichogrammatidae have a paltry 17.6% divergence between 33 genera. There is also little correspondence of rates at the generic level. Divergence rates in *Encarsia* (29.0%) and *Aphytis* (23.1%) are equivalent to family-level divergence in other groups, and vastly exceed the low rate of divergence in Eucharitini (9.6%), which, although

Table 3.1. Maximum percentage sequence divergence for the 28S-D2 region within various taxonomic groups of Chalcidoidea. Numbers in parentheses after subfamily or family names are the number of genera compared; those after generic names are the number of species compared.

Taxon	Divergence
Aphelinidae (11)[a]	28.8*
Coccophaginae (4)[b]	29.0
Encarsia (30)[b]	29.0
Aphelininae (6)[a]	25.8
Aphytis (30)[a]	23.1
Aphelinus (7)[a]	3.0
Azotinae (1)[c]	8.0
Chalcididae (7)[d]	25.9
Encyrtidae (12)[c]	28.6
Eucharitidae (31)[b]	21.5
Oraseminae (3)[b]	5.8
Orasema (21)[b]	5.0
Psilocharitini (2)[b]	12.0
Eucharitini (25)[b]	9.6
Eupelmidae (13)[a]	18.4
Eulophidae (51)[e]	17.4
Eulophidae (51)[e]	17.4
Eurytomidae (3)[d]	16.1
Mymaridae (2)[d]	18.2
Perilampidae (7)[b]	14.2
Chrysolampinae (3)[b]	5.7
Perilampinae (4)[b]	14.0
Pteromalidae (11)[d]	36.2
Cleonyminae (3)[d]	7.9
Pteromalinae (3)[b]	3.3
Signiphoridae (3)[c]	14.8
Trichogrammatidae (33)[g]	17.6
Paracentrobia (6)[g]	4.9
Trichogramma (34)[g]	8.1
Chalcidoidea (73)[d]	38.5

a = J.-W. Kim, UCR, unpublished; b = J. Heraty and D. Hawks, UCR, unpublished; c = J. Munro, UCR, unpublished; d = Babcock *et al.*, 2001; e = Gauthier *et al.*, 2000; g = J. George and A. Owen, UCR, unpublished.
* Not all species of *Encarsia* cited below are included in this estimate.

based on only 25 genera, is a representative sample of the entire tribe, with 47 morphologically distinct genera! Clearly, there is no single gene region that can be used for universal comparison. Each group and gene will need to be evaluated for its utility to answer particular questions.

Rates of divergence have a direct effect on the methods employed to analyse the data (Brower and DeSalle, 1994; Swofford *et al.*, 1996; Unruh and Woolley, 1999). Within the zone of optimal divergence, parsimony and likelihood approaches, with the latter compensating for estimated rates of change, should provide similar, if not the same, results. This has been the case for almost all studies within Chalcidoidea. The results from parsimony, distance (neighbour-joining) and likelihood methods of analysis are virtually the same (Rasplus *et al.*, 1998; Kerdelhue *et al.*, 1999; Babcock *et al.*, 2001; Lopez-Vaamonde *et al.*, 2001). In only one case was maximum likelihood the only method employed (Machado *et al.*, 2001), for an analysis of 20 genera of fig wasps using COI, but these results were not compared with a parsimony analysis of the same data. When saturation is a problem, analysing the data in combination with more conserved sequences to stabilize the deeper nodes is an option. For example, analyses of relationships in eucharitids and perilampids, which are reaching high levels of saturation (> 25%) for interfamilial divergences (Fig. 3.2), are relatively stable when analysed in combination with 28S-D3 and 18S-E23 (Fig. 3.4). As yet, none of the genes provide enough information on their own to indicate a single comprehensive set of relationships.

The stability of phylogenetic trees, and hence the confidence of predictions using those hypothesized relationships, is based on: (i) resolution of the resulting trees; (ii) statistical support; and (iii) comparison with previous taxonomic hypotheses. Resolution is determined by the number of trees generated and similarity of branch points across all of these trees, as determined by a consensus of all the resulting trees (Swofford, 1991). In some cases, there may be very different sets of solutions or 'islands' of tree topologies, with each island possibly representing a different evolutionary scenario for a trait of interest (Maddison, 1991). The goal of any study in phylogenetics is a single island of one or more trees with the highest possible resolution. Statistical support is measured in a variety of ways, but the most common are bootstrap analyses, decay indices and successive approximations character weighting (cf. Carpenter, 1988; Swofford *et al.*, 1996). Bootstrap analyses are based on iterative resampling and reanalysis of the data, with the percentage value indicating the proportion of iterations in which a particular node is supported. Decay or Bremer indices refer to the number of extra steps required for a branch not to be present in the resulting consensus of trees. Both are directly dependent on the number of synapomorphic changes of a character state (nucleotide change) on a given branch. Successive approximations involves reweighting the character data based on the fit of each character to a particular set of most-parsimonious tree topologies, followed by successive reanalyses of the trees and reweighted data until the final tree or trees stabilize to the same weighted length. Tree branches supported in both weighted and unweighted analyses are considered to be well supported. If one of the most-parsimonious trees is retrieved after reweighting, then this is considered to be the most favoured hypothesis of relationships. The last measure of stability, comparison with previous taxonomic hypotheses, is often not considered when evaluating trees; although it is probably the most important of considerations. In its simplest form, do species of the same genus, or higher-level taxon, cluster together in the final set of relationships? If not, why not? Admittedly, the taxonomy could be wrong, but then it should be backed with overwhelming support to the contrary, or a least a re-evaluation of the taxa in question.

In the following discussion, I will be focusing on various techniques of analysis, and the evaluation and interpretation of sequence data that have been applied within Chalcidoidea. There has been a significant change in the types of analyses being conducted since the recent review by Unruh and Woolley (1999). These following discussions are meant to build on their foundation, which focused largely on techniques other than sequencing. New examples have come to light within the parasitic wasps in the past few years.

Identification

To specialists in biological control, the recognition of units of interest (demes, races, populations, species) is initially the item of most interest. Systematics/taxonomy has always been regarded as important for providing names of both native and introduced species. An initial assessment, based solely upon morphological distinctness, is often later corroborated by information on degree of reproductive isolation or some other behavioural data, often in collaboration with biological control specialists working with these populations. The differentiation of these units has benefited from various molecular methods of analysis, ranging from allozyme profiles to the more recent use of molecular markers such as random amplified fragment polymorphism (RAPD), restriction fragment length polymorphism (RFLP) and amplified fragment length polymorphism (AFLP) analyses (Vos *et al.*, 1995; Unruh and Woolley, 1999). However, these methods are useful only for the differentiation of known populations or species, because of numerous analytical problems. They have little utility for achieving the second objective of systematics: the understanding of relationships and phylogenetic history between different groups. Decreasing costs and ease of use have enabled molecular sequencing to be more readily applied to the diagnosis of species and to understanding their relationships.

Aphelinidae and Trichogrammatidae are minute, often less than 2 mm in size, with some species of less than 0.5 mm overall body length. Morphological features in these tiny wasps (or flying bacteria as coined by Bruce Campbell) are often very reduced, and among closely related species the features used to recognize reproductively isolated species can sometimes be very minor and difficult to measure, or even absent (Pinto *et al.*, 2002a,b). With the need for accurate identification and a lack of available expertise, for both precision mounting of specimens and assignment to the correct species, alternative methods using molecular techniques have been developed (cf. Unruh and Woolley, 1999).

Fragments of DNA, once isolated, identified and sequenced, can be applied to the differentiation of species in three ways. First, is the difference in length of amplified gene fragments. Length differences are typical for either of the ITS1 or ITS2 fragments, which can differ even among closely related species (Sappal *et al.*, 1995; Stouthamer *et al.*, 1998). However, fragment length cannot be used as a measure of homology between different taxa, as fragments of similar length can be derived through different evolutionary pathways; thus taxa with vastly different sequences can have a similar length. Secondly, sequences compared from different species may exhibit fixed differences for either base changes (mutations) or insertion/deletion events (indels) that serve to differentiate groups (Fig. 3.3). It is essential for these differences to be assessed for different individuals within a population, or, if the comparison is to be made at the species level, amongst

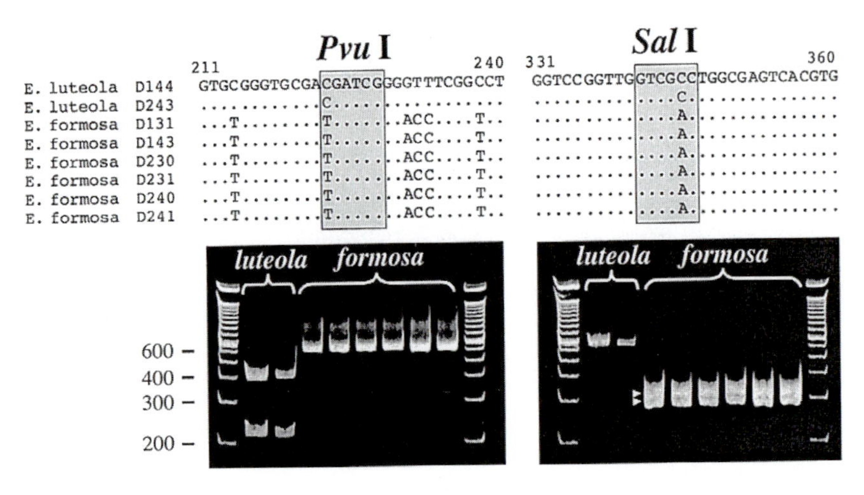

Fig. 3.3. Restriction enzyme recognition sites and digests for recognizing two closely related species of *Encarsia, E. formosa* and *E. luteola,* using the 635 base pair region of 28S-D2 rDNA (Babcock and Heraty, 2001). *Pvu*I cuts the fragment at the six-base recognition site (CGATAG), resulting in two disproportionate fragments, whereas *Sal*I cuts at another site (GTCGAC), resulting in two equal size fragments. (.) indicates sequence identity with *E. luteola*.

different individuals within geographically isolated populations. Thirdly, the identification of sequence differences that are fixed in the groups (populations or species) of interest, such that subsequent identification can be made using specific restriction enzymes that digest a fragment after polymerase chain reaction (PCR) amplification. For example, Fig. 3.3 shows partial sequences of 28S-D2 for eight populations (strains) of *Encarsia luteola* Howard and *Encarsia formosa* Gahan. The sequences for *E. formosa* are 592 bases long, whereas those of *E. luteola* are 590, with a sequence divergence of 2.8–3.3% between species (Babcock and Heraty, 2000). Both *E. luteola* and *E. formosa* had a within-species sequence diversity of 0.2% (three bases each between strains); however, each is fixed for the changes illustrated. Of eight populations of these two species that were examined, two restriction enzymes were found to cut six-base recognition sites unique to each of the species: *E. luteola* (*Pvu*I) and *E. formosa* (*Sal*I). When used in combination, the sites discriminate the two species accurately (Fig. 3.3). While sequencing is more accurate for species recognition, the use of restriction enzymes on targeted PCR products can be a rapid and cost-effective means for assessing species under certain conditions, especially if all of the expected species are known. Either approach is considered useful for the accurate identification of these two species of *Encarsia*, which differ by only minor morphological characters, which are difficult to observe even on slide-mounted specimens (Babcock and Heraty, 2000). Studies using restriction enzymes to separate species focus on Chalcidoidea that are traditionally very difficult to identify, such as *Encarsia* (Babcock and Heraty, 2000; Schmidt *et al.*, 2001), *Aphelinus* (Zhu and Greenstone,

1999; Zhu *et al.*, 2000; Prinsloo *et al.*, 2002; K. Hopper, Delaware, 2002, personal communication) and various species of *Trichogramma* (Stouthamer *et al.*, 1998).

De Barro *et al.* (2000) analysed the sequences of different populations of three species of *Eretmocerus* for differences in 28S-D2 and D3, COII, ITS1 and ITS2. Of these five genes, 28S-D3 was fixed at the species level, 28S-D2 was fixed for species with a few changes that differed between but not within populations, and both COII and the ITS regions demonstrated nucleotide variation between individuals within a population. At least for ITS1, an analysis of relationships grouped the various populations into the three distinct groups (100% bootstrap) that were recognized as morphologically distinct species. ITS1 had low levels of polymorphism within gene copies from a single specimen, which occurred at about the same levels as between individuals at different localities, but even with this demonstrated paralogy, the species grouped appropriately. Similar variation occurs in species of *Trichogramma*, which can result in multiple ITS2 digest bands for a single individual; however, all individuals possess at least some copies with the species-specific restriction enzyme cut sites (R. Stouthamer, California, 2002, personal communication). For purposes of identification, these examples emphasize the need for sampling of multiple individuals and populations to assure that enzyme recognition sites will be consistently diagnostic.

The application of sequencing as an identification tool for the separation of populations and species of *Trichogramma* using ITS1 was first applied by Orrego and Agudelo-Silva (1993). The authors identified two strains within one Californian culture of *Trichogramma pretiosum* that differed by three base substitutions and eight insertion/deletion events. These results were based on the comparison of only two individuals, making it difficult to determine whether the variation was real or a sequencing artefact. However, they did find a 1.1–4.1% sequence divergence among four Californian populations (=individuals) as compared with 27% with *Trichogramma dendrolimi* Matsumura from China. RFLP analyses were used on amplified ITS1, ITS2, 28S, and 18S to differentiate *Trichogramma minutum* Riley, *Trichogramma brassicae* Bezdenko and *Trichogramma* near *sibiricum* Sorokina (Sappal *et al.*, 1995). The ITS sequences showed substantial differences in both cut sites and length, whereas the 28S fragments were identical. The 18S fragment had a single cut site for *Bam*HI, which served to differentiate only one of the three species. Subsequent studies of *Trichogramma* were focused entirely on sequencing the ITS2 region and developing restriction enzyme assays (van Kan *et al.*, 1996, 1997; Pinto *et al.*, 1997, 2002a; Stouthamer *et al.*, 1998; Silva *et al.*, 1999; Ciociola *et al.*, 2001a,b). These studies have culminated in the development of dichotomous molecular keys that differentiate species on the basis of fragment length and cuts by specific restriction enzymes (Ciociola *et al.*, 2001a; Pinto *et al.*, 2002a). Beyond the identification of adults, length differences and restriction digests of ITS2 have been applied for the recognition of parasitism in the eggs of *Helicoverpa* by *Trichogramma australicum* (Amornsak *et al.*, 1998), and species of *Aphelinus* within their aphid host (Zhu *et al.*, 2000).

In *Trichogramma*, not all species can be differentiated using a single universal gene region. Two species pairs, *T. minutum*/*T. platneri* and *T. sibericum*/*T. alpha*, could not be differentiated using ITS2 alone (Pinto *et al.*, 2002; Stouthamer *et al.*, 2000a,b). Some base differences and indels were found in disjunct populations of *T. minutum* and *T. platneri*, but these were not fixed for either species. *T. minutum* and *T. platneri* are reproductively isolated, sympatric in the north-western USA, possess distinct non-overlapping sets of alleles at the phosphoglucomutase (PGM) enzyme locus, and possess very minor morphological differences (Nagarkati, 1975; Pinto *et al.*, 1991, 2002a,b; Pinto, 1999; Stouthamer *et al.*, 2000b; Burks and Heraty, 2002). A subsequent study found two fixed differences in the COI region that would discriminate both species unequivocally (R. Stouthamer, California, 2002, personal communication).

Not all species that express fixed behaviour differences have been shown to have detectable genetic differences. The differential host choice and isolation of populations of *Encarsia formosa* attacking *Bemisia* on poinsettia suggest that they should be genetically distinct from other populations, and yet no fixed differences were found for ITS or in a broader survey of AFLPs in several populations (Y. Gai and R. Stouthamer, unpublished). A similar case occurs in *Encarsia sophia*, which exhibit no fixed genetic differences for 28S-D2 between widely separated geographic localities, but populations from Spain and Pakistan exhibit mating incompatibilities and slight morphometric differences (Heraty and Polaszek, 2000; Babcock *et al.*, 2001; Hernández-Suárez *et al.*, 2003). Whether the correct genetic region needs to discovered or whether behavioural differences can accrue at a faster rate than molecular differences remains to be tested. Furthermore, no rules can be applied to correlate the amount of genetic divergence associated with speciation. Some reproductively isolated and partially sympatric species of *Trichogramma* can be recognized discretely by only a few bases of COI, and yet populations of *Megastigmus* differing by as much as 4.0% were not interpreted as different species (Fig. 6 of Scheffer and Grissell, 2003). For two sister species of *Encarsia*, *E. luteola* and *E. formosa*, sequence divergence ranged from 3.0 to 6.1% for the more conserved 28S-D2 gene (Babcock and Heraty, 2000). As with all taxonomic information, species boundaries must be determined by a summation of evidence from all sources of data, including geographic, morphological, behavioural and genetic.

Phylogenetics and their Applications

Relationships of Chalcidoidea

Chalcidoidea are recognized to contain somewhere in the order of 21,000 described species, distributed in 19 families and 89 subfamilies (Gibson *et al.*, 1999; Gauthier *et al.*, 2000). Whereas subfamily groups are relatively easy to define, the monophyly of many of the higher taxonomic groups, including larger family groups such as Aphelinidae, Pteromalidae and Eupelmidae, have not been determined (Gibson *et al.*, 1999; Campbell *et al.*, 2000). Above the family level, whether or not to include Mymarommatidae or Mymaridae in Chalcidoidea has been debated (Kozlov and Rasnitsyn, 1979; Gibson, 1986; Rasnitsyn, 1988), with Mymarommatidae currently excluded as a separate superfamily (Goulet and Huber, 1993). A sister-group relationship between Platygastroidea (Scelionidae + Platygastroidea) and Mymarommatoidea + Chalcidoidea has been proposed, based on morphology (Ronquist *et al.*, 1999), although the relationship with Platygastroidea was disputed by Gibson (1999).

Molecular studies of the relationships among families of Hymenoptera using either the 16S region or combined 16S+28S+COI regions all propose support for a sister-group relationship between Platygastroidea and Chalcidoidea (Dowton and Austin, 1994, 1998, 2001; Dowton *et al.*, 1998). Although Dowton and Austin (2001) favoured the Platygastroidea + Chalcidoidea sister-group relationship, other rearrangements were included in a variety of their analyses, including Chalcidoidea + Cynipoidea and Chalcidoidea + Proctotrupoidea. These latter hypotheses were based on what the authors considered to be unstable data, which included the hypervariable third base position of COI. When Mymaridae are included in molecular analyses, they are placed with other Chalcidoidea in a monophyletic group in all analytical results (Campbell *et al.*, 2000; Dowton and Austin, 2001), although they are placed unequivocally as a sister group to the remaining Chalcidoidea (excluding Mymaridae) only when numerous chalcidoid groups are included (Campbell *et al.*, 2000). If the Platygastroidea, which include two egg-parasitic families, are the sister group of Chalcidoidea, and in turn Mymaridae, another group of egg parasitoids, are the sister group of the remaining Chalcidoidea, then it seems most likely that egg parasitism is an ancestral behaviour for this lineage, and subsequent associations with larval or pupal parasitism must be derived (Dowton and Austin, 2001). However, both Mymarommatidae and Rotoitidae (basal member of Chalcidoidea) have unknown behaviour and are not yet included in any of the molecular analyses.

Relationships within Chalcidoidea

The relationships at the family level within Chalcidoidea have been addressed in only one study, which was based on 103 species from 39 of the 89 subfamilies (Campbell *et al.*, 2000). Whereas the relationships of some family-level taxa were supported (Encyrtidae, Eucharitidae, Eulophidae, Trichogrammatidae), many others were not, and it is far too early, in terms of both taxonomic and gene sampling, to draw any major conclusions from these data. Additional molecular studies using sequence data have focused on the within-family relationships of Eulophidae (Gauthier *et al.*, 2000), Aphelinidae (Babcock and Heraty, 2000; Babcock *et al.*, 2001; Manzari *et al.*, 2002), Agaonidae (Herre *et al.*, 1996; Machado *et al.*, 1996; Rasplus *et al.*, 1998; Kerdelhue *et al.*, 1999; Lopez-Vaamonde *et al.*, 2001), Pteromalidae (Campbell *et al.*, 1993), Torymidae (Scheffer and Grissell, 2003), and Trichogrammatidae (Stouthamer *et al.*, 1998). The remaining discussions are meant to focus attention on phylogenies of interest to biological control, even though some of the examples are not taxa commonly used in biological control programmes.

Eucharitidae: Competing Morphological and Molecular Trees

Eucharitidae are parasitoids of ants; adults deposit their eggs in or on vegetation and the active first-instar larvae are responsible for gaining access to the ant host (Heraty, 1994, 2002). Eucharitids are known to attack five subfamilies of ants, Myrmicinae, Formicinae, Ponerinae and Myrmeciinae, with one, probably erroneous, record from Ecitoninae (Heraty, 2000). Because of a large number of informative morphological characters, the proposed phylogenies of Eucharitidae are fairly well resolved and concordant in different studies with different characters and taxa (Heraty, 1994, 2000, 2002). The most recent analysis of morphological data encompasses all of the genera (Heraty, 2002), and supports a monophyletic Oraseminae that is a sister group to the Eucharitinae, which is comprised of the monophyletic Psilocharitini and Eucharitini (Fig. 3.4). Importantly, the genera *Gollumiella* and *Anorasema* are placed basally within the tribe Eucharitini, and another genus, *Tricoryna*, is included with *Pseudometagea* in a group basal to the major radiation of genera within the tribe.

The results from a molecular analysis of the 28S-D2, 28S-D3 and 18S-E23 regions (Fig. 3.4) are almost identical to the morphological analyses (Heraty and Hawks, unpublished), but with three important differences found in both independent and combined analyses of the three gene regions: (i) *Gollumiella* and *Anorasema* are highly supported (bootstrap values of 100%) as the sister group of Oraseminae and Eucharitinae; (ii) Psilocharitini are a paraphyletic group (bootstrap support of 80%); and (iii) *Tricoryna*, which are parasitoids of *Rhytidoponera* (large Ponerinae in Australia), is consistently placed as the sister group of *Austeucharis*, which are parasitoids of *Myrmecia* (Myrmeciinae; in Australia) and nested within a group that are all parasitic on large ponerine ants of the tribes Ectatommini or Ponerini, or Myrmeciinae (morphologically and behaviourally similar to large Ponerinae) (Heraty, 2002). A re-evaluation of morphological data led to the discovery of a unique ovipositor bulb in *Tricoryna* and *Austeucharis*, which places the two as sister taxa, but inclusion of this character in the morphological analyses did not change the position of this group. Morphological features were found that supported the monophyly of *Gollumiella* and *Anorasema*, but not their exclusion from Eucharitinae. There is no morphological support for paraphyly of Psilocharitini.

Why are these differences relevant? *Gollumiella* and the eucharitine genus, *Pseudometagea*, which is placed unequivocally within Eucharitinae in both analyses (Fig. 3.4), are both parasitoids of members of the tribe Lasiini within the ant subfamily Formicinae. This places Lasiini as the potential ancestral host for Eucharitidae. Under the morphological hypothesis, the ancestral host could be Myrmicinae, Ponerinae (host for *Neolosbanus* and other derived Eucharitini) or Formicinae. Furthermore, internal parasitism of the host ant larva by the

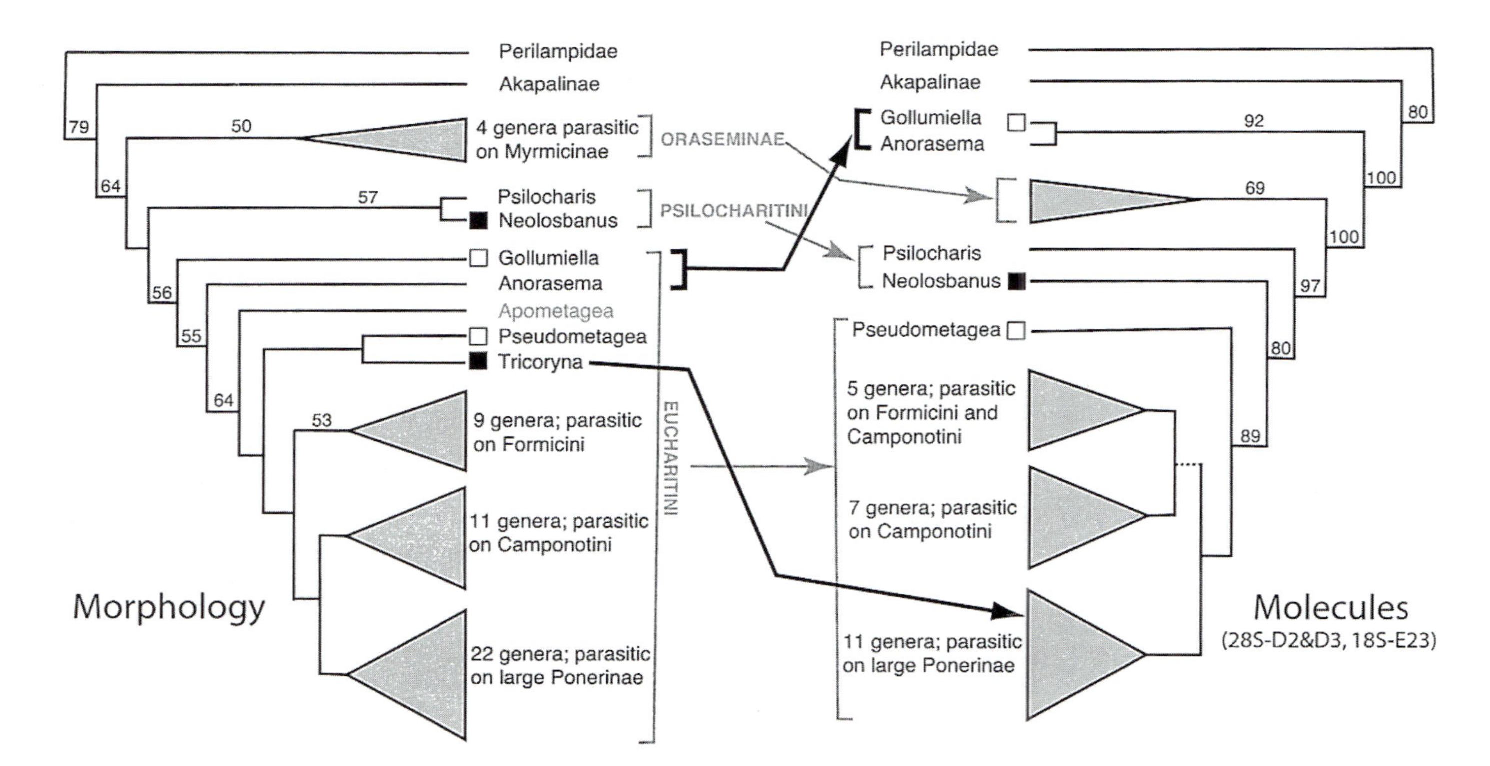

Fig. 3.4. Comparative phylogenies of Eucharitidae based on morphology (Heraty, 2002) and a parsimony analysis of the 28S-D2, -D3 and 18S-E23 gene regions (Heraty and Hawks, unpublished). Bootstrap values are indicated above branches. Ant hosts that are Formicinae (open boxes) or Ponerinae (black boxes) are indicated.

first-instar eucharitid larva occurs in *Gollumiella*, Oraseminae and *Pseudometagea*, but not *Neolosbanus* or other Eucharitini in which all larval stages are external parasites. The molecular hypothesis supports internal parasitism as an ancestral behaviour, whereas in the morphological hypothesis it is equivocal and either internal or external parasitism is possible.

With regard to the Psilocharitini, if the group is monophyletic, as suggested by the morphological hypothesis, then parasitism of small Ponerinae and external parasitism of the host ant larvae, as found in *Neolosbanus*, would be predicted for *Psilocharis*. However, no such assumptions can be made under the molecular hypothesis for *Psilocharis*. Finally, the new placement of *Tricoryna* in the group which is parasitic on large Ponerinae makes greater biological sense and leads to a single host shift in the common ancestor of this group, as compared with two independent acquisitions of a large ponerine host. When strongly supported by multiple genes, different analytical procedures, adequate taxonomic sampling, and after comparison with traditional morphology-based phylogenies, molecular hypotheses can be used to examine evolutionary change in a new perspective. It cannot, however, be done without first adopting a pessimistic view in which the molecular data are assumed to be wrong.

Encarsia: Unchallenged Trees and the Interpretation of Change

Species of *Encarsia* (Aphelinidae) are a diverse group of minute wasps, less than 2 mm in size. Immature stages usually develop as endoparasitoids of whiteflies and armoured scales, and perhaps less commonly in immatures of Hormaphididae, themselves, or eggs of Lepidoptera (Williams and Polaszek, 1996; Hunter and Woolley, 2001). *Encarsia* is one of the most important parasitic groups being exploited in biological control, and various species are currently being collected as part of foreign exploration efforts to search for biological control agents (Noyes and Hayat, 1994). Many species have attributes that allow them to be placed into discrete groups. These species groups are often defined by combinations of characters, many of which are characteristic of one or more species placed in other species groups. Importantly, these loosely defined groups are our first approximation of the phylogenetic relationships of species. However, species grouped arbitrarily on the basis of overall similarity can lead to misconceptions about behaviour and host associations. If we can prove that their defining features and behavioural traits have an evolutionary basis, then we can enhance our ability to place, and rapidly evaluate, the potential effectiveness of new species for use in biological control programmes.

Within *Encarsia*, 25 species groups are currently recognized, with 60 of the 273 described species unplaced (Heraty and Woolley, 2002). Because of their small size and general reduction or loss of morphological features, these groups are difficult to characterize on morphological data alone (Babcock *et al.*, 2001).

What appear to be obvious group characteristics can be found in unrelated groups of species; for example, the close placement of scutellar sensilla, which were considered diagnostic of the strenua group, are now known to be convergent and found in several very unrelated species groups (Heraty and Polaszek, 2000). Even the obvious characteristic of a reduction in number of tarsomeres from five to four was regarded as a poor character for defining species groups by Hayat (1998). Various analyses of morphological characters have led to differing opinions regarding the relationships, composition and placement of species into groups of *Encarsia* (Hayat, 1998; Huang and Polaszek, 1998). Trying to analyse these morphological traits within a phylogenetic framework yields little resolution of relationships (Babcock *et al.*, 2001).

The relationships of species within *Encarsia* were analysed in two papers using 28S-D2 rDNA (Babcock *et al.*, 2001; Manzari *et al.*, 2002). The species in the two data sets were re-analysed along with new sequence data for a total of 31 species of *Encarsia* and two outgroup genera (Fig. 3.5; Heraty *et al.*, 2003). Parsimony analysis resulted in only three competing tree topologies, with support for the inaron, luteola and strenua species groups, but not the parvella species group. The relationships between groups is not highly supported, and one of the outgroup genera, *Encarsiella*, occurs within *Encarsia*. The inclusion of this genus within *Encarsia* is strongly supported, with only an extra 3 steps required to force *Encarsia* to be monophyletic. These results were supported in the earlier studies (Babcock *et al.*, 2001; Manzari *et al.*, 2002). This odd generic placement could be an artefact of the 28S-D2 gene region (Fig. 3.5); however, studies with other genes (ITS2, COI) also support these conclusions. Quite possibly, the minor morphological differences between groups of *Encarsia* are representative of deeper phylogenetic differences that have not been recognized by taxonomists. Clearly, with only 31 of 273 described species represented in the molecular analyses, it is too early to make any formal conclusions from the data.

Each of the three studies reaches the same general conclusions, and if we have faith in the strongly supported species groups, then we can use these results to examine certain morphological and behavioural features for their ability to define monophyletic groups. The close placement of scutellar sensillae is present in all members of the strenua group and its putative sister group, *Encarsia quercicola* (Se; Fig. 3.5). A reduction to a four-segmented tarsus is found in all members of the luteola group, and separately in *Encarsia nigricephala* of the cubensis group (4; Fig. 3.5). Based on whitefly parasitism, which is characteristic of the outgroup taxa, a shift to parasitism of Diaspididae was derived, possibly as many as three times (D; Fig. 3.5). There does not appear to be any phylogenetic component to the parasitism of various whitefly genera (*Bemisia* [B], *Trialeurodes* [T], other whitefly genera [Ot]; Fig. 3.5). The association of thelytoky (Th) with sex-ratio-distorting bacteria (*Encarsia* bacterium [EB] and *Wolbachia* [W]; Fig. 3.5) also does not have a phylogenetic component, although *Wolbachia* is known only in *E. formosa* (Zchori-Fein *et al.*, 2001). In these cases, the lack of phylogenetic constraint affects how we approach our understanding of the results. For example, if the EB bacteria are associated with different lineages of *Encarsia*, are the EB

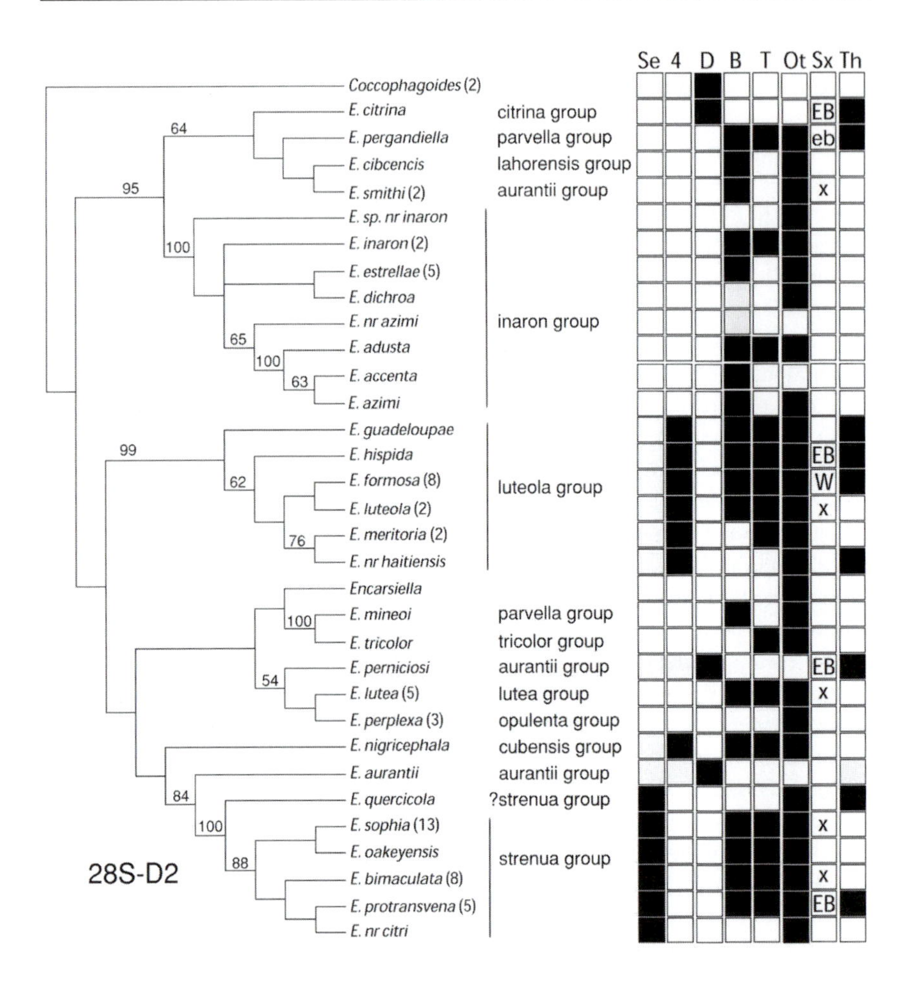

Fig. 3.5. Strict consensus of three trees of 844 steps recovered from a parsimony analysis of 28S-D2 rDNA from 31 species of *Encarsia* and two closely related genera (*Coccophagoides* and *Encarsiella*) (Heraty *et al.*, 2003). Data were analysed with PAUP 4.0*b9 using 100 random addition sequences and TBR branch swapping. Bootstrap proportions greater than 50% are shown above branches. The same results were obtained when additional populations, as identified by the numbers in parentheses, were added for a total of 80 terminal taxa (sequences from Babcock *et al.* (2001) and Manzari *et al.* (2002); some names corrected from Schmidt *et al.* (2001)), but with five trees of 893 steps. These were identical to the three trees after pruning out the extra populations. Both analyses were stable to successive approximations character weighting (Babcock *et al.*, 2001). Behavioural attributes indicated: Se = scutellar sensillae closely placed; 4 = mid-tarsus four-segmented; D = Diaspididae host; B = *Bemisia* host; T = *Trialeurodes vaporariorum* host; Ot = other whitefly host; S = symbiotic association (EB = *Encarsia* bacterium (eb is variable within species); W = *Wolbachia*; x = tested but none found (Zchori-Fein *et al.*, 2001)); Th = parthenogenetic species.

lineages concordant, indicating coevolution, or random, indicating horizontal transfer?

Phylogeography and Invasive Agents

Phylogeography is the study of the geographic variation within and between closely related species, and it is an effective research area to study population origins and the effect of colonization events (Avise, 2000). For biological control, this is important not only for understanding the nature of the introduced pest, but also their imported predators and parasitoids. For example, native species that have a broad geographic range can be expected to have high levels of genetic diversity, whereas introduced species should have decreased diversity due to a genetic bottleneck. This was well illustrated in a study of a torymid wasp, *Megastigmus transvaalensis* (Hussey), that feeds on the seeds of *Schinus* and *Rhus* (Anacardiaceae) in Africa, South America and North America (Scheffer and Grissell, 2003). *Schinus* is considered to have been introduced into Africa and North America, whereas *Rhus* is native to Africa and introduced into North America. *M. transvaalensis* exhibited a large degree of individual variation in populations from across continental Africa on both *Schinus* (S) and *Rhus* (R), with a total of 29 haplotypes discovered for 49 individual wasps (Fig. 3.6). Individuals from North and South America shared a single haplotype with an individual from the island of Réunion (which is just off of the coast of Kenya), and in their molecular analysis, these are also the most closely related populations. Pepper seed is a major export product from Réunion, and this is a likely explanation as to how the *Megastigmus* were distributed worldwide. Not only were the authors able to identify the source of the introduced population, but they were also able to identify a potential untapped source of genetic variation in the populations across Africa. These populations warrant consideration for programmes on the biological control of *Schinus terebinthifolius*, which is an invasive species found in Florida and Hawaii (Scheffer and Grissell, 2003).

It is interesting to note that *M. transvaalensis* sampled from different African countries had moderate or strong bootstrap support, with as many as 15 shared nucleotide differences separating individuals from Morocco. In other Chalcidoidea, this high level of fixed variation would be the equivalent of at least species-level, or even generic-level, differentiation. No morphological differences are apparent between these populations, and the authors are certain this is only a single genetically diverse species (E. Grissell, Washington, 2002, personal communication).

Cospeciation

The fig-pollinating wasps, Agaonidae, appear to be a poster child for the study of cospeciation in the Chalcidoidea using molecular data. Figs are dependent on

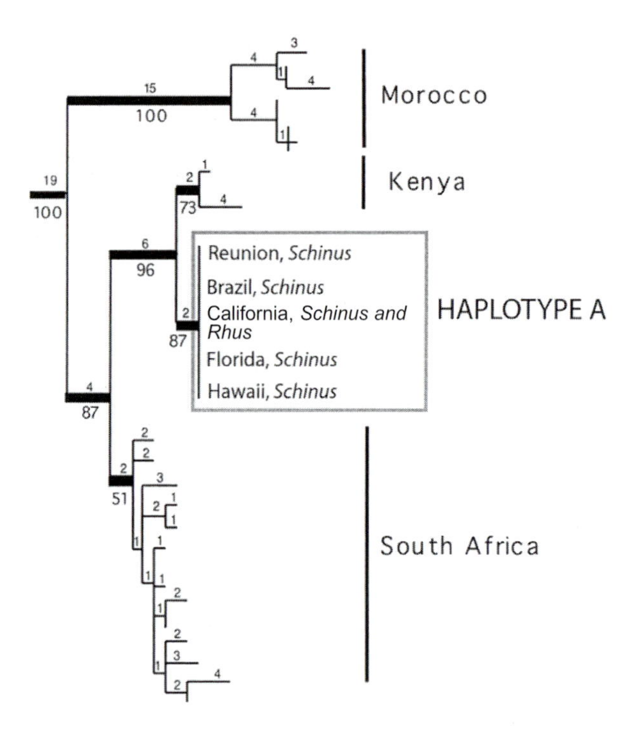

Fig. 3.6. Phylogram of *Megastigmus transvaalensis* from parsimony analysis of COI (Scheffer and Grissell, 2003). Thicker branches recovered in all 90 equally parsimonious trees. Branches are proportional to the number of nucleotide changes (indicated above branch), and bootstrap support is shown below branches. Of the species attacked, *Schinus* (S) is native to South America (Brazil) and introduced in all other areas, whereas *Rhus* (R) is native to Africa.

species of Agaoninae for pollination, and the wasps in turn require the seeds of the *Ficus* fruit for development. Studies at the generic level, using 12S and 28S rDNA (Herre *et al.*, 1996, Machado *et al.*, 1996; Rasplus *et al.*, 1998), are less conclusive than studies of more closely related species. Nevertheless, they do support the idea that the non-pollinating subfamilies traditionally placed in Agaonidae are not part of the same lineage, and probably became secondarily adapted to figs. This allows for comparison of not only the agaonid–fig pollinator association, but also the cospeciation of the associated inquilines and parasitoids.

Two studies initially focused on different aspects of obligate mutualism for Agaonidae involving the fig–pollinator and the inquiline–pollinator association. Herre *et al.* (1996) used mitochondrial cytochrome oxidase II (COII) to study the association between genera of Agaoninae and their hosts in the genus *Ficus* (*Moraceae*). The corresponding phylogeny of *Ficus* species was based on the chloroplast *rbcL* gene and tRNA spacer sequences. The generic-level divergences within Agaoninae were reasonably well correlated with the phylogeny for the species of *Ficus*. However, a parsimony-based analysis of the species in two genera of Agaoninae, *Tetrapus* and *Pegoscapus*, were identically correlated with a phylogeny of the species of *Ficus* in two subgenera, *Urostigmus* and *Pharmacosycea.* Their results provided strong evidence for a one-to-one coevolution between the pollinating wasps and their host figs. Machado *et al.* (1996) compared the phylogenies of non-pollinating species of *Idarnes* (Sycophaginae) and *Critogaster* (Sycoryctinae). *Idarnes* are either inquiline competitors or gall-formers in fruit pollinated by *Pegoscapus*. *Critogaster* are direct competitors, along with some *Idarnes*, in fruit that is pollinated by species of *Tetrapus*. The non-pollinating species oviposit through the fruit wall and compete directly with the pollinator for host resources, whereas the gallers develop in galls developed in the flowers or fruit walls and do not use the syconium for development. Again there was almost a one-to-one correspondence of the pollinating and non-pollinating species, with limited host switching and radiation on to closely related species of pollinators by the non-pollinated species. Nearly identical results were found in a study of species pairs of pollinators (*Pleistodontes*, Agaonidae: Agaoninae) and their parasitoids or possibly phytophagous cleptoparasites (*Sycoscapter*, Pteromalidae: Sycoryctinae) (Lopez-Vaamonde *et al.*, 2001) using 28S-D2 and -D3, cytochrome *b* (Cyt *b*) and ITS2. This is a clear test of Farenholz's rule, in which host and parasite phylogenies develop as mirrors of each other through vicariant cospeciation. Of the three gene regions analysed for *Pleistodontes*, only 28S + ITS2 provided clear and well-supported results with a single most-parsimonious tree. There was substantial incongruence with Cyt *b*, and these results were abandoned for between-group comparisons. In contrast, the combined 28S and Cyt *b* data for *Sycoscapter* were concordant and highly resolved. The authors found a high degree of correspondence and hence cospeciation indicated between the two groups. There was significant cospeciation (50–60% of nodes), but also a relatively high degree of host switching, with plant–host associations less likely to be constrained for the parasitoids. The cospeciation levels for the study discussed above for pollinating and non-pollinating fig wasps by Machado *et al.* (1996) was 62.5%. Although host switching by *Sycoscapter* is very likely, there is no explanation for the absence of any *Pleistodontes* with more than one parasitoid species, which might be expected under this scenario.

Ceratosolen (Agaonidae) are normally pollinators of several species of *Ficus*. Many of the species are associated with a single host fig species. However, a few pollinate the same species of *Ficus*, and one of these species, *C. arabicus*, does not pollinate its plant host and occurs as an inquiline with *C. galili*. The relationships of 13 species of *Ceratosolen* from the Old World tropical regions were studied using Cyt *b* (Kerdelhue *et al.*, 1999). Three groups were supported in the analysis and potentially two distinct origins of sympatry were found (Fig. 3.7). *Ficus sur* is pollinated by three species of *Ceratosolen*. Two of these pollinating species found on *F. sur* were well supported as sister taxa in all analyses. These two species are broadly sympatric in western Africa, although *C. silvestrianus* is specialized for open habitats, whereas *C. flabellatus* is found more in dense, closed-canopy forests. The close relationships were interpreted as a speciation event occurring in allopatry on the same host. Sympatry occurs only in mosaic forest–savannah habitats.

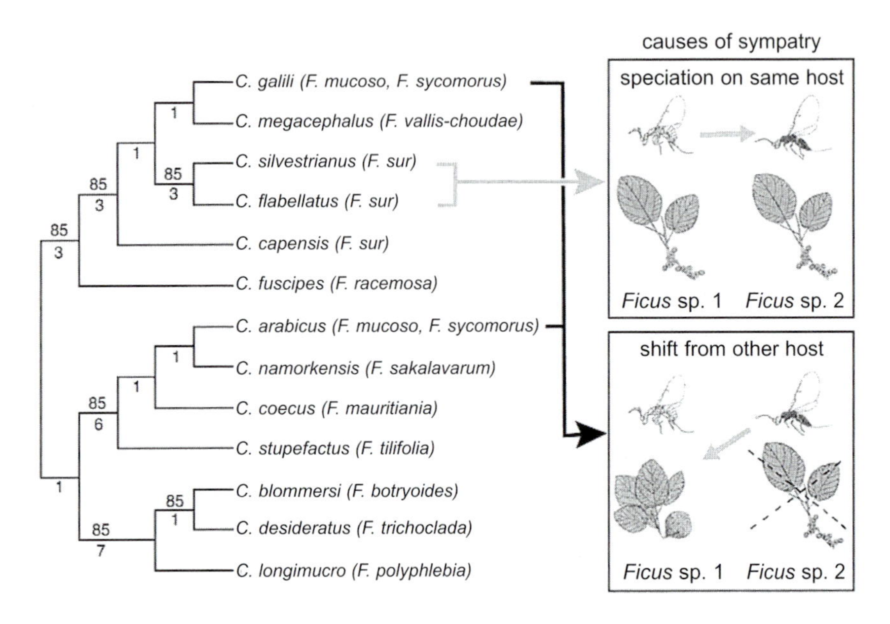

Fig. 3.7. Phylogenetic tree for *Ceratosolen* (Agaonidae) from an analysis of Cyt *b* sequences, with their associated host plants in the genus *Ficus* (*Moraceae*) (Kerdelhue *et al.*, 1999). Bootstrap values are shown above the branches and Bremer support values below. Two models of speciation that differ from a one-to-one host correspondence are indicated.

The relationships of these two species with *C. capensis* are less certain and vary with different analytical methods; however, the pattern of speciation can be explained by an allopatric vicariance across the African Rift (Kerdelhue *et al.*, 1999). The non-pollinating *C. galili* and pollinating *C. arabicus* occur together on *F. mucoso* and *F. sycomorus* in Africa and Madagascar. The two species belong to different lineages and must have colonized these hosts independently (Fig. 3.7; Kerdelhue *et al.*, 1999; Machado *et al.*, 2001). *C. galili* is the only species of its clade found in both Madagascar and Africa, with the other species being either African or Australasian. *C. arabicus* is also found in both Madagascar and Africa, but with the other species in the same clade known from Madagascar or Réunion Island. The authors propose a horizontal transfer for African *C. galili* to an introduced host, *F. sycomorus,* from Madagascar, with the eventual extinction of its original host species in Africa and the reverse colonization of Madagascar. Both species of *Ceratosolen* break the rule of host specificity and occur on two species, with *C. arabicus* probably the sole pollinator of both *F. mucoso* and *F. sycomorus*. This is regarded as a phylogenetic mystery that remains to be resolved!

Machado *et al.* (2001) undertook a broader study of the relationships of the fig-pollinating Agaonidae, with a study of 15 genera, using mitochondrial COI. Their results supported the monophyly of groups pollinating the *Ficus* subgenera *Urostigma, Pharmacosycea* and *Sycidium*, but not *Ficus* or *Sycomorus*. *Ceratosolen* was the only paraphyletic group of species, but, judging by the long branches represented on their phylogram, taxonomic sampling for this genus could be increased and ultimately improve the resolution and potential monophyly of this genus. There is almost no overlap of species sequenced by Kerdelhue *et al.* (1999). Regardless, Machado *et al.* (2001) removed these and other taxa that failed the relative rate test (probably due to a poor representation of species), and for the remaining 10 genera of Agaoninae proposed a molecular clock calibrated from *Drosophila*, which dates the oldest divergence for fig mutualism at 87.5 MYA and the most recent divergence of species of *Pegoscapus* at 21 MYA. The earliest know fossil of *Ficus* is from the early Eocene, about 50 MYA (Collinson, 1989). Although a bit young, the geographic disjunction of generic groups across the southern hemisphere suggests a Gondwanan distribution pattern during the late Cretaceous, which would predate these fossils by at least 30 MYA (Machado *et al.*, 2001). This most recent work points to a coordinated approach to understanding the evolution of host associations, biogeographic evaluations and dating of events that may eventually shape our understanding of many of the associations within Chalcidoidea and other insects.

Conclusion

Over the past few years, molecular sequencing has become cheaper, more accurate, more accessible and easier to do. Most laboratories are now doing only the extraction and PCR and sending the templates to centralized facilities for analysis, for a cost of only US$3–10 per sample, or US$6–20 per gene per species, if correctly sequenced in both directions. Verification of sequences and alignment is not trivial, but overall this is far easier than it was just a few years ago. Traditional taxonomy is not being replaced. Species are still primarily evaluated based on morphological characters. The tremendous variation in sequence divergence for species of different genera or family groups is well illustrated in the previous examples. For example, the divergence of COI in populations of *Megastigmus transvaalensis* is roughly equivalent to the divergence between generic groups of Trichogrammatidae and subfamily groups of Eucharitidae. Decisions of taxonomic status will still need to be based on an evaluation of overall evidence from morphology, behaviour and distribution, but now some of these data can be better evaluated with the added genetic component available from sequence information.

Importantly, as more taxa are sequenced for more genes, it will become easier to check new sequences as they are obtained, align them with existing data sets and retest previous hypotheses. The addition of new sequences to the *Encarsia* data set was relatively trivial and required the insertion of only one new site to the alignment to accommodate a single base insertion found in all five populations of *Encarsia estrellae*. The earlier molecular phylogenetic hypotheses were supported with this additional data, and the resulting trees much more resolved. Questions regarding the high level of genetic diversity within the genus will gradually be addressed with the addition of new taxa, more genes, and a better resolution of the placement of Coccophaginae within Chalcidoidea.

Molecular data also provide an independent test for observing our hypotheses, or lack of hypotheses, based on morphological evidence. Molecular and morphological studies have not supported a monophyletic Aphelinidae (Campbell *et al.*, 2000; Babcock *et al.*, 2001). Biologically, we are finding some extreme differences between Aphelininae and Coccophaginae, such as heteronomy, thelytoky, and a lack of an association with *Wolbachia*, all of which suggest that the subfamilies of Aphelinidae may indeed be very divergent lineages. Also, none of the molecular analyses for *Encarsia* support the monophyly of the group with *Encarsiella* included as an outgroup. The monophyly of *Encarsia* has been questioned (DeBach and Rose, 1981). Perhaps, the taxonomic status of the species groups within *Encarsia* will need to be re-examined in light of the disproportionate sequence divergence between species groups.

Some of our taxonomic concepts require revision. The placement of *Gollumiella* in Eucharitidae is very different in morphological and molecular analyses, and changes drastically how we interpret behavioural and host changes in the family. The molecular data is very strongly supported by three different gene regions, all of which have high levels of bootstrap and Bremer support for this hypothesis. A re-evaluation of some of the morphological features is now lending credence to the molecular hypothesis.

Studies of genetic levels of divergence in native and introduced populations of both pests and parasites will have significant applications to the study of biological control agents. This can only help to build on our understanding of the success or failure of certain agents, which often relies upon the 'do-or-die' strategy. Several parasite–host relationship rules are typically presented as founding theories for biological control (Farenholtz's rule of concordant phylogenies and Szidat's rule of the association between primitive hosts and parasites), but these have rarely been tested within a phylogenetic framework, or at least verified with an independent data set. Tests of cospeciation between *Wolbachia* and other sex-ratio-distorting organisms, which can often only be detected with sequencing technology, with their hosts will probably become a dynamic area of research.

With increased taxonomic sampling, a better understanding of the properties of the genes being applied to these problems, and a greater sampling of new genes in the future, molecular systematics will make further important contributions to biological control. Several years ago, Patterson *et al.* (1993; p. 179) made the statement 'As morphologists with high hopes of molecular systematics, we end this survey with our hopes dampened. Congruence between molecular phylogenies is as elusive as it is in morphology and as it is between molecules and morphology'. Many of the results presented herein are supported by morphological data, and sometimes a re-evaluation of morphology has led to better concordance with the molecular data. The times they are a changing!

Acknowledgements

This work was supported in part by National Science Foundation Research Grant DEB 0108245. Parts of Fig. 3.7 are reprinted from Kerdelhue *et al.* (1999) with permission from Elsevier. I would also like to thank James Munro and Albert Owen (UCR) and Phil Ward and Les Ehler (University of California, Davis) for comments on an earlier draft of the manuscript.

References

Amornsak, W., Gordh, G. and Graham, G. (1998) Detecting parasitised eggs with polymerase chain reaction and DNA sequence of *Trichogramma australicum* Girault (Hymenoptera: Trichogrammatidae). *Australian Journal of Entomology* 37, 174–179.

Antolin, M.F., Guertin, D.S. and Petersen, J.J. (1996) The origin of gregarious *Muscidifurax* (Hymenoptera: Pteromalidae) in North America: an analysis using molecular markers. *Biological Control* 6, 76–82.

Avise, J.C. (2000) *Phylogeography, the History and Formation of Species.* Harvard University Press, Cambridge, Massachusetts.

Babcock, C.S. and Heraty, J.M. (2000) Molecular markers distinguishing *Encarsia formosa* and *Encarsia luteola* (Hymenoptera: Aphelinidae). *Annals of the Entomological Society of America* 93, 738–744.

Babcock, C.S., Heraty, J.M., De Barro, P.J., Driver, F. and Schmidt, S. (2001) Preliminary phylogeny of *Encarsia* Forster (Hymenoptera: Aphelinidae) based on morphology and 28S rDNA. *Molecular Phylogenetics and Evolution* 18, 306–323.

Belshaw, R. and Quicke, D.L.J. (1997) A molecular phylogeny of the Aphidiinae (Hymenoptera: Braconidae). *Molecular Phylogenetics and Evolution* 7, 281–293.

Belshaw, R. and Quicke, D.L.J. (2002) Robustness of ancestral state estimates: evolution of life history strategy in ichneumonoid parasitoids. *Systematic Biology* 51, 450–477.

Belshaw, R., Fitton, M., Herniou, E., Gimeno, C. and Quicke, D.L.J. (1998) A phylogenetic reconstruction of the Ichneumonoidea (Hymenoptera) based on the D2 variable region of 28S ribosomal RNA. *Systematic Entomology* 23, 109–123.

Belshaw, R., Lopez-Vaamonde, C., Degerli, N. and Quicke, D.L.J. (2001) Paraphyletic taxa and taxonomic chaining: evaluating the classification of braconine wasps (Hymenoptera: Braconidae) using 28S D2–3 rDNA sequences and morphological characters. *Biological Journal of the Linnaean Society* 73, 411–424.

Brower, A.V.Z. and DeSalle, R. (1994) Practical and theoretical considerations for choice of a DNA sequence region in insect molecular systematics, with a short review of published studies using nuclear gene regions. *Annals of the Entomological Society of America* 87, 702–716.

Burks, R.A. and Heraty, J.M. (2002) Morphometric analysis of four species of *Trichogramma* Westwood (Hymenoptera: Trichogrammatidae) attacking codling moth and other tortricid pests in North America. *Journal of Hymenoptera Research* 11, 167–187.

Cameron, S.A., Derr, J.N., Austin, A.D., Woolley, J.B. and Wharton, R.A. (1992) The application of nucleotide sequence data to phylogeny of the Hymenoptera: a review. *Journal of Hymenoptera Research* 1, 63–79.

Campbell, B.C., Steffen-Campbell, J.D. and Werren, J.H. (1993) Phylogeny of the *Nasonia* species complex (Hymenoptera: Pteromalidae) inferred from an internal transcribed spacer (ITS2) and 28S rDNA sequences. *Insect Molecular Biology* 2, 225–237.

Campbell, B.C., Heraty, J.M., Rasplus, J.Y., Chan, K., Steffen-Campbell, J.D. and Babcock, C.S. (2000) Molecular systematics of the Chalcidoidea using 28S-D2 rDNA. In: Austin, A.D. and Dowton, M. (eds) *Hymenoptera, Evolution, Biodiversity and Biological Control.* CSIRO, Collingwood, Australia, pp. 59–73.

Carpenter, J.C. (1988) Choosing among equally parsimonious cladograms. *Cladistics* 4, 291–296.

Caterino, M.S., Cho, S. and Sperling, F.A.H. (2000) The current state of insect molecular systematics: a thriving tower of Babel. *Annual Review of Entomology* 45, 1–54.

Ciociola, A.I., Querino, R.B., Zucchi, R.A. and Stouthamer, R. (2001a) Molecular tool for identification of closely related species of *Trichogramma* (Hymenoptera: Trichogrammatidae): *T. rojasi* Nagaraja and Nagarkatti and *T. lasallei* Pinto. *Neotropical Entomology* 30, 575–578.

Ciociola, A.I. Jr, Zucchi, R.A. and Stouthamer, R. (2001b) Molecular key to seven Brazilian species of *Trichogramma* (Hymenoptera: Trichogrammatidae) using sequences of the ITS2 region and restriction analysis. *Neotropical Entomology* 30, 259–262.

Collinson, M.E. (1989) The fossil record of the Moraceae. In: Crane, P.R. and Blackmore, S. (eds) *Evolution, Systematics and Fossil History of the Hamamelidae.* Clarendon Press, Oxford, vol. 2, pp. 319–339.

Danforth, B.N. (1999) Phylogeny of the bee genus *Lasioglossum* (Hymenoptera: Halictidae) based on mitochondrial COI sequence data. *Systematic Entomology* 24, 377–393.

Danforth, B.N. and Ji, S. (1998) Elongation factor-1alpha occurs as two copies in bees: implications for phylogenetic analysis of EF-1alpha sequences in insects. *Molecular Biology and Evolution* 15, 225–235.

DeBach, P. (1971) *Biological Control by Natural Enemies.* Cambridge University Press, London.

DeBach, P. and Rose, M. (1981) A new genus and species of Aphelinidae with some synonymies, a rediagnosis of *Aspidiotiphagus* and a key to pentamerous and heteromerous Prospaltellinae (Hymenoptera: Chalcidoidea: Aphelinidae). *Proceedings of the Entomological Society of Washington* 83, 658–679.

De Barro, P.J., Driver, F., Naumann, I.D., Schmidt, S., Clarke, G.M. and Curran, J. (2000) Descriptions of three species of *Eretmocerus* Haldeman (Hymenoptera: Aphelinidae) parasitising *Bemisia tabaci* (Gennadius) (Hemiptera: Aleyrodidae) and *Trialeurodes vaporariorum* (Westwood) (Hemiptera: Aleyrodidae) in Australia based on morphological and molecular data. *Australian Journal of Entomology* 39, 259–269.

Derr, J.N., Davis, S.K., Woolley, J.B. and Wharton, R.A. (1992a) Reassessment of the 16S rRNA nucleotide sequence from members of the parasitic Hymenoptera. *Molecular Phylogenetics and Evolution* 1, 338–341.

Derr, J.N., Davis, S.K., Woolley, J.B. and Wharton, R.A. (1992b) Variation and the phylogenetic utility of the large ribosomal subunit of mitochondrial DNA from the insect order Hymenoptera. *Molecular Phylogenetics and Evolution* 1, 136–147.

Dowton, M. and Austin, A.D. (1994) Molecular phylogeny of the insect order Hymenoptera: Apocritan relationships. *Proceedings of the National Academy of Sciences USA* 91, 9911–9915.

Dowton, M. and Austin, A.D. (1995) Increased genetic diversity in mitochondrial genes is correlated with the evolution of parasitism in the Hymenoptera. *Journal of Molecular Evolution* 41, 958–965.

Dowton, M. and Austin, A.D. (1997) Evidence for AT-transversion bias in wasp

(Hymenoptera: Symphyta) mitochondrial genes and its implications for the origin of parasitism. *Journal of Molecular Evolution* 44, 398–405.

Dowton, M. and Austin, A.D. (1998) Phylogenetic relationships among the microgastroid wasps (Hymenoptera: Braconidae): combined analysis of 16S and 28S rDNA genes and morphological data. *Molecular Phylogenetics and Evolution* 10, 354–366.

Dowton, M. and Austin, A.D. (2001) Simultaneous analysis of 16S, 28S, COI and morphology in the Hymenoptera: Apocrita: evolutionary transitions among parasitic wasps. *Biological Journal of the Linnaean Society* 74, 87–111.

Dowton, M., Austin, A.D. and Antolin, M.F. (1998) Evolutionary relationships among the Braconidae (Hymenoptera: Ichneumonoidea) inferred from partial 16S rDNA gene sequences. *Insect Molecular Biology* 7, 129–150.

Dowton, M., Belshaw, R., Austin, A.D. and Quicke, D.L.J. (2002) Simultaneous molecular and morphological analysis of braconid relationships (Insecta: Hymenoptera: Braconidae) indicates independent mt-tRNA gene inversions within a single wasp family. *Journal of Molecular Evolution* 54, 210–226.

Gauthier, N., LaSalle, J., Quicke, D.L.J. and Godfray, H.C.J. (2000) Phylogeny of Eulophidae (Hymenoptera: Chalcidoidea), with a reclassification of Eulophinae and the recognition that Elasmidae are derived eulophids. *Systematic Entomology* 25, 521–539.

Gibson, G.A. (1986) Evidence for monophyly and relationships of Chalcidoidea, Mymaridae, and Mymarommatidae (Hymenoptera: Terebrantes). *Canadian Entomologist* 118, 205–240.

Gibson, G.A. (1999) Sister group relationships of the Platygastroidea and Chalcidoidea (Hymenoptera) – an alternative hypothesis to Rasnitsyn (1988). *Zoologica Scripta* 28, 125–138.

Gibson, G.A., Heraty, J.M. and Woolley, J.B. (1999) Phylogenetics and classification of Chalcidoidea and Mymarommatoidea: a review of current concepts (Hymenoptera, Apocrita). *Zoologica Scripta* 28, 87–124.

Gimeno, C., Belshaw, R. and Quicke, D.L.J. (1997) Phylogenetic relationships of the Alysiinae/Opiinae (Hymenoptera: Braconidae) and the utility of cytochrome b, 16S and 28S D2 rRNA. *Insect Molecular Biology* 6, 273–284.

Goodman, M., Czelusniak, J., Moore, G.W., Romero-Herrera, A.E. and Matsuda, G. (1979) Fitting the gene lineage into the species lineage, a parsimony strategy illustrated by cladograms constructed from globin sequences. *Systematic Zoology* 28, 132–163.

Gordh, G. (1979) Superfamily Chalcidoidea. In: Krombein, K.V., Hurd, P., Smith, D.R. and Burks, B.D. (eds) *Catalog of Hymenoptera in America North of Mexico*. Smithsonian Institution Press, Washington, DC, vol. 1, pp. 743–748.

Goulet, H. and Huber, J. (eds) (1993) *Hymenoptera of the World: an Identification Guide to Families*. Agriculture Canada Research Branch Publication, Ottawa.

Greathead, D.J. (1986) Parasitoids in classical biological control. In: Waage, J.K. and Greathead, D. J. (eds) *Insect Parasitoids*. Academic Press, London, pp. 287–318.

Hayat, M. (1998) Aphelinidae of India (Hymenoptera: Chalcidoidea): a taxonomic revision. *Memoirs of Entomology International* 13, 1–416.

Heraty, J.M. (1994) Classification and evolution of the Oraseminae in the Old World, including revisions of two closely related genera of Eucharitinae (Hymenoptera: Eucharitidae). *Royal Ontario Museum Life Sciences Contributions* 174, 1–176.

Heraty, J.M. (2000) Phylogenetic relationships of Oraseminae (Hymenoptera: Eucharitidae). *Annals of the Entomological Society of America* 93, 374–390.

Heraty, J.M. (2002) A revision of the Eucharitidae (Hymenoptera: Chalcidoidea) of the World. *Memoirs of the American Entomological Institute* 68, 1–359.

Heraty, J.M. and Polaszek, A. (2000) Morphometric analysis and descriptions of selected species in the *Encarsia strenua* group (Hymenoptera: Aphelinidae). *Journal of Hymenoptera Research* 9, 142–169.

Heraty, J.M. and Woolley, J.B. (2002) A catalogue of the world species of *Encarsia* Förster. http://cache.ucr.edu/~heraty/Aphelinidae.html.

Heraty, J.M., Polaszek, A. and Schauff, M. (2003) Systematics and biology of *Encarsia*. In: Gould, J. and Hoelmer, K. (eds) *Whiteflies and their Parasites*. Academic Press, New York.

Hernández-Suárez, E., Carnero, A., Aguiar, A., Prinsloo, G., LaSalle, J. and Polaszek, A. (2003) Whitefly parasitoids (Hemiptera: Aleyrodidae; Hymenoptera: Aphelinidae, Eulophidae, Platygastridae) from the Macaronesian archipelagos of the Canary Islands, Madiera and the Azores. *Systematics and Biodiversity* 1, 55–108.

Herre, E.A., Machado, C.A., Bermingham, E., Nason, J.D., Windsor, D.M., McCafferty, S.S., Van Houten, W. and Bachmann, K. (1996) Molecular phylogenies of figs and their pollinator wasps. *Journal of Biogeography* 23, 521–530.

Hillis, D.M. (1994) Homology in molecular biology. In: Hall, B.K. (ed.) *Homology: the Hierarchical Basis of Comparative Biology*. Academic Press, New York, pp. 102–151.

Huang, J. and Polaszek, A. (1998) A revision of the Chinese species of *Encarsia* Forster (Hymenoptera: Aphelinidae): parasitoids of whiteflies, scale insects and aphids (Hemiptera: Aleyrodidae, Diaspididae, Aphidoidea). *Journal of Natural History* 32, 1825–1966.

Hunter, M.S. and Woolley, J.B. (2001) Evolution and behavioral ecology of heteronomous aphelinid parasitoids. *Annual Review of Entomology* 46, 251–290.

Kambhampati, S., Volkl, W. and Mackauer, M. (2000) Phylogenetic relationships among genera of Aphidiinae (Hymenoptera: Braconidae) based on DNA sequence of the mitochondrial 16S rRNA gene. *Systematic Entomology* 25, 437–445.

Kerdelhue, C., Le Clainche, I. and Rasplus, J.-Y. (1999) Molecular phylogeny of the *Ceratosolen* species pollinating *Ficus* of the subgenus *Sycomorus sensu stricto*: biogeographical history and origins of the species-specificity breakdown cases. *Molecular Phylogenetics and Evolution* 11, 401–414.

Kozlov, M.A. and Rasnitsyn, A.P. (1979) On the limits of the family Serphitidae (Hymenoptera, Proctotrupoidea). *Entomologischekoye Obozreniye* 58, 402–416.

LaSalle, J. (1993) Parasitic Hymenoptera, biological control and biodiversity. In: LaSalle, J. and Gauld, I.D. (eds) *Hymenoptera and Biodiversity*. CAB International, Wallingford, UK, pp. 197–216.

LaSalle, J. and Gauld, I.D. (1992) Parasitic Hymenoptera and the biodiversity crisis. *Redia* 74, 315–334.

Landry, B.S., Dextraze, L. and Boivin, G. (1993) Random amplified polymorphic DNA markers for DNA fingerprinting and genetic variability assessment of minute parasitic wasp species (Hymenoptera: Mymaridae and Trichogrammatidae) used in biological control programs of phytophagous insects. *Genome* 36, 580–587.

Lopez-Vaamonde, C., Rasplus, J.Y., Weiblen, G.D. and Cook, J.M. (2001) Molecular phylogenies of fig wasps: partial cocladogenesis of pollinators and parasites. *Molecular Phylogenetics and Evolution* 21, 55–71.

Machado, C.A., Herre, E.A., McCafferty, S. and Bermingham, E. (1996) Molecular phylogenies of fig pollinating and non-pollinating wasps and the implications for the origin and evolution of the fig–fig wasp mutualism. *Journal of Biogeography* 23, 531–542.

Machado, C.A., Jousselin, E., Kjellberg, F., Compton, S.G. and Herre, E.A. (2001) Phylogenetic relationships, historical biogeography and character evolution of fig-pollinating wasps. *Proceedings of the Royal Society Biological Sciences Series B* 268, 685–694.

Maddison, D. (1991) The discovery and importance of multiple islands of most-parsimonious trees. *Systematic Biology* 40, 315–328.

Manzari, S., Polaszek, A., Belshaw, R. and Quicke, D.L.J. (2002) Morphometric and molecular analysis of the *Encarsia inaron* species-group (Hymenoptera: Aphelinidae), parasitoids of whiteflies (Hemiptera: Aleyrodidae). *Bulletin of Entomological Research* 92, 165–175.

Mardulyn, P. and Cameron, S.A. (1999) The major opsin in bees (Insecta: Hymenoptera): a promising nuclear gene for higher level phylogenetics. *Molecular Phylogenetics and Evolution* 12, 168–176.

Mardulyn, P. and Whitfield, J.B. (1999) Phylogenetic signal in the COI, 16S, and 28S genes for inferring relationships among genera of Microgastrinae (Hymenoptera; Braconidae): evidence of a high diversification rate in this group of parasitoids. *Molecular Phylogenetics and Evolution* 12, 282–294.

Nagarkati, S. (1975) Two new species of *Trichogramma* from the USA. *Entomophaga* 20, 245–248.

Nelson, G. (1994) Homology and systematics. In: Hall, B.K. (ed.) *Homology: the Hierarchical Basis of Comparative Biology.* Academic Press, New York, pp. 102–151.

Noyes, J.S. (1978) On the numbers of genera and species of Chalcidoidea (Hymenoptera) in the world. *Entomologist's Gazette* 29, 163–164.

Noyes, J.S. (1990) The number of described chalcidoid taxa that are currently regarded as valid. *Chalcid Forum* 13, 9–10.

Noyes, J.S. (2000) Encyrtidae of Costa Rica (Hymenoptera: Chalcidoidea), 1. The subfamily Tetracneminae, parasitoids of mealybugs (Homoptera: Pseudococcidae). *Memoirs of the American Entomological Institute* 62, 1–355.

Noyes, J.S. and Hayat, M. (1994) *Oriental Mealybug Parasitoids of the Anagyrini (Hymenoptera: Encyrtidae).* CAB International, Wallingford, UK.

Orrego, C. and Agudelo-Silva, F. (1993) Genetic variation in the parasitoid wasp *Trichogramma* (Hymenoptera: Trichogrammatidae) revealed by DNA amplification of a section of the nuclear ribosomal repeat. *Florida Entomologist* 76, 519–524.

Patterson, C., Williams, D.M. and Humphries, C.J. (1993) Congruence between molecular and morphological phylogenies. *Annual Review of Ecology and Systematics* 24, 153–188.

Pinto, J.D. (1999) Systematics of the North American species of *Trichogramma* Westwood (Hymenoptera: Trichogrammatidae). *Memoirs of the Entomological Society of Washington* 22, 1–287.

Pinto, J.D., Stouthamer, R., Platner, G.R. and Oatman, E.R. (1991) Variation in reproductive compatibility in *Trichogramma* and its taxonomic significance (Hymenoptera: Trichogrammatidae). *Annals of the Entomological Society of America* 84, 37–46.

Pinto, J.D., Stouthamer, R. and Platner, G.R. (1997) A new cryptic species of *Trichogramma* (Hymenoptera: Trichogrammatidae) from the Mojave Desert of California as determined by morphological, reproductive and molecular data. *Proceedings of the Entomological Society of Washington* 99, 238–247.

Pinto, J.D., Koopmanschap, A.B., Platner, G.R. and Stouthamer, R. (2002a) The North American *Trichogramma* (Hymenoptera: Trichogrammatidae) parasitizing certain Tortricidae (Lepidoptera) on apple and pear, with ITS2 DNA characterizations and description of a new species. *Biological Control* 23, 134–142.

Pinto, J.D., Platner, G.R. and Stouthamer, R. (2002b) The systematics of the *Trichogramma minutum* species complex (Hymenoptera: Trichogrammatidae), a group of important North American biological control agents: the evidence from reproductive compatibility and allozymes. *BioControl* 23, 134–142.

Prinsloo, G., Chen, Y., Giles, K.L. and Greenstone, M.H. (2002) Release and recovery in South Africa of the exotic aphid parasitoid *Aphelinus hordei* verified by the polymerase chain reaction. *BioControl* 47, 127–136.

Rasnitsyn, A.P. (1988) An outline of the evolution of the hymenopterous insects. *Oriental Insects* 22, 115–145.

Rasplus, J.-Y., Kerdelhue, C., Le Clainche, I. and Mondor, G. (1998) Molecular phylogeny of fig wasps; Agaonidae are not monophyletic. *Comptes Rendus de l'Academie des Sciences Serie III Sciences de la Vie* 321, 517–527.

Rokas, A., Nylander, J.A.A., Ronquist, F. and Stone, G.N. (2002) A maximum-likelihood analysis of eight phylogenetic markers in gallwasps (Hymenoptera: Cynipidae): implications for insect phylogenetic studies. *Molecular Phylogenetics and Evolution* 22, 206–219.

Ronquist, F., Rasnitsyn, A.P., Roy, A., Eriksson, K. and Lindgren, M. (1999) Phylogeny of the Hymenoptera: a cladistic reanalysis of Rasnitsyn's (1988) data. *Zoologica Scripta* 28, 13–50.

Sanchis, A., Latorre, A., Gonzalez-Candelas, F. and Michelena, J.M. (2000) An 18S rDNA-based molecular phylogeny of Aphidiinae (Hymenoptera: Braconidae). *Molecular Phylogenetics and Evolution* 14, 180–194.

Sappal, N.P., Jeng, R.S., Hubbes, M. and Liu, F. (1995) Restriction fragment length polymorphisms in polymerase chain reaction amplified ribosomal DNAs of three *Trichogramma* (Hymenoptera: Trichogrammatidae) species. *Genome* 38, 419–425.

Scheffer, S.J. and Grissell, E.E. (2003) Tracing the origin of *Megastigmus transvaalensis* (Hymenoptera: Torymidae): and African wasp feeding on a South American plant in North America. *Molecular Ecology* 12, 415–421.

Schmidt, S., Naumann, I.D. and De Barro, P.J. (2001) *Encarsia* species (Hymenoptera: Aphelinidae) of Australia and the Pacific Islands attacking *Bemisia tabaci* and *Trialeurodes vaporariorum* (Hemiptera: Aleyrodidae): a pictorial key and descriptions of four new species. *Bulletin of Entomological Research* 91, 369–387.

Silva, I.M.M.S., Honda, J., van Kan, F., Hu, J., Neto, L., Pintureau, B. and Stouthamer, R. (1999) Molecular differentiation of five *Trichogramma* species occurring in Portugal. *Biological Control* 16, 177–184.

Simon, C., Frati, F., Beckenbach, A., Crespi, B., Liu, H. and Flook, P. (1994) Evolution, weighting, and phylogenetic utility of mitochondrial gene sequences and a compilation of conserved polymerase chain reaction primers. *Annals of the Entomological Society of America* 87, 651–701.

Smith, S.M. (1996) Biological control with *Trichogramma*: advances, successes, and potential of their use. *Annual Review of Entomology* 41, 375–406.

Stouthamer, R., Hu, J., Van Kan, F.J.P.M., Platner, G.R. and Pinto, J.D. (1998) The utility of internally transcribed spacer 2 DNA sequences of the nuclear ribosomal gene for distinguishing sibling species of *Trichogramma*. *Biocontrol* 43, 421–440.

Stouthamer, R., Gai, Y., Koopmanschap, A.B., Platner, G.R. and Pinto, J.D. (2000a) ITS-2 sequences do not differ for the closely related species *Trichogramma minutum* and *T. platneri*. *Entomologia Experimentalis et Applicata* 95, 105–111.

Stouthamer, R., Jochemsen, P., Platner, G.R. and Pinto, J.D. (2000b) Crossing incompatibility between *Trichogramma minutum* and *T. platneri* (Hymenoptera:

Trichogrammatidae): implications for application in biological control. *Environmental Entomology* 29, 832–837.

Swofford, D. (1991) When are phylogeny estimates from morphological and molecular data incongruent? In: Miyamoto, M.M. and Cracraft, J. (eds) *Phylogenetic Analysis of DNA Sequences.* Oxford University Press, New York, pp. 295–333.

Swofford, D., Olsen, G.J., Waddell, P. and Hillis, D.M. (1996) Phylogenetic inference. In: Hillis, D.M., Moritz, C. and Mable, B.K. (eds) *Molecular Systematics.* Sinauer Associates, Massachusetts, pp. 407–514.

Unruh, T.R. and Woolley, J.B. (1999) Molecular methods in classical biological control. In: Bellows, T.S. and Fisher, T.W. (eds) *Handbook of Biological Control.* Academic Press, San Diego, pp. 57–85.

van Kan, F.J.P.M., Silva, I.M.M.S., Schilthuizen, M., Pinto, J.D. and Stouthamer, R. (1996) Use of DNA-based methods for the identification of minute wasps of the genus *Trichogramma. Proceedings of Experimental and Applied Entomology, N.E.V. Amsterdam* 7, 233–237.

van Kan, F.J.P.M., Honda, J., Pinto, J.D. and Stouthamer, R. (1997) Molecular based techniques for *Trichogramma* identification. *Proceedings of Experimental and Applied Entomology, N.E.V. Amsterdam* 8, 59–62.

Vanlerberghe-Masutti, F. (1994) Molecular identification and phylogeny of parasitic wasp species (Hymenoptera: Trichogrammatidae) by mitochondrial DNA RFLP and RAPD markers. *Insect Molecular Biology* 3, 229–237.

Viggiani, G. (1984) Bionomics of Aphelinidae. *Annual Review of Entomology* 29, 257–276.

Vos, P., Hogers, R., Blecker, M., Reijans, M., Van de Lee, T. and Hornes, M. (1995) AFLP: a new technique for DNA fingerprinting. *Nucleic Acids Research* 23, 4407–4414.

Whitfield, J.B. and Cameron, S.A. (1998) Hierarchical analysis of variation in the mitochondrial 16S rRNA gene among Hymenoptera. *Molecular Biology and Evolution* 15, 1728–1743.

Wiegmann, B.M., Mitter, C., Regier, J.C., Friedlander, T.P., Wagner, D.M. and Nielsen, E.S. (2000) Nuclear genes resolve Mesozoic-aged divergences in the insect order Lepidoptera. *Molecular Phylogenetics and Evolution* 15, 242–259.

Williams, T. and Polaszek, A. (1996) A re-examination of host relations in the Aphelinidae (Hymenoptera: Chalcidoidea). *Biological Journal of the Linnaean Society* 57, 35–45.

Yoshimoto, C. (1975) Cretaceous chalcidoid fossils from Canadian Amber. *Canadian Entomologist* 107, 499–528.

Zchori-Fein, E., Gottlieb, Y., Brown, J.K., Wilson, J.M., Karr, T.L. and Hunter, M.S. (2001) A newly discovered bacterium associated with parthenogenesis and a change in host selection behavior in parasitoid wasps. *Proceedings of the National Academy of Sciences USA* 98, 12555–12560.

Zhu, Y.-C. and Greenstone, M.H. (1999) Polymerase chain reaction techniques for distinguishing three species and two strains of *Aphelinus* (Hymenoptera: Aphelinidae) from *Diuraphis noxia* and *Schizaphis graminum* (Homoptera: Aphididae). *Annals of the Entomological Society of America* 92, 71–79.

Zhu, Y.-C., Burd, J.D., Elliott, N.C. and Greenstone, M.H. (2000) Specific ribosomal DNA marker for early polymerase chain reaction detection of *Aphelinus hordei* (Hymenoptera: Aphelinidae) and *Aphidius colemani* (Hymenoptera: Aphididae) from *Diuraphis noxia* (Homoptera: Aphididae). *Annals of the Entomological Society of America* 93, 486–491.

4

Genetic Markers in Rust Fungi and their Application to Weed Biocontrol

K.J. Evans[1] and D.R. Gomez[2]*

[1]Cooperative Research Centre for Australian Weed Management, Tasmanian Institute of Agricultural Research, New Town Research Laboratories, 13 St Johns Avenue, New Town, Tasmania 7008, Australia; [2]Department of Applied and Molecular Ecology, University of Adelaide, Waite Campus, PMB 1, Glen Osmond, South Australia 5064, Australia

Introduction

The deliberate release of plant pathogens as biocontrol agents for weeds has a short history, dating back to the 1970s (Evans *et al.*, 2001a). For classical biocontrol of alien weeds, genetic markers provide tools for (i) investigating relationships between weed diversity and pathogen variation; (ii) identifying and characterizing the released pathogen strain or strains with certainty; (iii) monitoring the fate of pathogen strains in the environment; and (iv) investigating coevolution of the pathogen and target weed in both their native and alien environments. Depending on the evolutionary time scale studied, centres of diversity may be located or apparent shifts in host range explained. Among other things, the knowledge generated by such studies may provide clues as to how to develop search and selection strategies and improve the composition of pathogen strains released on target weed populations for effective, long-term biocontrol.

In this chapter, we focus on rust fungi (order *Uredinales*, division *Basidiomycota*) of the type that produce discrete lesions in annual hosts or in perennial species in which the site of infection has an annual habit (Burdon *et al.*, 1996). In particular, we will define and evaluate genetic tools that can be used to characterize rust strains or populations and when these tools can be applied during the course of weed biocontrol research. Specific examples of the practical application of markers will be given where possible.

*See Contributors list for new address.

Rust Fungi as Biocontrol Agents of Weeds

The rust fungus, *Puccinia chondrillina*, was released in 1971 against skeleton weed (*Chrondrilla juncea*) in wheat crops in south-eastern Australia (Cullen *et al.*, 1973). This appears to be the first deliberate introduction of a pathogen for weed control anywhere in the world. The release of one strain of the rust controlled one biotype of the weed spectacularly, whereas the agent had no impact on at least five other biotypes of *C. juncea* in Australia and the USA (Burdon *et al.*, 1981; Tisdell, 1990; Hasan *et al.*, 1995). Most importantly, it raised awareness of the importance of considering intraspecific variation in plant resistance and fungal virulence in designing successful biocontrol programmes. Genetic markers are essential for characterizing the plant–pathogen interaction, a process we shall describe in detail later.

The initial success of the rust on a single biotype of skeleton weed led to a rapid expansion in the use of rust fungi for weed biocontrol (Julien and Griffiths, 1998). In 2002, at least 11 rust fungi had been released legally into Australia, including nine species listed by Evans (2000) plus *Prospodium tuberculatum* for *Lantana camara* (Lantana) and *Puccinia myrsiphylli* for *Asparagus asparagoides* (bridal creeper, L. Morin, Canberra, 2002, personal communication). Descriptions of a number of biocontrol programmes around the world using rust fungi can be found in Evans (2000) and Evans *et al.* (2001a). With the exception of skeleton weed, rust fungi are mostly targeted at weeds of pasture, forestry, rangeland habitats and conservation areas. The peculiarities of using rust fungi for classical biocontrol of weeds are now described.

The rust fungi represent a very ancient group of microorganisms that have coevolved intimately with their plant hosts (Savile, 1971). As such, a level of host specificity suitable for biological control is often satisfied. As obligate biotrophs, rust fungi derive their energy by penetrating the plant cell wall and interfacing with the plant-cell plasma membrane via a feeding structure known as a haustorium. This allows the fungus to parasitize the living plant cells for weeks prior to, and during, reproduction. The continuous diversion of assimilates from the host plant has a negative impact on plant growth and reproduction. Severe disease results in leaf death and significant defoliation of the plant canopy. Ideally, plant biomass and/or weed density is reduced to some predetermined threshold level where the weed is no longer an economic or environmental problem (Briese, 2000). For weeds with perennial root systems or large reserves of tubers underground, repeated defoliation over a period of 5–10 years, sometimes sooner, results in a gradual reduction in plant biomass and a reduced rate of vegetative spread (Mahr and Bruzzese, 1998). In natural ecosystems, biocontrol is often the only weed management option for remote and inaccessible locations. Biocontrol is successful when natural succession of native vegetation resumes, because the weed is no longer the dominant plant type.

Most rust fungi used for weed biocontrol are macrocyclic and autoecious (Fig. 4.1). Although the life history is complex, it is only the repeating uredinial phase that results in multiple generations and damaging levels of plant disease.

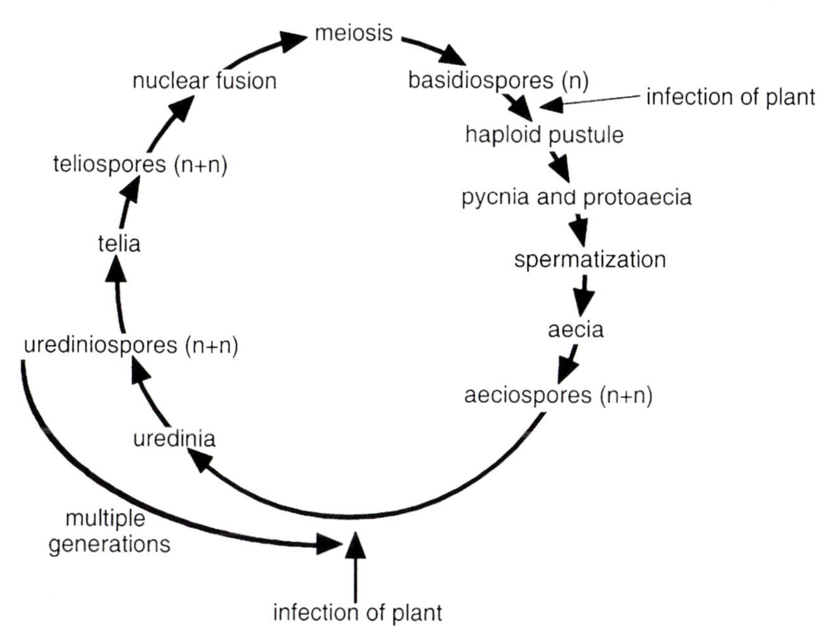

Fig. 4.1. Schematic representation of the life cycle of a macrocyclic, autoecious rust fungus (Order *Uredinales*). (Modified from Hawksworth *et al.*, 1995.)

Long-distance dispersal is achieved by wind-borne urediniospores, which are capable of aerial dispersal over 500 km or more (Brown and Høvmoller, 2002). After *P. chondrillina* was introduced to Australia, the fungus was found 80, 160 and 320 km from the initial release site at approximately the eighth, tenth and twelfth generations, respectively, after release (Cullen *et al.*, 1973). *Puccinia carduorum* was introduced into Virginia, USA, in 1987 for biological control of musk thistle, *Carduus thoermeri*. Five years later this rust pathogen had spread more than 500 km from the release site (Baudoin and Bruckart, 1996). Another rust biocontrol agent, *Phragmidium violaceum*, was first reported to occur on European blackberry in Australia in 1984 (Marks *et al.*, 1984). Wind-borne dispersal of urediniospores across the Tasman Sea, a distance of more than 1000 km, may explain the first report of *P. violaceum* in New Zealand in 1990 (Anon., 1990). Unlike insect biocontrol agents, redistribution of rusts following initial establishment is rarely necessary. In summary, rust fungi have a high degree of host specificity, widespread dispersal in a short space of time and a huge potential to suppress weed infestations.

Genetic Markers in Rust Fungi

We define a rust strain as a mitotic cell lineage, which constitutes all fungal structures derived from an identical genetic source. In practice, a rust strain is derived by isolating a single uredinium and subsequent transfer of its urediniospores to a healthy leaf for propagation. Given that rust strains are morphologically identical, they must be distinguished by markers that are stable and correlated to genetic differences.

Pathotyping

Following the identification of the host–pathogen 'gene-for-gene' interaction by Flor (1955), avirulence genes have been used extensively for characterizing rust strains (Crute *et al.*, 1997). Each rust strain is bioassayed to determine its virulence or avirulence on a given host clone, cultivar or biotype. The strain is designated a physiological 'race' or 'pathotype' based on the pattern of disease expression across a well-characterized 'differential set' of host lines carrying different resistance genes. Unlike the well-characterized cereal rust pathosystems, a set of differential host lines must be developed for each weed rust pathosystem.

Pathotyping by bioassay can provide important biological information when evaluating individual rust strains for their range of virulence within a genetically diverse weed. In the cereal rust fungi at least, avirulence is potentially a very mutable character. One estimate of the change of virulence in *Puccinia graminis* f. sp. *tritici* was 8.3 spores per million spores per generation (Schafer and Roelfs, 1985). The variability for virulence may differ greatly from that of molecular markers that are phenotypically neutral. An example of how rapidly new pathotypes can evolve in a rust population that has a low genetic diversity can be found in the *Puccinia striiformis* f. sp. *tritici* wheat pathosystem (Wellings and McIntosh, 1990; Hovmøller, 2001). As will be described below, there are a number of applications where molecular data can be combined with pathotyping to provide greater insight into the maintenance of intraspecific variation when compared with using each technique in isolation.

Selecting molecular markers for population-genetic studies

Comprehensive reviews of the basic properties of a range of genetic markers and their selection for a given research question in population biology are presented by McDonald and McDermott (1993), Brown (1996), McDonald (1997), Parker *et al.* (1998) and Sunnucks (2000). Rust fungi contain DNA both in the nuclei and in the cytoplasm, with extrachromosomal DNA occurring in the mitochondria. Extrachromosomal genetic elements such as double-stranded RNAs (dsRNAs, Pryor *et al.*, 1990) may also act as genetic markers if these elements are retained during asexual propagation. The utility of dsRNAs as genetic markers has not yet been demonstrated.

Before selecting a molecular marker, genome organization, quantity of rust DNA available for assay, the likely resolution of a particular method at the intraspecific level and costs of marker development should be considered. Of critical importance for population-genetic studies is how the marker fits the genetic assumptions that underlie analyses of the data (Brown, 1996).

Relative utility of some common DNA markers

Isozymes are simple and reliable genetic markers, although very little isozyme variation exists at the rust intraspecific level (Burdon *et al.*, 1982; Burdon and Roelfs, 1985; Newton *et al.*, 1985; Burdon and Roberts, 1995). Analysis of restriction fragment length polymorphisms (RFLPs) involving hybridization of anonymous or generic DNA probes, while technically demanding, remains a robust and useful technique. In rust fungi, it may be applied for preliminary analyses of intraspecific genetic variation (Evans *et al.*, 2000) and for discriminating races of rust species (Anderson and Pryor, 1992). No preliminary sequence information is required and, unlike generic polymerase chain reaction (PCR)-based techniques, contamination by small quantities of non-target DNA will not confound the result.

In rust fungi, randomly amplified polymorphic DNAs (RAPDs) have been applied to study gene flow (Braithwaite *et al.*, 1994) and the linkage of molecular markers to virulence phenotypes (Liu and Kolmer, 1998). Amplified fragment length polymorphisms (AFLPs) are widely adopted at present because they are highly polymorphic and reproducible (Majer *et al.*, 1996; Justesen *et al.* 2001). No prior sequence knowledge is required, ensuring minimal development time. For example, Steele *et al.* (2001) investigated AFLP variation among five geographically distinct isolates of *P. striiformis* f. sp. *tritici* collected from the United Kingdom, Denmark and Columbia. With each primer set tested, an average of 6.5 polymorphic bands were detected. All five isolates tested could be uniquely identified when compared with six primer combinations. A modification of the AFLP technique, known as selectively amplified microsatellites (SAM; for example, Witsenboer *et al.*, 1997), has the additional advantage that markers have the potential to be converted to locus-specific microsatellites.

Microsatellites (Olsen *et al.*, 1989) are considered an ideal single-locus marker for population studies (Blouin *et al.*, 1996; Jarne and Lagoda, 1996), and methods for identifying DNA clones with microsatellite sequences are rapidly becoming more efficient. Single-locus microsatellite markers are now being evaluated for studying the population genetics of some cereal rust fungi (R.F. Park, Sydney, 2003, personal communication). An important consideration in the research cost–benefit analysis for weed rust pathosystems is that microsatellite primers must be developed for each species investigated.

Sequences of mtDNA and rDNA may be combined with sequences from nuclear genetic regions (microsatellites and single-copy nuclear markers) to generate gene genealogies (Luikart and England, 1999). Only a few plant pathogens of economic importance have genomes that have been sequenced extensively. Nevertheless, DNA sequences relevant for basidiomycete fungi might be found on the expanding genomic databases that are facilitating this type of data exploration.

Emerging technologies

High-quality sequence data are likely to be a prerequisite for emerging technologies that utilize high-density DNA arrays or microarrays that merge DNA and silicon chips (Lipshutz *et al.*, 1999). The so-called 'lab on a chip' (Lévesque, 2001) is a substrate of thousands of 'spotted' oligonucleotides that can be hybridized to labelled DNA (Maughan *et al.*, 2001). The detection of multiple DNA hybrids can be used, among many other things, for genotyping and simultaneous monitoring of multiple genetic variants (Dalma-Weiszhausz *et al.*, 2002). This technology was applied recently to examine relationships among closely related microbial species (Murray *et al.*, 2001). The ability to genotype thousands of individuals will generate overwhelming amounts of data, which will require appropriate analytical skills for making valid conclusions about complex population dynamics. As always, adoption of new technology will depend on accuracy, sensitivity, reproducibility and cost effectiveness.

Application of Molecular Markers in Weed Biocontrol

Investigating relationships between weed diversity and pathogen variation

Beginning with the search phase of a biocontrol programme, a reliable taxonomy of the weed will assist identification of the plant in its native range. This is not a trivial task, considering that the taxonomic treatment of the plant genus in the native range is often incomplete and, depending on the time since introduction, the weed may have evolved significantly in its introduced range. Given the tight coevolution between rusts and their hosts, delineation of weed biotypes or apomictic clones at the sub-species level has become an integral component of many biocontrol programmes using rust fungi. Weed diversity can be determined using isozyme markers (Chaboudez, 1994) or by DNA-based marker systems (Nissen *et al.*, 1995). In the case of skeleton weed, *C. juncea*, isozyme markers detected a large amount of clonal variation in this apomict, even though the technique may not have separated the entire range of genetic variation (Chaboudez, 1994).

The issue of biocontrol agent diversity appears to be of greater significance for plant pathogens than for insect biocontrol agents, where a genetically variable founding population of the insect is anticipated (Hopper *et al.*, 1993). Failure of an insect agent can result from inadequate genetic composition, but this is sometimes the result of genetic deterioration during culture maintenance in quarantine. Identifying genetic diversity of the weed in its introduced range is a first step in defining what is an 'adequate' genetic composition for the introduced rust pathogen.

The skeleton rust story

The skeleton rust biocontrol programme provides an instructive example of the impact of genetic diversity in an apomictic weed on the success of biocontrol. It also provides a conceptual framework for the selection of rust strains. Despite early success in controlling the 'narrow-leaf' form of *C. juncea* in Australia, the other two clones have since colonized wheat crops in areas previously occupied by the 'narrow-leaf' form. The origin of *C. juncea* clones in Australia is unknown, although the centre of origin of the genus *Chrondrilla* was assumed to be eastern Europe. Following the identification of the isozyme variants of *C. juncea* (Burdon *et al.*, 1980), the first strategy employed to find suitable strains of *P. chondrillina* for the other forms of *C. juncea* relied on the existence of a strong correlation between isozyme variants and plant disease resistance phenotypes. Subsequently, Chaboudez (1989, 1994) used isozyme marker systems to identify western Turkey as the source of the diploid progenitor of the triploid apomicts of *C. juncea*. Not surprisingly, there is a concentration of triploid apomicts in this region (Chaboudez, 1994). In a parallel evolutionary story, sexual stages of *P. chondrillina* are known to occur in the general region of Turkey where apomicts of *C. juncea* arise (Hasan and Wapshere, 1973). Climatic conditions in western Europe are generally unfavourable for the sexual stage, consequently clonal reproduction is the principal mode of reproduction. The current hypothesis is that virulent forms of *P. chondrillina* originate in eastern Europe and, by wind dispersal of asexual spores, infect susceptible clones of *C. juncea* as they migrate away from the centre of origin. In western Europe, where *P. chondrillina* appears to be predominantly clonal, the outcome of this selection process is that clones of the plant are matched to clones of the pathogen (Chaboudez, 1989). The practical outcome of this research was that Hasan *et al.* (1995) established a field planting or 'trap garden' of the Australian and North American forms of *C. juncea* in Turkey, the 'centre of diversity', in the hope that suitable pathotypes of *P. chondrillina* would infect the target weed biotypes. Suitable pathotypes were found for clones of *C. juncea* in the USA (Hasan *et al.*, 1995) and potential matches were found for the 'intermediate form' in Australia. As the CSIRO's work on skeleton weed ceased in 1996, the impact of the release of additional strains of *P. chondrillina* on *C. juncea* in Australia has never been assessed.

In a later study, Espiau *et al.* (1998) determined that there was only 58% congruence between host resistance phenotype and multi-locus isozyme variants in a population of *C. juncea* from Turkey. Assuming sexual recombination in *P. chondrillina* is common in this region, linkages among virulence loci would not always be maintained. Therefore, the disease response of an individual clone of *C. juncea* would be dependent on the genetic structure of the local pathogen population at the time of infection. This information, plus the high level of isozyme variability in Turkey, adds to the complexity of recognizing forms of *C. juncea* in Europe that might provide appropriate pathotypes of *P. chondrillina* for the introduced weedy biotypes. In summary, the population genetics of this pathosystem suggest that the selection of *P. chondrillina* strains by the trap garden approach is a more efficient strategy than searching for the weedy biotypes of *C. juncea* in the native range in the hope of collecting a virulent rust strain. If the source populations of the clones in Australia and the USA had become extinct in Europe, then the trap garden approach would offer the only hope of finding matching rust strains.

The blackberry rust story

The increasing availability of molecular tools and the lessons learned from research on skeleton weed were the main reasons for the revival of the biocontrol programme for weedy blackberry in Australia, some 12 years after *Phragmidium violaceum* was first reported in Australia in 1984. European blackberry, comprising closely related taxa of the *Rubus fruticosus* aggregate, is an important weed of agriculture, forestry and natural ecosystems in Australia. Identification of the many apomictic taxa can be difficult, as phenotypic plasticity may be high and morphological variants can arise by hybridization between taxa. A consensus in the taxonomic treatment of *Rubus* in Europe, let alone in Australia, has not been reached.

A strain of *P. violaceum*, F15, was released as a biological control agent in Australia in 1991 and 1992. Despite the spectacular success of biocontrol in a number of blackberry infestations, there appear to be some blackberry biotypes that are escaping severe disease in locations where the weather is mostly favourable for the development of rust disease. Clones of the *R. fruticosus* agg., a facultative apomict, were collected for identification of biotypes resistant to disease. By using M13 DNA phenotyping, we were able to identify each *Rubus* clone propagated for use in pathogenicity studies with *P. violaceum* (Evans *et al.*, 2000, 2001b). A timely interaction with *Rubus* taxonomists, D.E. Symon (Australia), H.E. Weber (Germany) and A. Newton (UK), enabled us to widen the application of this DNA marker to clarify some taxonomic problems in the *R. fruticosus* agg.

Evans *et al.* (1998 and unpublished data) identified 33 M13 DNA phenotypes that were correlated to 13 taxa of the *R. fruticosus* agg. and one undetermined taxon. A further 16 DNA phenotypes were undetermined, based on morphology, or determined with only a moderate level of confidence. These undetermined DNA phenotypes are new biotypes, biotypes that have not yet been recognized and characterized in Europe, or biotypes that no longer exist in Europe. Exotic *Rubus* spp. have had over 150 years to evolve in Australia, and it is conceivable that new biotypes may have arisen by hybridization or somatic mutation.

An unexpected outcome of the taxonomic research was the identification of the most common and widespread weedy blackberry in Australia. This taxon, previously misnamed *Rubus procerus*, *Rubus discolor* or *Rubus* affin. *armeniacus*, appears to exist as a clonal lineage with greater than 97% of samples (n = 76) collected across Australia representing a single DNA phenotype. With relatively little effort we were able to sample the same DNA phenotype from England among a population of *Rubus anglocandicans* that is morphologically similar to the Australian material and also uniform in DNA phenotype. The weedy taxon in Australia has been renamed *R. anglocandicans* and can now be distinguished from *R. armeniacus*, the common weedy European blackberry in the pacific north-western region of the USA and in some parts of New Zealand (Evans and Weber, 2003). It has long been assumed by weed managers that Australia, New Zealand and north-western America shared their most widespread biotype of the *R. fruticosus* agg.: this new finding may have implications for the selection of biocontrol agents in the respective countries.

Disease resistance in the *R. fruticosus* agg. and physiological specialization among three Australian isolates of *P. violaceum* was identified in pathogenicity assays of 26 *Rubus* clones representing 17 DNA phenotypes and 14 taxa (Table 4.1). Physiological specialization in *P. violaceum* was detected readily, as it was for seven *P. chondrillina* isolates tested over six populations of *C. juncea* from the USA (Emge *et al.*, 1981). M13 DNA phenotyping of the rust strains used in the blackberry bioassays confirmed that genetically different rust strains were being tested (Fig. 4.2).

Table 4.1. Physiological specialization in *Phragmidium violaceum.* Twenty-six clones of the *Rubus fruticosus* agg. (European blackberry) were grouped according to patterns of susceptibility (S) or resistance (R) when inoculated with each of four strains of *P. violaceum.* Isolates V1, V2 and SA1 were collected in Australia between 1997 and 1999.

Strain of *P. violaceum*	Group 1 (n = 22)[a,b]	Group 2 (n = 1)[a,c]	Group 3 (n = 3)[a,d]
F15, France	S	R	S
V1, western Victoria	S	S	S
V2, eastern Victoria	S	S	R
SA1, Adelaide Hills	S	S	R

[a] n is the number of *Rubus* clones identified in each group.
[b] Group 1 includes the most common weedy taxon of blackberry in Australia, *R. anglocandicans.*
[c] A clone of *R. laciniatus.*
[d] Clones representing taxa of *R. erythrops*, *R. leucostachys* and *R.* sp. (not determined). In other tests, a clone of *R. cissburiensis* was found to be resistant to strain SA1.

P. violaceum strain V1, isolated from western Victoria, produced a susceptible disease response in all *Rubus* clones tested (Table 4.1) but many questions remain as to why some blackberry biotypes are escaping severe disease at some locations.

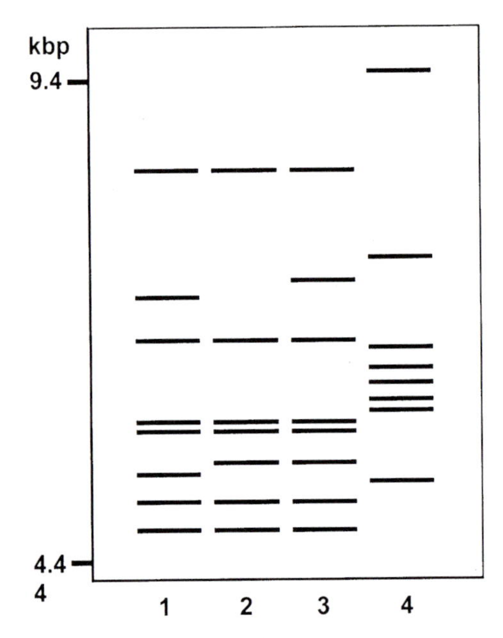

Fig. 4.2. Schematic representation of Southern hybridization of probe M13 to total *Phragmidium violaceum* DNA, digested with the restriction enzyme *Hae*III: lane 1, strain SA1, Adelaide Hills; Lane 2, strain V1, western Victoria; Lane 3, strain V2, eastern Victoria; Lane 4, strain F15 from France.

It may be that a rust strain with the corresponding virulence does arrive on the 'resistant' host biotype, but that it arrives too late (Burdon *et al.*, 1996); this would delay the initiation of the epidemic and reduce disease levels at critical times in the growing season. Another explanation relates to the fact that blackberry plants exhibit leaf-age-related disease resistance (Evans and Bruzzese, 2003). Two blackberry biotypes susceptible to a particular strain of *P. violaceum* and growing adjacent to each other may have different growth rates and/or cane densities (Amor, 1975). Different growth characteristics result in blackberry canopies with different leaf-age profiles and differences in the proportion of the canopy that is susceptible to disease at any given time. Indeed, *P. violaceum* strain V1 was isolated from *Rubus* clone EB19 growing adjacent to *Rubus* clone EB18, and clone EB19 appeared more severely diseased than clone EB18. Both of these *Rubus* clones were characterized as 'susceptible' when inoculated with strain V1 under controlled-environment conditions (Evans, 2001b, and unpublished data), which suggests that leaf-age-related disease resistance might have been the factor most limiting rust disease on *Rubus* clone EB18 *in situ*.

While this research was being conducted, a complementary research programme was initiated by CSIRO and the CRC for Australian Weed Management to search for additional strains of *P. violaceum* in Europe for release in Australia. DNA phenotyping of *Rubus* clones enabled characterized plant material to be shipped to Europe as *in vitro* cultures. These pest- and disease-free clones were established in an outdoor trap garden at the CSIRO laboratory in Montpellier. Urediniospores of *P. violaceum*, in air currents over the Montpellier region, were trapped on the plants and incited disease in all 21 *Rubus* clones, representing 19 DNA phenotypes, planted in the garden. Rust strains were then isolated and multiplied from the infected plants. Prior to importation in Australia, strains of *P. violaceum* collected from the Montpellier trap garden are being evaluated for their genetic similarity to the existing population of *P. violaceum* in Australia, using DNA and virulence phenotyping (J.K. Scott, M. Jourdan and K.J. Evans, unpublished data).

This approach demonstrates the use of DNA typing in streamlining the collection of rust strains through a well-characterized trap garden. Unlike the skeleton rust pathosystem, lack of knowledge about the origin of *P. violaceum* in Europe or the Middle East meant that we were unable to determine the best location to place a stationary trap garden. If necessary, additional trap gardens could be placed in regions of high diversity of *Rubus* species, for example southern England, where the clone of *R. anglocandicans* was found. Alternatively, a mobile trap garden could be developed, whereby target plants are placed on the roof-rack of a car so that urediniospores are sampled from a wide range of air currents. Infection of trap plants would be promoted by covering plants in moist chambers during overnight stopovers at blackberry-infested but tourist-friendly destinations!

Identifying the released pathogen with certainty

Responsible risk assessment associated with releasing an exotic organism into the environment should include identification, with certainty, of the released agent. Separation of closely related rust species can sometimes be problematic, as illustrated by the development of rust fungi as biocontrol agents for musk thistle (*Carduus thoermeri*).

The musk thistle rust story

In North America, seven species of *Carduus* (thistles) are introduced weeds (McCarty, 1982). Rust fungi of the genus *Puccinia* were evaluated as candidates for biological control of the widespread musk thistle (*C. thoermeri*), but the selection process was confounded by unclear relationships and morphological similarity among *Carduus* rusts. Urediniospore morphologies, host ranges and isozyme patterns of rust isolates from *C. thoermeri*, *C. tenuiflorus* and *C. pycnocephalus* were compared (Bruckart and Peterson, 1991). The isozyme data supported the hypothesis that all rust isolates were *P. carduorum* and could be distinguished from *P. carthami*, a morphologically similar rust species which infects the economically important crop, safflower (*C. tinctorius*). Morphological differences and differences in host preference among isolates of *P. carduorum* from the different *Carduus* hosts were observed. These relationships were explored further by RFLP analysis of internal transcribed spacer (ITS) regions of rDNA (Berthier *et al.*, 1996), which allowed *P. carduorum* to be separated into two groups, each group being correlated to particular *Carduus* hosts. Eight years following this release of an isolate of *P. carduorum* in the USA, eight samples of urediniospores were collected across the USA from infected *C. thoermeri*. DNA was extracted from each sample of urediniospores, and the ITS2 region of rDNA was amplified. Given that each collection of urediniospores might have been genetically heterogeneous, Luster and Bruckart (1998) omitted the time-consuming step of isolating single rust pustules prior to extraction of DNA. The population of amplified products was cloned and positive clones purified before being sequenced. Therefore, each DNA clone represented an individual in the population of amplified products. The sequences of clones representing the field samples were compared with ITS2 DNA sequences of the original rust isolate released in Virginia and those of the closely related rust species, *P. cyani, P. jaceae* and *P. chondrillina.* All rust isolates from infected *C. thoermeri* had identical ITS2 nucleotide sequences and could be distinguished from the other rust species. This information, combined with analysis of urediniospore morphology, confirmed the identity of the pathogen on musk thistle as *P. carduorum.*

Identifying and monitoring the fate of the released pathogen strain with certainty

Following identification of the released pathogen strain, its fate should be monitored, primarily to assess the impact of the release. If the pathogen strain is being released in an existing or resident population of the pathogen species, then genetic markers are essential for distinguishing the 'immigrant' from the resident population. DNA markers can provide sufficient resolution to separate fungal genets. Characterization of a rust strain at the level of genet should provide the most robust 'fingerprint' of the foreign entity being released into the environment. Microsatellite loci would be ideal for this purpose, but their development is time consuming. DNA fingerprints generated with multiple AFLP primer sets should provide enough polymorphic loci to reduce the chance of detecting the same DNA phenotype by chance. Even higher levels of variation might be detected by techniques such as single-strand conformation polymorphism (Orita *et al.*, 1989) or by heteroduplex mobility assay (Wang and Hiruki, 2000). It is unlikely that such a high genetic resolution would be required unless the rust strain being released was clonally related to the resident pathogen population.

In the past, thorough characterization of pathogen strains prior to their release as a biocontrol agent has rarely been attempted. In the case of the blackberry rust pathosystem, the virulence phenotype of strain F15 of *P. violaceum* was determined partially during tests for host specificity (Bruzzese and Hasan, 1986). The establishment and fate of strain F15 of *P. violaceum* was not monitored immediately after it was released in 1991. In the late 1990s, M13 DNA phenotyping was used to identify genetic variation in the Australian population of *P. violaceum* (Evans *et al.*, 2000). Thirteen DNA phenotypes of *P. violaceum* were identified among 18 strains collected from various locations in mainland Australia between 1997 and 1999. The restriction fragment patterns and DNA band-sharing indices of similarity suggest that the so-called 'illegal' strains predominate in the population of *P. violaceum*, and that strain F15 did not become well established in Australia. In order to verify the paucity of strain F15 or its descendants in Australia, DNA markers developed for the mitochondrial genome could define lines of descent of *P. violaceum*, regardless of its mode of reproduction. Otherwise, larger sample sizes, additional DNA markers and phylogeographic analyses could be used to separate current population structure from historical events.

Morin *et al.* (2001) combined RAPD data with sequence determination of the ITS region of rDNA to examine genetic variation among a collection of *P. myrsiphylli* strains isolated in South Africa for the biocontrol of *Asparagus asparagoides* (bridal creeper) in Australia. Three rust strains collected from the winter rainfall region of Western Cape Province were found to be genetically distinct from three rust strains collected from the 'even' rainfall region of the same province. Given that the sample size is small, this genetic difference appears to be correlated to the susceptibility of Australian accessions of bridal creeper to disease caused by strains of *P. myrsiphylli* from each region. One strain of *P. myrsiphylli* from the winter rainfall region was released in Australia in 2000 and severe rust disease has been observed near release sites in the 2002 growing season (Morin *et al.*, 2003). This work provides one of the first examples where a rust biocontrol agent has been characterized molecularly before being released in a new environment.

Population genetics of rust fungi in relation to strain selection

Having highlighted the importance of matching weed and pathogen diversity, the question remains about what constitutes an 'adequate' genetic composition of the introduced agent. More precisely, what is the appropriate genetic structure of the introduced pathogen that will have the greatest impact on the target weed?

The genetic structure of a population refers to the amount and distribution of genetic variation within and between populations (McDonald, 1997). The genetic structure of a population reflects its evolutionary history and its potential to evolve. Studies of genetic diversity as described previously are an obvious starting point for exploring the question of whether or not the search strategy has exhausted the potential genetic variation available in the native range. How the genetic diversity is distributed within and between populations, and on what spatial scale, is the next question. For weed biocontrol, we might ask how the genetic structure and evolution of the pathogen population varies from populations in their native range. An understanding of the evolutionary processes in each range might provide clues as to how to change the genetic structure in the introduced range for improved biocontrol. Studying the pathogen in its introduced range may even provide a simplified model of what is happening in the native range (Chaboudez and Burdon, 1995). We shall now focus on reproductive mode as an example of how to use knowledge of an evolutionary process in selecting plant pathogens for weed biocontrol.

Many rust fungi undergo alternating cycles of sexual and asexual reproduction, but the relative contribution of each type of reproduction to the genetic structure of populations is often poorly understood. The use of DNA markers in exploring questions of clonality versus recombination depends on setting appropriate and testable hypotheses (Milgroom, 1996). Evidence for clonal reproduction can be obtained by a simple criterion such as genotype over-representation (Pei and Ruiz, 2000). The patterns of genetic variation among individuals sampled may also suggest the presence of a clonal lineage, which is a subset of a population in which all genotypes originate from the same clone (Anderson and Kohn, 1995). When a clonal pathogen is released in a new environment there may be a close relationship between pathotype and clonal lineage. Over time there may be diversification of pathotypes within the lineage (Zeigler *et al.*, 1995).

Evidence for clonal reproduction does not mean that recombination is absent, but, rather, that it may be infrequent (Förster *et al.*, 1994). A potentially robust way of detecting recombination that is rare or episodic is by testing whether reconstructed phylogenies of different genes are congruent. Milgroom (1996) outlines this approach and its application to the human fungal pathogen, *Coccidioides immitis* (Burt *et al.*, 1996).

Rust fungi undergo clonal propagation during the uredinial phase of the life cycle, so rather than undertaking analyses of gametic disequilibrium or other tests for random mating, it would be more pertinent to obtain evidence for recombination. Evidence for recombination, like clonality, can be obtained by examining genotype diversity. Samils *et al.* (2001) used AFLPs to study the genetic structure of populations of *Melampsora epitea* in Swedish willow plantations. In a sample of 197 isolates collected from three locations, two AFLP primer combinations revealed 83 loci, 80 of which were polymorphic. Most isolates (97%) had unique DNA phenotypes. The fact that a single DNA phenotype was not represented in excess over other genotypes is evidence that recombination is an important factor in the creation and maintenance of genetic variation among Swedish willow leaf rust populations.

If sexual structures are found in nature, then the next question is whether the fungus outcrosses in nature. Sometimes it is possible to compare sexual and asexual populations of the same species to demonstrate that the sexual population has a greater diversity of genotypes. Burdon and Roelfs (1985) used isozyme markers to demonstrate that a sexually reproducing population of *P. graminis* f. sp. *tritici* was consistently more diverse than the asexually reproducing population. Furthermore, there was a complete association between virulence phenotype and isozyme pattern in the asexual population. Segregation of single-locus DNA markers among progeny from naturally occurring teliospores is another way of demonstrating recombination in nature.

Having obtained evidence of recombination, estimating the proportion of fungal inoculum that is sexual or asexual at any given time may provide an explanation for changes in population structure throughout the course of a fungal epidemic. Brown and Wolfe (1990) demonstrated how this might be done for the mildew fungus *Erysiphe graminis* f. sp. *hordei*. The application of such knowledge to biocontrol may be relevant when establishing additional strains of a plant pathogen in an existing population. The release could be timed to coincide with the likely window for recombination with the existing population. The additional virulence genes can then be introgressed into the resident population before selection allows one or more fit clones to dominate the clonal phase of the epidemic.

Establishment of additional pathogen strains in an existing population of the biocontrol agent is, in population-genetic terms, the same as the arrival of an immigrant. If a pathogen is not strictly clonal, then a means for detecting the introgression of 'foreign' genes in the resident population will provide clues as to how the introduced strain is changing the genetic structure of the existing population. Identifying introgression of new genes (gene flow) into the resident pathogen population could be achieved by first discriminating the introduced strain from the resident population, using multiple, single-locus DNA markers. The appearance of 'intermediate' genotypes that contain loci from both the introduced strain and the resident population would suggest that gene flow is occurring.

The relative importance of reproductive mode and gene flow has implications for the selection of rust strains for weed biocontrol. If the introduced pathogen has a strictly clonal mode of reproduction, then weedy biotypes that are disease resistant can be identified by assaying them across the range of introduced pathogen strains. These resistant weedy biotypes can then be used to identify additional and virulent rust strains in the native range that will produce a susceptible disease response. Ideally, a virulence gene tightly linked to a fitness gene would create a high potential for severe disease. Conversely, if the mode of pathogen reproduction is both sexual and asexual, then gene linkages can be disrupted during recombination, including virulence genes that might be linked with genes for fitness. The introduction of a rust strain with additional virulence genes has the potential to recombine genetically with the existing rust population to create new combinations of virulence genes. The selection objective for additional rust strains then becomes increasing the genetic diversity of the pathogen population; that is, the population should be given the potential to evolve and infect the entire range of weedy biotypes. In this situation the use of a differential set of host genotypes during strain selection can aid the identification of additional virulence phenotypes.

An example of the consequence of pathogen reproductive mode in biocontrol can be found in the skeleton rust pathosystem. *P. chondrillina* in Australia does not reproduce sexually in the Australian climate, unlike populations in Turkey

(Chaboudez and Sheppard, 1995). If the diploid progenitor of *C. juncea* clones was introduced to Australia, then the *P. chondrillina* population might have a reduced capacity to evolve virulence phenotypes in the absence of sexual reproduction. Unlike Australasian populations of *P. striiformis* f. sp. *tritici* (Steele *et al.*, 2001), there is currently no evidence to suggest that *P. chondrillina* can generate new pathotypes by a high mutation rate combined with selection for new or altered virulences.

Host-mediated selection is another evolutionary force that might impact on biocontrol efficacy. If some weed biotypes escape infection, does the pathogen population have the capacity to evolve new virulence phenotypes? If so, will the new pathotypes reach their target and incite disease early enough to have an impact on weed biomass? How fast can the pathogen evolve in the introduced range when compared with the rate of evolution in the native range? McDonald and Linde (2002) list high mutation rate, large population sizes, high gene flow, a mixed reproduction system and efficient directional selection as factors that create high evolutionary potential in plant pathogens. Rust fungi potentially satisfy all these criteria in agricultural systems, but much less is known about these factors in natural rust pathosystems. Recently, there has been considerable theoretical treatment of factors that affect the evolution of virulence (Thrall and Burdon, 2002; Zhan *et al.*, 2002) and host resistance (Carlsson-Granér and Thrall, 2002). Genetic markers are essential in testing theoretical concepts as they evolve.

Locating centres of diversity and evolutionary new associations

A widely accepted premise for a classical biocontrol strategy is that the source of a virulent and host-specific agent lies in the centre of origin or diversification of the weed species. The alternative view is that new associations in the course of evolution may in fact increase the chance of successful control. Focusing on the former premise, multilocus DNA sequence data can be used to construct genealogical lineages or phylogenies. If the spatial distribution of phylogenies is investigated (phylo-geography), then 'old' populations can be distinguished from 'new' populations in combined studies of population splitting and speciation. For example, Carbone and Kohn (2001) determined the DNA haplotypes of 385 strains of *Sclerotinia sclerotiorum* (order *Leotiales*, division *Ascomycota*), a haploid, filamentous fungus reported from more than 400 native and agricultural plant species. Phillips *et al.* (2002) outline the basic analytical steps taken by Carbone and Kohn (2001) utilizing DNA sequence data for seven loci. A significant outcome of their work was that they were able to distinguish populations of *S. sclerotiorum* in North America that were relatively old and endemic to local areas versus populations that were more recently evolved and highly dispersed over the entire region. This approach could therefore be applied in the search for rust biocontrol strains, especially when the centre of diversity is ill-defined or 'somewhere in Europe or the Middle East'. The cost of sampling and sequencing a large number of individuals over a wide geographical area may necessitate collaborative type research.

The studies of Phillips *et al.* (2002) also have relevance for testing the notion that an evolutionary new association or 'new encounter disease' can cause high levels of disease. The hypothesis of Phillips *et al.* (2002) is that observed high yield losses in canola when it was introduced to south-eastern USA were caused by the endemic pathogen genotypes adapted to many hosts and able to infect over a wide range of environmental conditions. It is possible the genotypes of *S. sclerotiorum*, representing the population that was more recently evolved, were introduced with canola to the region. Unlike rust fungi, *S. sclerotiorum* is a generalist pathogen, in that it has a wide host range. It seems unlikely that an alien weed could succumb to an indigenous rust pathogen or any rust species for which there was no coevolutionary history. However, there are several examples where a host plant developed rust disease upon exposure to a pathogen with which it has had no previous contact. Species of the Australian plant genera, *Eucalyptus*, *Callistemon* and *Melaleuca*, when grown in South America, were exposed to the guava rust fungus *Puccinia psidii* for the first time and developed a 'new encounter' disease (Walker, 1996). It is thought that these host plants contain genes for compatibility with the pathogen that were inherited and maintained from ancestors. Walker (1996) suggests that the ancestors were present in East Gondwana before the continents of Australia and South America separated. With no previous exposure to the pathogen, the host plant has not developed any resistance and severe disease can occur.

A variation on host-range extension is when an exotic host and pathogen are moved to a new area, and other plants in the area develop disease even though they coexisted disease free with the pathogen in its original range. When the rust fungus *Uromyces minor* was introduced with clover (*Trifolium* spp.) to Chile, New Zealand and Tasmania, it subsequently infected peas (*Pisum sativum*) in those countries, despite having never been recorded as a pathogen of these hosts in the original range of the northern hemisphere (Wiberg and Walker, 1990). This surprising host jump might be explained by studies of molecular phylogeography.

Host-range expansion is a concern raised frequently by those who oppose the introduction of any exotic organism to the environment. In this context it is worth remembering that plant pathogenicity is the exception rather than the norm for microorganisms (Paxton and Groth, 1994). There are likely to be exceptions, however, and pre-release host-testing programmes are designed to minimize potential risks associated with the release of an exotic organism (Wapshere, 1974). For example, a species of *Puccinia melampodii*, evaluated as a biocontrol agent for parthenium weed (*Parthenium hysterophorus*) in Australia, was found to induce advanced symptoms of disease in several local cultivars of sunflower (*Helianthus annuus*), when grown and inoculated under artificial conditions (Evans, 2000). Whether or not this level of disease would occur in the field is unknown. At first glance, a risk assessor might have classed this information as a case of the rust fungus 'jumping hosts'. Upon further analysis, the rust was found to represent a *forma specialis* of *Puccinia xanthii* (Evans, 2000). *P. xanthii* was introduced accidentally to Australia: it causes severe disease in weedy *Xanthium* spp. but only a mild infection on sunflower growing in a field situation. Curiously, *P. xanthii* has not been recorded on species of *Helianthus* growing in their original northern hemisphere range (Walker, 1996). Phylogenetic analysis could be used to assess the relationships between members of *Asteraceae* and their rust pathogens as a way of evaluating if the sunflower infection was an unexpected case of host-range expansion.

In addition to understanding phylogenetic relationships among rust fungi, molecular markers are powerful tools for assessing the potential for hybridization between closely related, but previously geographically isolated, pathogens. The interaction of the introduced pathogen with a closely related species in the new environment may be unpredictable unless the breeding system of the pathogen is well understood (Brasier, 2001). In particular, the strength of barriers to gene flow between the two species should be studied. The potential for hybridization among closely related rust species is worth investigating.

Conclusions

Ultimately, it is the weed genotype × pathogen genotype × environment interaction that will determine the impact of rust disease on a weed population. Assuming the pathogen population has the appropriate virulence traits, the fitness of introduced strains in a new environment or a resident pathogen population is difficult to predict from controlled-environment experiments (Mundt, 1995). The release of a genetically diverse pathogen population, even for a genetically uniform weed, will allow natural selection of those pathotypes that are fit in the new environment.

The use of molecular markers in understanding the creation and maintenance of intraspecific variation in rust fungi and their plant hosts provides a basis for comparing pathogen populations in their native and introduced ranges. Central to the discussion is the amount of genetic diversity that should be present in the founding population of the biocontrol agent in relation to weed diversity and the capacity for the fungal population to evolve new pathotypes by mutation and recombination.

When the genetic composition of the founding population is found to be inadequate, then efficient strategies for selecting additional rust strains must be devised. The use of a trap garden of weedy biotypes planted in the centre of diversity of the rust pathogen is likely to provide an efficient means of selecting additional strains when the introduced rust population has a mixed reproductive mode. For rust fungi that are strictly clonal in their introduced and native range, then collection of rust strains from the infected weedy biotype in the native range may prove fruitful, providing the biotype still exists in the native range. Where phenotypic plasticity in the host plant is high, molecular markers can identify weed clones with certainty.

Evans *et al.* (2001a) used the term 'pathophobia' to describe the resistance of 'environmentalists' towards the release of exotic plant pathogens in the environment. These concerns have resulted in a tradition of releasing one pathogen strain or genet at a time, after satisfying quarantine authorities that the risk of the pathogen strain shifting its host specificity is minimal. This release strategy is possibly detrimental to the establishment of additional strains in populations where the resident pathogen is well established and well adapted to local conditions. Testing multiple rust strains for host specificity will continue to add significantly to the cost of research until regulatory authorities accept less rigorous procedures based on sound knowledge of rust phylogeny, phylogeography and host–pathogen coevolution.

Acknowledgements

We thank J.K. Scott and R.T. Roush for helpful discussions and comments on the manuscript. The blackberry rust biocontrol programme is supported by the Cooperative Research Centre for Australian Weed Management, CSIRO, and the State government of Western Australia.

References

Amor, R.L. (1975) Ecology and control of blackberry (*Rubus fruticosus* L. agg.) IV. Effect of single and repeated applications of 2,4,5–T, picloram and aminotriazole. *Weed Research* 15, 39–45.

Anderson, J.B. and Kohn, L.M. (1995) Clonality in soilborne, plant-pathogenic fungi. *Annual Review of Phytopathology* 33, 369–391.

Anderson, P.A. and Pryor, A.J. (1992) DNA restriction fragment length polymorphisms in the wheat stem rust fungus, *Puccinia graminis tritici*. *Theoretical and Applied Genetics* 83, 715–719.

Anon. (1990) Blackberry fungus disease arrives in NZ. *Straight Furrow* 46, 20.

Baudoin, A.B.A.M. and Bruckart, W.L. (1996) Population dynamics and spread of *Puccinia carduorum* in the eastern United States. *Plant Disease* 80, 1193–1196.

Berthier, Y.T., Bruckart, W.L., Chaboudez, P. and Luster, D.G. (1996) Polymorphic restriction patterns of ribosomal internal transcribed spacers in the biocontrol fungus *Puccinia carduorum* correlate with weed host origin. *Applied and Environmental Microbiology* 62, 3037–3041.

Blouin, M.S., Parsons, M., Lacaille, V. and Lotz, S. (1996) Use of microsatellite loci to classify individuals by relatedness. *Molecular Ecology* 5, 393–401.

Braithwaite, K.S., Manners, J.M., Irwin, J.A.G. and Maclean, D.J. (1994) DNA markers reveal hybrids between two diverse background genotypes in Australian collections of the bean rust fungus *Uromyces appendiculatus*. *Australian Journal of Botany* 42, 255–257.

Brasier, C.M. (2001) Rapid evolution of introduced plant pathogens via interspecific hybridization. *BioScience* 51, 123–133.

Briese, D.T. (2000) Classical biological control. In: Sindel, B. (ed.) *Australian Weed Management Systems*. R.G. and F.J. Richardson, Melbourne, Australia, pp. 161–192.

Brown, J.K.M. (1996) The choice of molecular marker methods for population genetic studies of plant pathogens. *New Phytologist* 133, 183–195.

Brown, J.K.M. and Hovmøller, M.S. (2002) Aerial dispersal of pathogens on the global and continental scales and its impact on plant disease. *Science* 297, 537–541.

Brown, J.K.M. and Wolfe, M.S. (1990) Structure and evolution of a population of *Erysiphe graminis* f.sp. *hordei*. *Plant Pathology* 39, 376–390.

Bruckart, W.L. and Peterson, G.L. (1991) Phenotypic comparison of *Puccinia carduorum* from *Carduus thoermeri*, *C. tenuiflorus* and *C. pycnocephalus*. *Phytopathology* 81, 192–197.

Bruzzese, E. and Hasan, S. (1986) Host specificity of the rust *Phragmidium violaceum*, a potential biological control agent of European blackberry. *Annals of Applied Biology* 108, 585–596.

Burdon, J.J. and Roberts, J.K. (1995) The population genetic structure of the rust fungus *Melampsora lini* as revealed by pathogenicity, isozyme and RFLP markers. *Plant Pathology* 44, 270–278.

Burdon, J.J. and Roelfs, A.P. (1985) The effect of sexual and asexual reproduction on the isozyme structure of populations of *Puccinia graminis*. *Phytopathology* 75, 1068–1073.

Burdon, J.J., Marshall, D.R. and Groves, R.H. (1980) Isozyme variation in *Chondrilla juncea* L. in Australia. *Australian Journal of Botany* 28, 193–198.

Burdon, J.J., Groves, R.H. and Cullen, J.M. (1981) The impact of biological control on the distribution and abundance of *Chondrilla juncea* in south-eastern Australia. *Journal of Applied Ecology* 8, 957–966.

Burdon, J.J., Marshall, D.R., Luig, N.H. and Gow, D.J.S. (1982) Isozyme studies on the origin and evolution of *Puccinia graminis* f.sp. *tritici* in Australia. *Australian Journal of Biological Science* 35, 231–238.

Burdon, J.J., Wennström, A., Elmquist, T. and Kirby, G.C. (1996) The role of race specific resistance in natural populations. *Oikos* 76, 411–416.

Burt, A., Carter, D.A., Koenig, G.L., White, T.J. and Taylor, J.W. (1996) Molecular markers reveal cryptic sex in the human pathogen *Coccidioides immitis*. *Proceedings of the National Academy of Sciences USA* 93, 770–773.

Carbone, I. and Kohn, L.M. (2001) A microbial population–species interface: nested cladistic and coalescent inference with multilocus data. *Molecular Ecology* 10, 947–964.

Carlsson-Granér, U. and Thrall, P.H. (2002) The spatial distribution of plant populations, disease dynamics and evolution of resistance. *Oikos* 97, 97–110.

Chaboudez, P. (1989) Modes de reproduction et variabilité génétique des populations de *Chondrilla juncea* L.: implications dans la lutte microbiologique contre cette mauvaise herbe. Thèse Doctorale, Université de Montpellier, Sciences et Techniques du Languedoc, Montpellier, France.

Chaboudez, P. (1994) Patterns of clonal variation in skeleton weed (*Chondrilla juncea*), an apomictic species. *Australian Journal of Botany* 42, 283–293.

Chaboudez, P. and Burdon, J.J. (1995) Frequency-dependent selection in a wild plant–pathogen system. *Oecologia* 102, 490–493.

Chaboudez, P. and Sheppard, A.W. (1995) Are particular weeds more amenable to biological control? – a reanalysis of mode of reproduction and life history. *Proceedings of the Eighth International Symposium on Biological Control of Weeds*. CSIRO, Melbourne, pp. 95–102.

Crute, I.R., Holub, E.B. and Burdon, J.J. (1997) *The Gene-for-Gene Relationship in Plant–Parasite Interactions*. CAB International, Wallingford, UK.

Cullen, J.M., Kable, P.F. and Catt, M. (1973) Epidemic spread of a rust imported for biological control. *Nature* 244, 462–464.

Dalma-Weiszhausz, D.D., Chicurel, M.E. and Gingeras, T.R. (2002) Microarrays and genetic epidemiology: a multipurpose tool for a multifaceted field. *Genetic Epidemiology* 23, 4–20.

Emge, R.G., Melching, J.S. and Kingsolver, C.H. (1981) Epidemiology of *Puccinia chondrillina*, a rust pathogen for the biological control of rush skeleton weed in the United States. *Phytopathology* 71, 839–843.

Espiau, C., Riviere, D., Burdon, J.J., Gartner, S., Daclinat, B., Hasan, S. and Chaboudez, P. (1998) Host–pathogen diversity in a wild system: *Chondrilla juncea–Puccinia chondrillina*. *Oecologia* 113, 133–139.

Evans, H.C. (2000) Evaluating plant pathogens for biological control of weeds: an alternative view of pest risk assessment. *Australasian Plant Pathology* 29, 1–14.

Evans, K.J. and Bruzzese, E. (2003) Life history of *Phragmidium violaceum* in relation to its effectiveness as a biological control agent of European blackberry. *Australasian Plant Pathology* 32, 231–239.

Evans, K.J. and Weber, H.E. (2003) *Rubus anglocandicans* (Rasaceae) is the most widespread taxon of European blackberry in Australia. *Australian Systematic Botany* 16 (in press).

Evans, K.J., Symon, D.E. and Roush, R.T. (1998) Taxonomy and genotypes of the *Rubus fruticosus* L. aggregate in Australia. *Plant Protection Quarterly* 13, 152–156.

Evans, K.J., Jones, M.K., Mahr, F.A. and Roush, R.T. (2000) DNA phenotypes of the blackberry biological control agent, *Phragmidium violaceum*, in Australia. *Australasian Plant Pathology* 29, 249–254.

Evans, H.C., Greaves, M.P. and Watson A.K. (2001a) Fungal biocontrol agents of weeds. In: Butt, T.M., Jackson, C. and Magan, N. (eds) *Fungi as Biocontrol Agents: Progress Problems and Potential.* CAB International, Wallingford, UK, pp.169–192.

Evans, K.J., Jones, M.K. and Roush, R.T. (2001b) Biological control of blackberry: characterization of host and pathogen. *Proceedings of the 13th Biennial Conference of the Australasian Plant Pathology Society*. Department of Primary Industries, Mareeba, Queensland, p. 344.

Flor, H.H. (1955) Host–parasite interaction in flax rust – its genetics and other implications. *Phytopathology* 45, 680–685.

Förster, H., Tyler, B.M. and Coffey, M.D. (1994) *Phytophthora sojae* races have arisen by clonal evolution and by rare outcrosses. *Molecular Plant–Microbe Interactions* 7, 780–791.

Hasan, S. and Wapshere, A.J. (1973) The biology of *Puccinia chondrillina*, a potential biological control agent of skeleton weed. *Annals of Applied Biology* 74, 325–332.

Hasan, S., Chaboudez, P. and Espiau, C. (1995) Isozyme patterns and susceptibility of North American forms of *Chondrilla juncea* to European strains of the rust fungus *Puccinia chondrillina*. In: Delfosse, E.S. and Scott, R.R. (eds) *Proceedings of the Eighth International Symposium on Biological Control of Weeds*. CSIRO, Melbourne, Australia, pp. 367–373.

Hawksworth, D.L., Kirk, P.M., Sutton, B.C. and Pegler, D.N. (1995) *Ainsworth and Bisby's Dictionary of the Fungi*, 8th edn. CAB International, Wallingford, UK.

Hopper, K.R., Roush, R.T. and Powell, W. (1993) Management of genetics of biological-control introductions. *Annual Review of Entomology* 38, 27–51.

Hovmøller, M.S. (2001) Disease severity and pathotype dynamics of *Puccinia stiiformis* f.sp. *tritici* in Denmark. *Plant Pathology* 50, 181–189.

Jarne, P. and Lagoda, P.J.L. (1996) Microsatellites, from molecules to populations and back. *Trends in Ecology and Evolution* 11, 424–429.

Julien, M.H. and Griffiths, M.W. (1998) *Biological Control of Weeds: a World Catalogue of Agents and their Target Weeds*, 4th edn. CAB International, Wallingford, UK.

Justesen, A.F., Ridout, C.J. and Hovmøller, M.S. (2001) The recent history of *Puccinia striiformis* f. sp. *tritici* in Denmark as revealed by disease incidence and AFLP markers. *Plant Pathology* 51, 13–23.

Lévesque, C.A. (2001) Molecular methods for detection of plant pathogens – What is the future? *Canadian Journal of Plant Pathology* 24, 333–336.

Lipshutz, R.J., Fodor, S.P.A., Gingeras, T.R. and Lockhart, D.J. (1999) High density synthetic oligonucleotide arrays. *Nature Genetics* 21, 20–24.

Liu, J.Q. and Kolmer, J.A. (1998) Molecular and virulence diversity and linkage disequilibria in asexual and sexual populations of the wheat leaf rust fungus, *Puccinia recondita*. *Genome* 41, 832–840.

Luikart, G. and England, P.E. (1999) Statistical analysis of microsatellite data. *Trends in Ecology and Evolution* 14, 253–256.

Luster, D.G. and Bruckart, W.L. (1998) DNA fingerprinting of weed pathogens for biological control. *Proceedings of the 7th International Congress of Plant Pathology*. British Society for Plant Pathology, Edinburgh, Abstract 3.5.9.

Mahr, F.A. and Bruzzese, E. (1998) The effect of *Phragmidium violaceum* (Schultz) Winter (Uredinales) on *Rubus fruticosus* L. agg in south-eastern Victoria. *Plant Protection Quarterly* 13, 182–185.

Majer, D., Mithen, R., Lewis, B.G. and Pieter, V. (1996). The use of AFLP fingerprinting for the detection of genetic variation in fungi. *Mycological Research* 100, 1107–1111.

Marks, G.C., Pascoe, I.G. and Bruzzese, E. (1984) First record of *Phragmidium violaceum* on blackberry in Victoria. *Australasian Plant Pathology* 13, 12–13.

Maughan, N.J., Lewis, F.A. and Smith, V. (2001) An introduction to arrays. *Journal of Pathology* 195, 3–6.

McCarty, M.K. (1982) Musk thistle (*Carduus thoermeri*) seed production. *Weed Science* 30, 441–445.

McDonald, B.A. (1997) The population genetics of fungi: tools and techniques. *Phytopathology* 87, 448–453.

McDonald, B.A. and Linde, C. (2002) The population genetics of plant pathogens and breeding strategies for durable resistance. *Euphytica* 124, 163–180.

McDonald, B.A. and McDermott, J.M. (1993) Population genetics of plant pathogenic fungi. *BioScience* 43, 311–319.

Milgroom, M.G. (1996) Recombination and the multilocus structure of fungal populations. *Annual Review of Phytopathology* 34, 457–477.

Morin, L., Armstrong, J. and Driver, F. (2001) The science underpinning the release of the bridal creeper rust in Australia. *Proceedings of the 13th Biennial Conference of the Australasian Plant Pathology Society*. Department of Primary Industries, Mareeba, Queensland, p. 78.

Morin, L., Armstrong, J. and Kriticos, D. (2003) Spread and epidemic development of *Puccinia myrsiphylli*, a rust released for weed control in Australia. *Proceedings of the 8th International Congress of Plant Pathology*, Vol. 2, Offered Papers. International Society for Plant Pathology, Christchurch, New Zealand, p. 49.

Mundt, C.C. (1995) Models from plant pathology on the movement and fate of new genotypes of microorganisms in the environment. *Annual Review of Phytopathology* 33, 467–468.

Murray, A.E., Lies, D., Li, G., Nealson, K., Zhou, J. and Tiedje, J.M. (2001) DNA/DNA hybridisation to microarrays reveals gene-specific differences between closely related microbial genomes. *Proceedings of the National Academy of Sciences USA* 98, 9853–9858.

Newton, A.E., Caten, C.E. and Johnson, R. (1985) Variation for isozymes and double-stranded RNA among isolates of *Puccinia striiformis* and two other cereal rusts. *Plant Pathology* 34, 235–247.

Nissen, S.J., Masters, R.A., Lee, D.J. and Rowe, M.L. (1995) DNA-based marker systems to determine genetic diversity of weedy species and their application to biocontrol. *Weed Science* 43, 504–513.

Olsen, M., Hood, L., Cantor, C. and Botstein, D. (1989) A common language for physical mapping of the human genome. *Science* 245, 1434–1435.

Orita, M., Suzuki, Y., Sekiya, T. and Hayashi, K. (1989) Rapid and sensitive detection of point mutations and DNA polymorphisms using the polymerase chain reaction. *Genomics* 5, 874–879.

Parker, P.G., Snow, A.A., Schug, M.D., Booton, G.C. and Fuerst, P.A. (1998) What molecules can tell us about populations: choosing and using a molecular marker. *Ecology* 79, 361–382.

Paxton, J.D. and Groth, J. (1994) Constraints on pathogens attacking plants. *Critical Reviews in Plant Sciences* 13, 77–95.

Pei, M.H. and Ruiz, C. (2000) AFLP evidence of distinct patterns of life-cycle in two forms of *Melampsora* rust on *Salix viminalis*. *Mycological Research* 104, 937–942.

Phillips, D.V., Carbone, I., Gold, S.E. and Kohn, L.M. (2002) Phylogeography and genotype-symptom associations in early and late season infections of canola by *Sclerotinia sclerotiorum*. *Phytopathology* 92, 785–793.

Pryor, A., Boelen, M.G., Dickinson, M.J. and Lawrence, G.J. (1990) Widespread incidence of double-stranded RNAs of unknown function in rust fungi. *Canadian Journal of Botany* 68, 669–676.

Samils, B., Lagercrantz, U., Lascoux, M. and Gullberg, U. (2001) Genetic structure of *Melampsora epitea* populations in Swedish *Salix viminalis* plantations. *European Journal of Plant Pathology* 107, 339–409.

Savile, D.B.O. (1971) Coevolution of the rust fungi and their hosts. *Quarterly Review of Biology* 46, 211–218.

Schafer, J.F. and Roelfs, A.P. (1985) Estimated relation between numbers of urediniospores of *Puccinia graminis* f. sp. *tritici* and rates of occurrence of virulence. *Phytopathology* 75, 749–750.

Steele, K.A., Humphreys, E., Wellings, C.R. and Dickinson, M.J. (2001) Support for a stepwise mutation model for pathogen evolution in Australasian *Puccinia striiformis* f.sp. *tritici* by use of molecular markers. *Plant Pathology* 50, 174–180.

Sunnucks, P. (2000) Efficient genetic markers for population biology. *Trends in Ecology and Evolution* 15, 199–203.

Thrall, P.H. and Burdon, J.J. (2002) Evolution of gene-for-gene systems in metapopulations: the effect of spatial scale of host and pathogen dispersal. *Plant Pathology* 51, 169–184.

Tisdell, C.A. (1990) Economic impact of biological control of weeds and insects. In: Mackauer, M., Ehler, L.E. and Roland, J. (eds) *Critical Issues in Biological Control.* Intercept Press, Andover, UK, pp. 301–316.

Walker, J. (1996) Biogeography of fungi with special reference to Australia. In: Grgurinovic, C. and Mallett, K. (eds) *Fungi of Australia*, Vol. 1A, Introduction–Classification. Australian Biological Resources Study/CSIRO, Canberra, Australia, pp. 263–320.

Wang, K. and Hiruki, C. (2000) Heteroduplex mobility assay detects DNA mutations for differentiation of closely related phytoplasma strains. *Journal of Microbiological Methods* 41, 59–68.

Wapshere, A.J. (1974) A strategy for evaluating the safety of organisms for biological weed control. *Annals of Applied Biology* 77, 201–211.

Wellings, C.R. and McIntosh, R.A. (1990) *Puccinia striiformis* f. sp. *tritici.* Australasia: pathogenic change during the first 10 years. *Plant Pathology* 39, 316–325.

Wiberg, L. and Walker, J. (1990) *Uromyces minor* on peas in Australia, with notes on other rusts of *Pisum. Australasian Plant Pathology* 19, 42–45.

Witsenboer, H., Vogel, J. and Michelmore, R.W. (1997) Identification, genetic localization, and allelic diversity of selectively amplified microsatellite polymorphic loci in lettuce and wild relatives. *Genome* 40, 923–936.

Zeigler, R.S., Cuoc, L.X., Scott, R.P., Bernado, M.A., Chen, D.H., Valent, B. and Nelson, R.J. (1995) The relationship between lineage and virulence in *Pyricularia grisea* in the Phillipines. *Phytopathology* 85, 443–451.

Zhan, J., Mundt, C.C., Hoffer, M.E. and McDonald, B.A. (2002) Local adaptation and effect of host genotype on the rate of pathogen evolution: an experimental test in a plant pathosystem. *Journal of Evolutionary Biology* 15, 634–647.

5

Tracing the Origin of Pests and Natural Enemies: Genetic and Statistical Approaches

G.K. Roderick

Environmental Science, Policy and Management, Division of Insect Biology, University of California, Berkeley, CA 94720-3112, USA

Introduction

Invasive species are now recognized as a major economic and environmental concern affecting nearly every aspect of life on Earth. Statistics concerning the impact of invasive species are staggering. For example, it is roughly estimated that costs associated with the damage caused by, and the control of, invasive species may total over US$100 billion annually (Pimentel *et al.*, 2000). Invasive species have been implicated as a significant factor in approximately half of all efforts to recover threatened and endangered species in the USA, with much higher impacts in other geographical regions (Wilcove *et al.*, 1998). This environmental impact from invasive species rivals that of impending climate change and clearly must be considered a significant form of global change.

Despite the impact of invasive organisms, and the fact that they have been recognized as an issue of scientific study for many decades (e.g. Elton, 1958), many aspects of the biology of invasions remain unknown (Williamson, 1996). One area in which knowledge is lacking is in the identification of the source and general pattern and spread of invasive populations. This information is of value both to basic researchers and to resource managers, whose goal it is to contain, control or eradicate invasive species. For example, determining the origins of a pest and understanding the role of natural enemies in its native habitat can aid in designing strategies for control.

Here, I examine the general issues associated with determining origins of non-indigenous species, including relevant basic principles of population biology and genetics, as well as novel molecular and statistical tools. It is hoped that this review can not only assist those interested in determining the origins of populations, but also prompt an interest in the collection of data that might facilitate the use of these methods in the future.

Problems and Approaches

Patterns of spread

The problem of determining origins of populations is greatly simplified if one can propose a general model for the pattern of spread. For example, for an invasive species that is currently restricted in its range, one might ask 'Has the species only recently arrived in this locality, perhaps through anthropogenic intervention, or has the species been present in the area for some time, though with restricted expansion in range?' While both scales of colonization are probably involved in most biological invasions, it may be difficult, in practice, to distinguish these patterns. For example, some invasive species may be present undetected for many generations at low population levels, before rapidly exploding in numbers to detectable levels (Kaneshiro, 1993; Williamson, 1996; Roderick and Howarth, 1999). In such cases it may be difficult or impossible to infer the invasion history.

Until recently, biological invasions have been modelled largely as diffusion processes – indeed, modelling expanding populations that spread in this fashion has been very successful (Elton, 1958; Andow, 1999). It has been more difficult to model the 'jump' type of dispersal events, in which populations are founded by colonists that do not come from neighbouring localities, but rather from longer distances, and in many instances from different continents. Such 'jump' types of dispersal appear to be the rule for many invasive species, particularly where association with humans is an important element of their lifestyle. Good examples of jump-type dispersal include Argentine ants (Suarez *et al.*, 2001) and Mediterranean fruit flies (Carey, 1996; Davies *et al.*, 1999a). For invasions both of Argentine ants and of Mediterranean fruit flies, spread within more-defined geographical localities has occurred subsequent to the initial colonization. It may be that most biological invasions consist of some form of jump dispersal associated with the initial colonization event, followed by more local spread. The discussion that follows will emphasize the importance of determining origins of jump-type dispersal events, though many of the same methods can also be used to track the spread of populations expanding more gradually.

Invasion genetics

Individuals that colonize a new habitat are, by definition, a genetic subset of their source population, representing a genetic bottleneck relative to the source population. Over time, the colonizing population will be expected to continue to diverge genetically from its source in allele frequency, as alleles are lost by chance (Fig. 5.1). The alleles most likely to be lost are the so-called 'rare alleles', or those in low frequency in the population. Given more generations, entire lineages of alleles will be

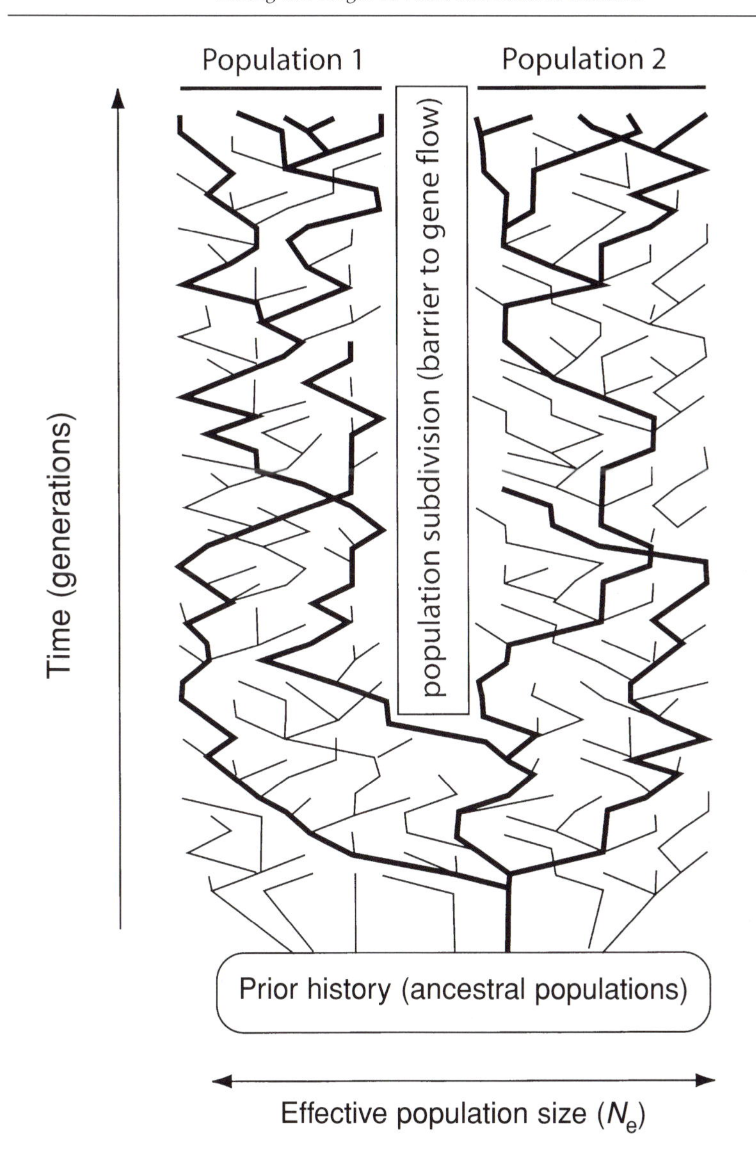

Fig. 5.1. Effective population size (horizontal axis) and number of generations (vertical axis) affect the loss of lineages following population subdivision (after Avise, 2000). A genealogy of haplotypes (or alleles) is shown. Dark lines represent ancestral lineages present in the populations at the time of subdivision; light lines are lineages that have been lost by chance through time.

lost, and with a sufficient number of generations in isolation, the alleles found in a new population will become monophyletic, that is, all alleles at a locus within the population will be descended from one ancestral allele. This scenario embodies the process of genetic drift, which in itself can be a powerful force of evolution.

The rate at which alleles at a locus within a population sort to monophyly depends on two parameters: the genetic effective size of the population (N_e) and the number of generations since separation. The larger the effective size of the founding population, the longer it will take for the population to diverge genetically from its source, and the longer it will take for lineages to be lost and for the existing lineages to 'sort' to monophyly. Indeed, very small founding populations will show a severe decrease in genetic diversity. The change in genetic composition will also depend on the number of generations since separation – more specifically, these processes are likely to take place after in the order of N_e number of generations (see discussion by Avise, 2000).

Through the colonization process described above, a founding population will be expected to lose alleles relative to its source population. This reduction will also be reflected in measures of heterozygosity, or the number of loci that are expected to be heterozygous. Initially, if the initial bottleneck at colonization is brief, and if the population can grow rapidly, there may be little reduction in heterozygosity. Under such conditions – a short-duration bottleneck followed by rapid population growth – heterozygosity can recover rather rapidly (Fig. 5.2),

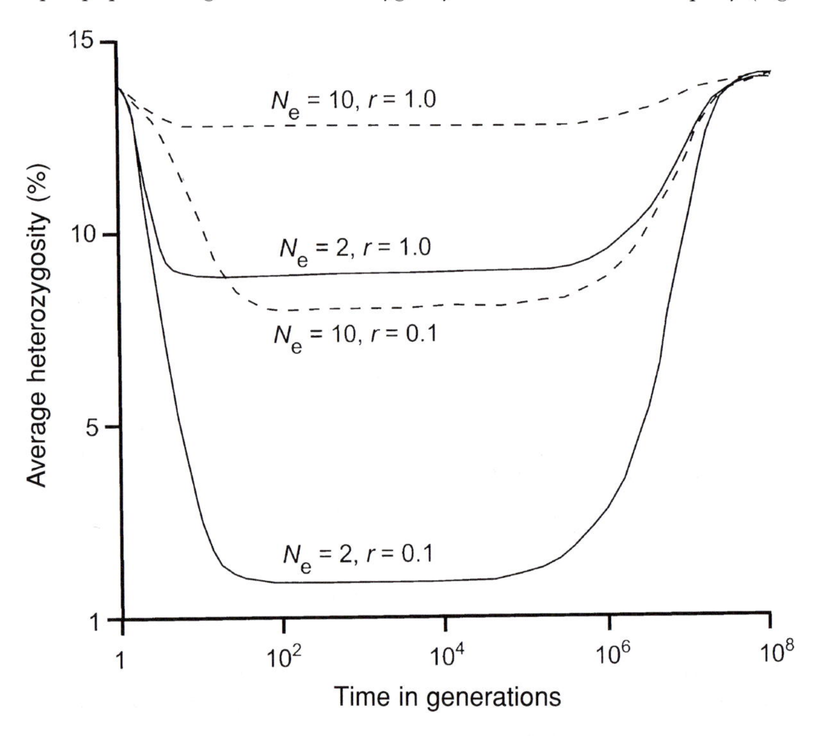

Fig. 5.2. The loss of heterozygosity following a bottleneck is influenced not only by the severity of the bottleneck, but also by the subsequent rate of population growth (after Nei *et al.*, 1975). Patterns of heterozygosity through time following a bottleneck are shown for a bottleneck of N_e = 2 (solid lines) and N_e = 10 (dashed lines), each for two growth rates, r = 1.0 and r = 0.1.

ameliorating the potentially negative genetic effects of colonization (Nei *et al.*, 1975). By contrast, the number of alleles in a population following a bottleneck may remain reduced for some time, because new alleles in the founding population can only arise through mutation or migration from other populations. Changes in both allele number and heterozygosity can be used to estimate the severity of a bottleneck (e.g. Luikart and Cornuet, 1998).

Genetic bottlenecks often characterize invasive populations – they usually show a reduction in allele number and genetic diversity compared with suspected source populations (Fig. 5.3). Good examples of this phenomenon can be found in Mediterranean fruit flies (Davies *et al.*, 1999b) and *Culex* mosquitoes (Fonseca *et al.*, 2000). As is usually the case, the genetic bottleneck is expected to be more severe in mitochondrial DNA (mtDNA), which is usually haploid (one copy), compared with diploid nuclear DNA. MtDNA therefore has a lower effective population size (N_e) than nuclear DNA.

Despite its ability to document a loss in genetic variability associated with colonization, there are relatively few cases in which this reduction in genetic variation can be associated directly with a reduction in invasive success. For example,

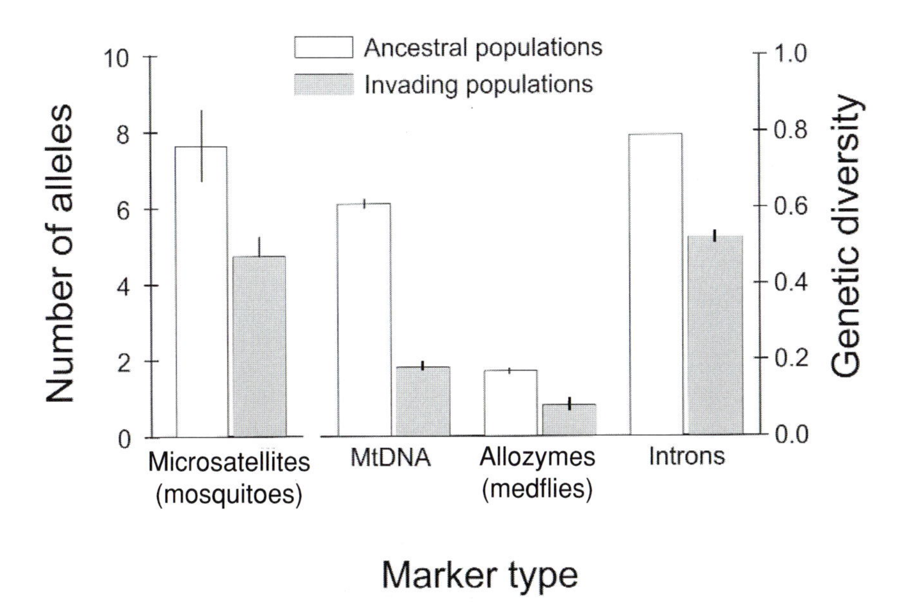

Fig. 5.3. Evidence for genetic bottlenecks following invasions can be seen in loss of alleles and decrease in genetic diversity. Reduction in numbers of alleles are shown for a study of microsatellites in *Culex* mosquitoes, comparing the presumed ancestral source population (Hawaii, Maui) with invading populations (other Hawaiian Islands) (data from Fonseca *et al.*, 2000). Decreases in genetic diversity are shown in analyses of mitochondrial DNA (mtDNA), allozymes and nuclear introns in medfly, comparing ancestral African populations with invading populations worldwide (after Davies *et al.*, 1999b).

in some plants loss of variability upon colonization may reduce the ability to outcross, thus limiting reproduction to vegetative means (Yano *et al.*, 1999). Yet, even plants that reproduce vegetatively can be effective invaders (e.g. knotweed; Hollingsworth and Bailey, 2000).

In some systems it may be that a lack of genetic variability actually *contributes* to success. A good example of this phenomenon is shown in invasive populations of the Argentine ant, which have probably lost genetic variation coding for external hydrocarbon differences associated with colony recognition (Holway *et al.*, 1998, 2002; Tsutsui *et al.*, 2000). The result has been that invasive colonies are able to merge and share queens, unlike colonies in their native range, where such cues are associated with inter-colony conflict, fighting and population control (Suarez *et al.*, 2002; Tsutsui *et al.*, 2003). Similar results have been found for Argentine ants elsewhere, and polygyne colonies appear to be the rule for invasive ant species when compared with their native counterparts (Holway *et al.*, 2002).

It is also likely that one may underestimate the effects of genetic loss associated with biological invasions, particularly if the negative impact of the loss is associated with initial colonization. Under this scenario, populations that were lost quickly as a result of low genetic variability no longer exist to be sampled and thus one has no way to assess the negative effects of loss of genetic variability.

Three useful conclusions follow from these general considerations:

1. As a result of sampling variation and isolation, different invading populations should be expected to differ genetically from one another, and to differ genetically from their source population. This phenomenon is seen in invading populations (e.g. different medfly populations in California; Meixner *et al.*, 2002). Unless they are actively exchanging genes, different invading populations of the same species should be expected to diverge genetically, regardless of their source of origin. Thus, the fact that invading populations differ from one another genetically cannot be used alone as evidence of multiple colonization events.
2. The importance of genetic effects, if they exist, will be influenced by the founding population size and subsequent rate of population growth.
3. Because of the period of time that a population must be isolated for unique genetic markers to develop (in the order of N_e generations), it may be difficult, if not impossible, to find population-specific markers that can be used, either to distinguish particular invasive populations or to identify source populations. Thus, finding diagnostic markers with which to study invasions will be problematic, unless the populations in question have been established and isolated for a very long time.

Clearly, a number of factors complicate matters for those desiring to use genetic markers to trace the history of populations. Not only can patterns of colonization be confusing, such as a mix of jump and diffusive types of dispersal, but founding populations are expected to have little genetic diversity that can be easily quantified. Further, most invasive populations have been founded relatively recently, violating the usual assumption of genetic equilibrium, under which an increase in numbers of alleles as a result of migration is balanced by a loss of alleles through genetic drift. This assumption is an important one in the traditional use of genetic data to infer past and current demographic processes. Fortunately, several methods of analysis have been developed recently that do not rely on the assumption of genetic equilibrium. These methods are discussed below, following a brief description of the population-genetic markers they employ.

Types of genetic markers

A diversity of genetic markers now exists that can be used for studies of populations and sub-species (Table 5.1). Allozymes (protein isozymes), developed in the 1960s, were among the first genetic markers used for the study of populations, and are still useful where access to fresh material is possible. Probably the most commonly used marker for animal studies is mitochondrial DNA, through either

Table 5.1. Types of genetic markers used in studies of population origins (see text for references providing more detailed information for each type of marker).

Marker	Effective population size	Co-dominant?	Notes
MtDNA sequences	N (haploid)	–	Construction of trees possible
Mitochondrial restriction fragment length polymorphisms (RFLPs)	N (haploid)	–	
Allozymes	$2N$ (diploid)	Yes	Requires fresh or frozen specimens
Microsatellites	$2N$ (diploid)	Yes	
Single nucleotide polymorphisms (SNPs)	$2N$ (diploid)	Yes	
Randomly amplified polymorphic DNA (RAPDs)	$2N$ (diploid)	No	Highly variable
Amplified fragment length polymorphisms (AFLPs)	$2N$ (diploid)	No	Highly variable
Introns (nuclear)	$2N$ (diploid)	Yes	
Intragenic transcribed spacers (ITS)	$2N$ (diploid)	Yes	Concerted evolution probably involved
Nuclear sequences	$2N$ (diploid)	Yes	Construction of trees possible
Nuclear RFLPs	$2N$ (diploid)	Yes	

sequencing or fragments, such as restriction fragment length polymorphisms (RFLPs) (Simon *et al.*, 1994; Caterino *et al.*, 2000). Markers that reflect genetic variation at nuclear loci include microsatellites, randomly amplified polymorphic DNA (RAPD), amplified fragment length polymorphisms (AFLPs), non-coding regions within genes (introns) of nuclear DNA, and single nucleotide polymorphisms (SNPs), among others (for description of markers and associated methods see Hillis *et al.*, 1996; Roderick, 1996, 2003; Bohonak, 1999). In plants, mitochondrial DNA does not display as much within-population variation as animal mitochondrial DNA, and studies at the level of populations have used sequence variation in chloroplast DNA and intragenic transcribed spacing regions (ITS), as well as nuclear introns (e.g. Gaskin and Schaal, 2002). Which marker to choose depends largely on the question asked as well as access to material. For questions of origins, simulation studies suggest that increased power comes with information from many loci (Boecklen and Howard, 1997; Rannala and Mountain, 1997), and since mitochondrial DNA is typically inherited as one locus, the need for multiple loci necessitates examination of nuclear DNA. Most population-level analyses also require information from both alleles at a nuclear locus, as can be determined by so-called co-dominant loci, such as microsatellites, allozymes, introns and SNPs. An alternative approach for dominant markers is to estimate heterozygote information from genotype frequencies and assuming Hardy–Weinberg equilibrium. Because this assumption may not be valid for recently founded populations, dominant markers such as RAPDs and AFLPs are not as well suited as dominant markers for studies of origins. Nevertheless, the accessibility and generally high levels of variability of both RAPDs and AFLPs make them ideal for other uses, such as quickly quantifying genetic variation or in providing markers for the mapping of genomes.

Genealogies

The genealogical history of life provides one set of approaches to estimating origins of populations. The idea is simple – provided sufficient genetic variation exists for study, one can simply trace the alleles of an individual back to their parents, grandparents and so on. This is the basis for the concept of *phylogeography*, in which the genealogical history of alleles is examined in the context of the geographical localities from which the alleles are sampled. It should be noted that this approach examines the history of the alleles present *within individuals* (i.e. such that haplotypes are at the tips of the tree), rather than using a composite of individuals within populations as the datum (i.e. with populations at the tips of the tree). The problem with the latter approach for determination of origins is that a single population may contain individuals that stem from more than one population, and by grouping these individuals, information about potential separate origins cannot be recovered by inspecting the tree.

Many commonly used methods of phylogenetic analysis can be used for this approach of tracing the history of alleles (see Swofford *et al.*, 1996; Hall, 2001).

These methods have been particularly useful in questions at the species level (e.g., cryptic species, closely related species) and above (reviewed in Chapter 6, this volume). However, as a group, these methods have three potential pitfalls for the study of origins of populations:

1. These methods usually assume that the branching points, or nodes, represent alleles within individuals that existed in the past and are therefore no longer represented in the population. However, it is quite likely that, within current populations, individuals exist with alleles that have given rise to other alleles that are also present in the population.
2. The emphasis in the development of these methods has been on the level of species and above (genera, families, etc.), rather than on populations.
3. The methods require that the history of alleles be bifurcating, an assumption which may be violated, due to recombination.

Despite these limitations, such phylogeographic methods have proven useful in determining origins of populations of some organisms, particularly when rapidly evolving loci are examined; a good example is the history of the West Nile virus that was traced from a likely source in the Middle East, probably Israel, to New York city (Lanciotti *et al.*, 1999).

A recent methodology based on the idea of genealogical nature of individuals is that of *nested clade analysis* (NCA; Templeton, 1998; Posada and Crandall, 2001). Unlike phylogenetic methods suited for understanding patterns in the distant past, this type of analysis allows both for extant alleles to be represented as historical nodes and for non-bifurcating branches, or networks. The method works by reconstructing the most recent history of individuals first, through procedures similar to parsimony, optimizing for the fewest changes. Subsequently, progressively deeper relationships are reconstructed, creating layers of groupings. The groupings are then examined to assess independence of genealogy and geography, as well as to infer aspects of population history, including the identification of ancestral genotypes, the possibility of population expansion and the existence of isolation by distance. The method can be implemented by the programs TCS (Clement *et al.*, 2000) and GEODIS (Posada *et al.*, 2000). While the general inference structure of this approach is only beginning to be tested in a rigorous statistical manner (e.g. Knowles and Maddison, 2002), the hypothesized history of alleles that results from TCS has been very useful in testing hypotheses concerning population histories (e.g. Templeton, 2002). For example, Gaskin and Schaal (2002) identified Eurasian origins for the invasive plant *Tamarix* in the USA and were able to document that most invasive populations were of hybrid origin. An example of this approach for insect populations is provided by J.-H. Mun and colleagues (2003) in an effort to determine whether the pumpkin fruit fly, *Bactrocera depressa* (Tephritidae), is a recent immigrant to Japan or Korea. In this study, both Japan and Korea were found to have ancestral populations of this fly, perhaps dating to a common land connection between them; however, there was also evidence of some recent movement of individuals from Korea to Japan.

Analysis of frequencies

While the genealogical approach can be very powerful, it is laborious and expensive to obtain DNA sequence information for many individuals and for many loci. Thus, analysis of frequencies of alleles remains the preferred method for determining origins of individuals within populations. One advantage of examining frequencies is that it is possible to obtain information from many loci, a prerequisite for any substantial statistical power (see Boecklen and Howard, 1997). Studies of frequencies of alleles fall into two general categories: assignment tests and, for lack of a better term, average estimates of gene flow.

In assignment tests, alleles at one or more loci are used to statistically assign the individual in question to one or more populations from which it could have originated. The general procedure is to compare the likelihood that an individual could have come from the locality in which it was found with the likelihood that it came from any of the other possible source populations. Paetkau *et al.* (1995) and Rannala and Mountain (1997) were among the first to adapt this more general statistical approach for the study of origins of biological populations, and it has been applied to many species since (Davies *et al.*, 1999b). Most recently, the method has been expanded to facilitate comparisons between many populations (Pritchard *et al.*, 2000), as implemented in the program STRUCTURE (Pritchard, 2003), which allows assessment of a number of associated hypotheses, including whether an individual is likely to have descended from a particular population in the past. This procedure has been used to determine the origins of populations of a variety of invasive species, including tephritid fruit flies, such as Mediterranean fruit fly (Davies *et al.*, 1999a; Bonizzoni *et al.*, 2001) and pumpkin fruit fly (J.-H. Mun *et al.*, 2003).

A more traditional approach to examining population origins has been the use of genetic data to estimate gene flow between populations, making use of summaries of population-genetic subdivision among populations, such as F_{ST}, R_{ST}, or ϕ_{ST} (see Slatkin, 1985, 1995; Excoffier *et al.*, 1992). The limitation of this approach has been that usually an average level of gene flow is calculated, rather than a more specific determination of the pathways of movement of individuals (for a discussion of this topic relevant to insects, see Roderick, 1996; Bohonak, 1999; Bohonak and Roderick, 2001). However, recent developments in both theory and computational power have provided more specific estimation of genetic isolation by distance (Slatkin, 1994), and the ability to estimate non-reciprocal gene flow among sets of populations (Beerli and Felsenstein, 2001). These approaches can be implemented by the programs IBD (Bohonak, 2002) and MIGRATE (Beerli, 2002), respectively.

Another approach to examining origins, again made possible by recent advances in computing power, is the use of simulation, such as Monte Carlo procedures, to distinguish between alternative scenarios concerning population history and demography. This method has many possible applications, as it can be tailored to test very specific hypotheses. The general idea is to simulate historical scenarios that might have to do with different sources, different patterns of introduction, and different population sizes. Parameters calculated from actual observed data from the field can then be compared with those calculated from simulated data, allowing one to ask, do the real data fall outside the confidence limits predicted by any particular scenario? If so, that scenario can be rejected. Recently, Bohonak and colleagues (2001) used this approach to distinguish between various scenarios for the invasion by medflies of the New World, particularly Central and South America. Under reasonable population sizes and levels of movement estimated from existing populations, existing genetic data are not consistent with multiple introductions of flies into these regions.

Bioinformatics

It has been noted that the numbers of loci are important in determining the power of the various statistical approaches, both to be able to distinguish sources, and to estimate population parameters with some confidence. One way to harness the power provided by different approaches employed by different research groups is to coordinate efforts, particularly with respect to analysis of genetic data. Such data can also be linked to geographical origin in a geographical information system (GIS) framework. Work in this area is currently in progress by a number of research groups, with the ultimate goal of providing electronically (over the Internet) real-time access to existing genetic data, including relevant statistics (e.g. likelihood of assignment). One example is the BioInvasion Genetic (BIG) database, currently under development by Bohonak and colleagues (San Diego State University) for the study of invasive tephritid fruit flies.

Other methods

While this review has focused on the use of genetic methods to trace the history of populations, it should be noted that other non-genetic methods might also be of use in determining origins. For example, ratios of stable isotopes have been used to examine the origins of red wine (Gimenez-Miralles *et al.*, 1999) and ratios of stable isotopes coupled with ratios of amino acids to examine the history of cheeses (Manca *et al.*, 2001). As for the genetic methods discussed, the usefulness of non-genetic methods in determining the histories of populations will depend largely on the availability of samples of potential source populations. Bioinformatic approaches would also be applicable to isotope data, provided that similar standards can be achieved among different laboratories.

Discussion

The study of origins of populations is currently expanding rapidly, facilitated by novel molecular methods and advances in computational ability. Computer programs for analysing population-genetic data for the study of population origins have already been developed and are available on the Internet. A number of case studies have demonstrated their usefulness. However, despite the ease of analysis, some limitations in the use of genetic data to determine population origins persist, particularly the difficulties involved in the development of molecular markers for novel taxa, as well as limited accessibility to material from potential source material worldwide. It is likely that, for both issues, coordinated international and multi-agency efforts will be needed for substantial progress.

The methods for the determination of population origins described here can be applied to the area of biological control in a number of ways, including: (i) the determination of origins to identify the pathways of invasion of non-indigenous species, as well as their subsequent spread in their new habitat; and (ii) understanding of the biology of pests in the habitats in which they are indigenous, in order to develop effective methods of control, particularly as it relates to the use of predators and parasites. Other applications of the methodologies show some promise for the future (Roderick and Navajas, 2003). For example, by following the fates of genetically or chemically marked individuals, it may be possible to address the issue of the importance of genetic variation in the success of colonization.

One might ask, 'Do multiple genotypes or individuals from multiple origins have a greater chance of success in new environments?' That is, do populations with greater genetic variability have a greater chance of establishment? The importance of particular traits can also be examined by studying individuals in the field that possess those traits. For example, are particular strains of parasitoids that have been selected for increased searching ability better able to persist in novel habitats? If and when genetically modified insects are released, methods described here can be used to trace their fate (and the fate of their genes) in natural and managed habitats. Finally, extensions of the methods used to determine origins can be applied in identification of cryptic species in all stages of the process of colonization and establishment (see Chapter 6, this volume).

Conclusions

1. The use of genetic and other methods for determining origins of non-indigenous species offers much promise.

2. The typical demographic processes associated with invasions necessitate the use of non-traditional methods in population genetics.

3. It is unlikely that population-specific genetic markers can be developed routinely for non-indigenous species.

4. There is little evidence that lack of genetic variation limits success in colonization, though more data are needed.

5. More genetic loci are nearly always more useful in tracing population histories; bioinformatics approaches can be used to combine information from the work of many research laboratories.

6. Tracing the fate of particular genotypes in the environment allows the study of a number of issues related to invasion biology, biological control and the use of genetically modified organisms.

Acknowledgements

I thank the editors of this volume and the symposium organizers for making both resounding successes! This work is supported by grants from the California Department of Food and Agriculture, US National Science Foundation, US Department of Agriculture, and the University of California Agricultural Experiment Station.

References

Andow, D. (1999) Spread of invading organisms. In: Yano, E., Matsuo, K., Shiyomi, M. and Andow, D. (eds) *Biological Invasions of Pests and Beneficial Organisms*. National Institute of Agro-Environmental Sciences, Tsukuba, Japan, pp. 66–77.

Avise, J.C. (2000) *Phylogeography. The History and Formation of Species*. Harvard University Press, Cambridge, Massachusetts.

Beerli, P. (2002) *Migrate: Documentation and Program*, Version 1.6.9. University of Washington, Seattle, Washington.

Beerli, P. and Felsenstein, J. (2001) Maximum likelihood estimation of a migration matrix and effective population sizes in *n* subpopulations by using a coalescent approach. *Proceedings of the National Academy of Sciences USA* 98, 4563–4568.

Boecklen, W.J. and Howard, D.J. (1997) Genetic analysis of hybrid zones: numbers of markers and power of resolution. *Ecology* 78, 2611–2616.

Bohonak, A.J. (1999) Dispersal, gene flow, and population structure. *Quarterly Review of Biology* 74, 21–45.

Bohonak, A.J. (2002) IBD (Isolation By Distance): a program for analyses of isolation by distance. *Journal of Heredity* 93, 153–154.

Bohonak, A.J. and Roderick, G.K. (2001) Dispersal of invertebrates among temporary ponds: are genetic estimates accurate? *Israel Journal of Zoology* 47, 367–386.

Bohonak, A.J., Davies, N., Villablanca, F.X. and Roderick, G.K. (2001) Invasion genetics of New World medflies: testing alternative colonization scenarios. *Biological Invasions* 3, 103–111.

Bonizzoni, M., Zheng, L., Guglielmino, C.R., Haymer, D.S., Gasperi, G., Gomulski, L.M. and Malacrida, A.R. (2001) Microsatellite analysis of medfly bioinfestations in California. *Molecular Ecology* 10, 2515–2524.

Carey, J.R. (1996) The incipient Mediterranean fruit fly population in California: implications for invasion biology. *Ecology* 77, 1690–1697.

Caterino, M.S., Cho, S. and Sperling, F.A.H. (2000) The current state of insect molecular systematics: a thriving tower of Babel. *Annual Review of Entomology* 45, 1–54.

Clement, M., Posada, D. and Crandall, K.A. (2000) TCS: a computer program to estimate gene genealogies. *Molecular Ecology* 9, 1657–1659.

Davies, N., Villablanca, F.X. and Roderick, G.K. (1999a) Bioinvasions of the medfly,

Ceratitis capitata: source estimation using DNA sequences at multiple intron loci. *Genetics* 153, 351–360.

Davies, N., Villablanca, F.X. and Roderick, G.K. (1999b) Determining the sources of individuals in recently founded populations: multilocus genotyping in non-equilibrium genetics. *Trends in Ecology and Evolution* 14, 17–21.

Elton, C. (1958) *The Ecology of Invasions by Animals and Plants.* Chapman & Hall, London.

Excoffier, L., Smouse, P. and Quattro, J. (1992) Analysis of molecular variance inferred from metric distances among DNA haplotypes: application to human mitochondrial DNA restriction data. *Genetics* 131, 479–491.

Fonseca, D.M., LaPointe, D.A. and Fleischer, R.C. (2000) Bottlenecks and multiple introductions: population genetics of the vector of avian malaria in Hawaii. *Molecular Ecology* 9, 1803–1814.

Gaskin, J.F. and Schaal, B.A. (2002) Hybrid *Tamarix* widespread in U.S. invasion and undetected in native Asian range. *Proceedings of the National Academy of Sciences USA* 99, 11256–11259.

Gimenez-Miralles, J.E., Salazar, D.M. and Solana, I. (1999) Regional origin assignment of red wines from Valencia (Spain) by ^{2}H NMR and ^{13}C IRMS stable isotope analysis of fermentative ethanol. *Journal of Agricultural and Food Chemistry* 47, 2645–2652.

Hall, B.G. (2001) *Phylogenetic Trees Made Easy: a How-to Manual for Molecular Biologists.* Sinauer, Sunderland, Massachusetts.

Hillis, D., Moritz, C. and Mable, B.K. (eds) (1996) *Molecular Systematics,* 2nd edn. Sinauer, Sunderland, Massachusetts.

Hollingsworth, M.L. and Bailey, J.P. (2000) Evidence for massive clonal growth in the invasive weed *Fallopia japonica* (Japanese knotweed). *Botanical Journal of the Linnean Society* 133, 463–472.

Holway, D.A., Suarez, A.V. and Case, T.J. (1998) Loss of intraspecific aggression underlies the success of a widespread invasive social insect. *Science* 282, 949–952.

Holway, D.A., Lach, L., Suarez, A.V., Tsutsui, N.D. and Case, T.J. (2002) The causes and consequences of ant invasions. *Annual Review of Ecology and Systematics* 33, 181–233.

Kaneshiro, K.Y. (1993) Introduction, colonization, and establishment of exotic insect populations: fruit flies in Hawaii and California. *American Entomologist* 39, 23–29.

Knowles, L.L. and Maddison, W.P. (2002) Statistical phylogeography. *Molecular Ecology* 11, 2623–2635.

Lanciotti, R.S., Roehrig, J.T., Deubel, V., Smith, J., Parker, M., Steele, K., Crise, B., Volpe, K.E., Crabtree, M.B., Scherret, J.H., Hall, R.A., MacKenzie, J.S., Cropp, C.B., Panigrahy, B., Ostlund, E., Schmitt, B., Malkinson, M., Banet, C., Weissman, J., Komar, N., Savage, H.M., Stone, W., McNamara, T. and Gubler, D.J. (1999) Origin of the West Nile virus responsible for an outbreak of encephalitis in the northeastern United States. *Science (Washington DC)* 286, 2333–2337.

Luikart, G. and Cornuet, J.M. (1998) Empirical evaluation of a test for identifying recently bottlenecked populations from allele frequency data. *Conservation Biology* 12, 228–237.

Manca, G., Camin, F., Coloru, G.C., Del Caro, A., Depentori, D., Franco, M.A. and Versini, G. (2001) Characterization of the geographical origin of Pecorino Sardo cheese by casein stable isotope (^{13}C/^{12}C and ^{15}N/^{14}N) ratios and free amino acid ratios. *Journal of Agricultural and Food Chemistry* 49, 1404–1409.

Meixner, M.D., McPheron, B.A., Silva, J.G., Gasparich, G.E. and Sheppard, W.S. (2002) The Mediterranean fruit fly in California: evidence for multiple introductions and persistent populations based on microsatellite and mitochondrial DNA variability. *Molecular Ecology* 11, 891–899.

Mun, G.-H., Bohonak, A.G. and Roderick, G.K. (2003) Invasion Genetics of the pumpkin fruit fly (*Bactrocera depressa*). *Molecular Ecology* (in press).

Nei, M., Maruyama, T. and Chakraborty, R. (1975) The bottleneck effect and genetic variability in populations. *Evolution* 29, 1–10.

Paetkau, D., Calvert, W., Stirling, I. and Strobeck, C. (1995) Microsatellite analysis of population structure in Canadian polar bears. *Molecular Ecology* 4, 347–354.

Pimentel, D., Lack, L., Suniga, R. and Morrison, D. (2000) Environmental and economic costs of nonindigenous species in the United States. *Bioscience* 50, 53–65.

Posada, D. and Crandall, K.A. (2001) Intraspecific gene genealogies: trees grafting into networks. *Trends in Ecology and Evolution* 16, 37–45.

Posada, D., Crandall, K.A. and Templeton, A.R. (2000) GeoDis: a program for the cladistic nested analysis of the geographical distribution of genetic haplotypes. *Molecular Ecology* 9, 487–488.

Pritchard, J.K. (2003) *Structure*, Version 2.0. University of Chicago, Chicago.

Pritchard, J.K., Stephens, M. and Donnelly, P. (2000) Inference of population structure using multilocus genotype data. *Genetics* 155, 945–959.

Rannala, B. and Mountain, J.L. (1997) Detecting immigration using multilocus genotypes. *Proceedings of the National Academy of Sciences USA* 94, 9197–9201.

Roderick, G.K. (1996) Geographic structure of insect populations: gene flow, phylogeography, and their uses. *Annual Review of Entomology* 41, 263–290.

Roderick, G.K. (2003) Genetic variation. In: Resh, V.H. and Cardé, R. (eds) *Encyclopedia of Insects*. Academic Press, San Diego, pp. 478–481.

Roderick, G.K. and Howarth, F.G. (1999) Invasion genetics: natural colonizations, nonindigenous species, and classical biological control. In: Yano, E., Matsuo, K., Shiyomi, M. and Andow, D. (eds) *Biological Invasions of Pests and Beneficial Organisms*. National Institute of Agro-Environmental Sciences, Tsukuba, Japan.

Roderick, G.K. and Navajas, M. (2003) Genetics and evolution in biological control: proving ground or predictive science? *Nature Reviews Genetics* (in press).

Simon, C., Fratti, F., Beckenbach, A., Crespi, B., Liu, H. and Flook, P. (1994) Evolution, weighting, and phylogentic utility of mitochondrial gene sequences and a compilation of conserved polymerase chain reaction primers. *Annals of the Entomological Society of America* 87, 651–701.

Slatkin, M. (1985) Gene flow in natural populations. *Annual Review of Ecology and Systematics* 16, 393–430.

Slatkin, M. (1994) Gene flow and population structure. In: Real, L.A. (ed.) *Ecological Genetics*. Princeton University Press, Princeton, New Jersey, pp. 3–17.

Slatkin, M. (1995) A measure of population subdivision based on microsatellite allele frequencies. *Genetics* 139, 457–462.

Suarez, A.V., Holway, D.A. and Case, T.J. (2001) Predicting patterns of spread in biological invasions dominated by long-distance jump dispersal: insights from Argentine ants. *Proceedings of the National Academy of Sciences USA* 98, 1095–1100.

Suarez, A.V., Holway, D.A., Liang, D., Tsutsui, N.D. and Case, T.J. (2002) Spatiotemporal patterns of intraspecific aggression in the invasive Argentine ant. *Animal Behaviour* 64, 697–708.

Swofford, D.L., Olsen, G.J., Waddell, P.J. and Hillis, D.M. (1996) Phylogenetic inference. In: Hillis, D.M., Moritz, C. and Mable, B.K. (eds) *Molecular Systematics*. Sinauer, Sunderland, Massachusetts.

Templeton, A.R. (1998) Nested clade analysis of phylogeographic data: testing hypotheses about gene flow and population history. *Molecular Ecology* 7, 381–397.
Templeton, A.R. (2002) Out of Africa again and again. *Nature* 416, 45–51.
Tsutsui, N.D., Suarez, A.V., Holway, D.A. and Case, T.J. (2000) Reduced genetic variation and the success of an invasive species. *Proceedings of the National Academy of Sciences USA* 97, 5948–5953.
Tsutsui, N.D., Suarez, A.V. and Grosberg, R.K. (2003) Genetic diversity, asymmetrical aggression, and recognition in a widespread invasive species. *Proceedings of the National Academy of Sciences USA* 100, 1078–1083.
Wilcove, D.S., Rothstein, D., Dubow, J., Phillips, A. and Losos, E. (1998) Quantifying threats to imperiled species in the United States. *BioScience* 48, 607–615.
Williamson, M. (1996) *Biological Invasions*. Chapman & Hall, London.
Yano, E., Matsuo, K., Shiyomi, M. and Andow, D. (eds) (1999) *Biological Invasions of Pests and Beneficial Organisms*. National Institute of Agro-Environmental Sciences, Tsukuba, Japan.

6

Tracing the Origin of Cryptic Insect Pests and Vectors, and their Natural Enemies

J.K. Brown

Department of Plant Sciences, University of Arizona, Tucson, AZ 85721, USA

Introduction

In the United Kingdom alone, 1700 arthropods have been introduced, with 1500 of them having an economic impact, and about 170 being considered alien pests. Thirty to forty per cent of crop losses are attributed to damage caused by introduced pests in the UK and USA (Pimentel, 2001). More than 50,000 species of arthropods, birds, mammals, microbes and plant weeds have been introduced into the USA since North America was discovered by Columbus. Damage attributed to these introductions has been estimated at US$137 billion. Worldwide it is estimated that over 400,000 non-indigenous species exist, and it has become increasingly clear that many introduced species negatively affect the 'global ecological integrity', and are major factors in the extinction of native species within the habitat (Pimentel *et al.*, 2000; Craig, 2001; Pimentel, 2001; Ebert *et al.*, 2002). Most mammals, birds and plants have been introduced intentionally, whereas arthropods and microbes are typically accidental introductions (Pimentel, 2001). As such, arthropods and microbial pathogens are amongst the most difficult to discover in a timely manner and, hence, to restrict their movement and/or eradicate them.

In addition, upsurgent insects from both local and exotic locations are important pests and vectors of plant pathogens. Insect pests that occur unexpectedly frequently become problematic because they have developed resistance to insecticides or have overcome genetic (introgressed) or engineered resistance in the host plant. In the case of insect vectors of transmissible plant pathogens, virus–vector specificity may shift; an introduced, or upsurgent, new vector biotype may occur unexpectedly; or new, more virulent viral or microbial variants may emerge as pathogens, causing unprecedented epidemics or pandemics. In such situations, it is essential to act quickly to provide a rapid and accurate

 Genetics, Evolution and Biological Control (eds L.E. Ehler, R. Sforza and T. Mateille)

assessment of the situation, so that remedies may be put into place as quickly and effectively as possible. Often, tools for rapid identification and assessment of plant viruses and microbes, and their vectors, are unavailable during such times of crisis.

The ability to accurately identify and differentiate invasive insect species, among the most difficult being those referred to as 'cryptic species', and to discriminate between indigenous and introduced species or strains, is essential both to crop security and to the ecological health of the environment. Molecular markers have become important tools for differentiating 'cryptic' insect pests (Stern *et al.*, 1997; Hoy *et al.*, 2000), vector (Brown, 2000) and natural enemy (Manzari *et al.*, 2002) species, and for identifying microbial and viral pathogens transmitted by invasive vector species, which can potentially affect agricultural and non-cultivated habitats.

Some of the most important cryptic insect species relevant to agricultural introductions and subsequent outbreaks belong to the Order Hemiptera. Among those that present particular challenges to systematists and evolutionary biologists are aphids, leafhoppers, planthoppers, mealybugs, psyllids and whiteflies. Development of molecular tools for identifying or distinguishing difficult to identify biotypes or variant hemipterans are thus particularly in demand. Such tools can be applied for accurate identification, and, together with knowledge of the biology of the organism, are essential requirements at major ports of entry, where quarantine is an issue. Accurate identification of such variants is crucial to the early recognition of invasive species, or when outbreaks of a previously unidentified insect strain or biotype occur. Molecular markers are also of use to facilitate the coevolutionary 'matching' of biological control agents with target versus non-target hosts. Finally, accurate identification is also necessary when discrete host races, such as aphid biotypes, develop in response to plant host resistance (or other mechanisms) or, similarly, when insecticide resistance evolves and alleles comprise a significant frequency in an insect pest or vector population (Costa *et al.*, 1993; Coats *et al.*, 1994; Anthony *et al.*, 1995; Morin *et al.*, 2002).

Genetic surveys that employ molecular markers have been used for some time to study intraspecific genetic variation (genealogy) and its relationships to geographical or spatial distribution of species (Templeton, 1998), for which the term 'phylogeography' was ultimately coined and thereby established as a discipline more than 10 years ago (Avise, 2000). Phylogeographic studies have typically demonstrated an association between geographic space and the population demography for a particular species or species complex, based on the analysis of particular genetic loci in a non-equilibrium (historical) nature. This concept has become an integral component of biodiversity analysis and is useful in studying the diversity and dispersal of economically important insects (Avise, 2000).

Microevolutionary processes that involve chromosomal mutation, recombination, genetic drift, and natural selection and speciation, operating within a species, are thought, in part, to explain macroevolutionary differences among species (Avise *et al.*, 1987; Templeton, 1998). Several theoretical definitions of species exist, but the three most widely used are: the 'reproductive community', or the biological species (Mayr, 1992); the evolutionary, or phylogenetic, species; and the cohesion species concept (see Avise *et al.*, 1987). An evolutionary lineage, which is more like a biological species, is used by evolutionary and phylogenetic concepts (Templeton, 1998).

Molecular-genetic approaches are considered by many to be best served by the biological species concept, which involves community boundaries defined by reproductively isolating mechanisms that are subsequently reflected in an evolutionary lineage. In the cohesion concept of species, the derived adaptations and shared ecological attributes that constrain the reproductive demographics define the boundaries of 'interchangeability'. Thus a cohesion species is a distinct evolutionary lineage and a reproductive community, with respect to genetic processes or ecological adaptiveness (Templeton, 1998). The monitoring of such processes (heredity and evolution thereof) and the phylogeographical interpretation of molecular sequence data sets, together with within and between population variation, are therefore useful for predicting macroevolutionary patterns, in relation to evolutionary history, through phylogenetic analysis of informative gene sequences (Avise *et al.* 1987; Avise, 2000; Huelsenbeck *et al.*, 2000; Gaunt and Miles, 2002).

Application of phylogeographical analysis of insect pests and vectors, and natural enemies

The more recent application of systematics and population-genetic analysis to the study of insect pest/vector–plant host–natural enemy biosystems strives to apply unified concepts (Matsuda and Ishii, 2001; Gaunt and Miles, 2002) and to explain the dynamics and evolution of multitrophic-level interactions. In this context, certain population-genetic approaches are increasingly applicable to investigating the composition, distribution, dispersal and population dynamics of insects that colonize agricultural cropping systems as pests, and/or those that vector damaging plant virus pathogens (Mun *et al.*, 1999; Brown, 2000, 2001; Legg *et al.*, 2002), in relation to the natural enemy fauna and the ensuing complex tritrophic interactions that influence gene flow, and the natural history and ecology of agricultural biosystems. As such, it behoves entomologists to utilize all available and applicable tools to address both theoretical and practical questions, employing the same principles to different, but fundamentally similar, questions. The application of genetics tools to insect pest and vector population biology questions that are of practical import will serve to test the underlying theoretical basis of the approaches, and further their application.

An important application of molecular phylogeographic analysis in the study of insect pest and vector populations includes the identification of the predominant genotype/haplotype or groups in a population or community. To accomplish this, DNA sequences that cluster in particular phylogenetic groupings are obtained, and additional sequences for the same locus can be used to assess the abundance and distribution of that locus or gene in the ecosystem, or even globally. Phylogenetic analysis of DNA sequence data can also be utilized to detect previously unidentified variants in local or invasive populations, identify insects to species/biotype by sequence comparisons, assess the efficacy of natural enemy releases in target population species (Kirk *et al.*, 2000), and even to catalogue the associated endosymbionts (Buchner, 1965; Costa *et al.*, 1995; Caballero *et al.*, 2001; Zchori-Fein and Brown, 2002) for certain insects (including all members of the Sternorrhyncha), which can further be studied to determine prospective contributions to the host insect fitness and fecundity, reproductive compatibility, and specialist or generalist host range behaviours, among other phenotypes.

Molecular-phylogenetic analysis begins with the extraction of DNA directly from individual insects, or lysis of whole insects, such that DNA can be amplified when template DNA is aliquoted to tubes containing all other reagents necessary for the polymerase-catalysed chain reaction (PCR) (Saiki *et al.*, 1985). A typical process involves extraction or lysis of DNA from the sample, PCR amplification of a specific gene, gene fragment or intergenic region, such as an intron (in nuclear DNA) (VillaBlanca *et al.*, 1998), using sequence-specific or degenerate primers, or universal primers that base pair to a sequence conserved across numerous species or even orders. Numerous primers have been designed that permit the targeting of sequences from either the nuclear or the mitochondrial genome of insects. Once obtained, amplicons are cloned, particular clones selected, based on the expected insert size or following digestion with selected restriction enzymes, and the nucleotide (nt) sequence is determined. Relationships are predicted using various algorithms (Avise, 2000; see Von Haeseler, 1999), which permit the reconstruction of a phylogenetic history based on the aligned sequences. Typically, more than one dominant locus is targeted in search of concordance between trees. Often the tree can be associated reasonably well with the extant geographical distribution and dispersal of a particular insect, or 'phylogeography', which, subsequently, can provide insights and clues to the history and formation of species (Avise, 2000), although, clearly, pitfalls may arise in which it is not readily possible to distinguish between population structure and population history (Templeton, 1998; Avise, 2000). Finally, oligonucleotide sequences or cloned amplicons can be used subsequently as hybridization probes to investigate occurrence, abundance and diversity (e.g. fluorescence-labelled PCR-amplified sequences (oligonucleotides) or randomly labelled probes).

Genetic markers

DNA-based genetic markers that have been useful, specifically for examining nuclear loci, rely on PCR (Saiki *et al.*, 1985; Mullis and Faloona, 1987), and include microsatellites (Massonnet *et al.*, 2001, 2002; Sloane *et al.*, 2001; Bonizzoni *et al.*, 2002; Estoup *et al.*, 2002; Meixner *et al.*, 2002), randomly amplified polymorphic DNA-PCR (RAPD-PCR) (Black *et al.*, 1992; Figueroa *et al.*, 1999, 2002;

Moya *et al.*, 2001), restriction fragment length polymorphisms (RFLPs), amplified fragment length polymorphisms (AFLPs), sequence characterized amplified region (SCAR) markers (Agusti *et al.*, 2000), single nt polymorphisms (SNPs) (see Black *et al.*, 1992; Avise, 2000) and analysis of gene or intergenic sequences (Kwon and Ishikawa, 1992; Campbell *et al.*, 1995). These approaches have also been referred to by Roderick (Chapter 5, this volume), and specific methodologies, applications and interpretations of the different approaches have been reviewed widely by others (see references in Avise, 2000).

Mitochondria are widely used as molecular markers, to infer robust genealogical inference for intraspecific organisms (Moritz *et al.*, 1987; Slatkin and Hudson, 1991; Simon *et al.*, 1994; Figueroa *et al.*, 1999; Roehrdanz *et al.*, 2002). This is due primarily to the rapid accumulation of nt substitutions occurring in the mitochondrial, compared with the nuclear, genome, the (assumed) neutral effects of drift and mutation, and its maternal inheritance (generally) (Kondo *et al.*, 1990), which eliminates confounding interpretations of nuclear sequences due to recombining genes (Saccone *et al.*, 2000). Thus, the mitochondrion has been widely explored as a molecular marker to study the population genetics of many insects, including agriculturally important insect pests and vectors (Brown *et al.*, 1995, 2000; Lunt *et al.*, 1996; Nikolaidis and Scouras, 1996; Zhang and Hewitt, 1996; Stern *et al.*, 1997; Frohlich *et al.* 1999; Aikhionbare and Mayo, 2000; Medina and Walsh, 2000; Anstead *et al.*, 2002; Viscarret *et al.*, 2003) and natural enemies (Kirk *et al.*, 2000). Female-mediated (or asexually transmitted) inheritance of mitochondrial cytochrome oxidase I (mtCOI) is employed to examine population-level processes, including matrilineal gene flow, founder events and population bottlenecks, the dynamics of hybrid zones, island constraints and the histories of hybrid taxa (Gillespie and Roderick, 2002).

Studies involving mitochondrial markers indicate that, in general, a high level of polymorphisms typically corresponds to definitive geographical separation, despite the fact that different regions of the molecule evolve at different rates (Lunt *et al.*, 1996). High diversity is interpreted to indicate a matrilineal relationship within a population, whereas low divergence may occur due to frequent dispersal of females, or to recent bottlenecks in populations. Owing to the 'self-pruning' of mtDNA evolutionary trees due to stochastic lineage extinction, which is expected to occur rapidly, it is predicted that the mtDNA will display limited sequence divergence values within local populations, or within entire species that undergo historically high levels of gene flow and/or recent expansion from a single founder event (McCauley *et al.*, 1995; Templeton, 1998). However, other processes are thought to come into play that resist extinction of certain mtDNA clades, resulting in greater than expected divergence within populations, particularly when they are allopatric. It is thought that this condition results from long-term reproductive isolation. Avise *et al.* (1987) describe five categories of possible interpretations of phylogenetic analyses for population studies using mitochondrial molecular markers: discontinuity due to (i) spatial separation or (ii) lack of spatial separation; phylogenetic continuity but (iii) spatial separation or (iv) lack of spatial separation, and (v) partial spatial separation.

In this overview, I have included representative examples (by no means all-inclusive) that illustrate the exciting potential for the use of molecular markers for identification, genetically based tracking and biodiversity studies of cryptic hemipteran pest– and virus vector–host plant complexes in agricultural (and non-cultivated) settings.

Studies in which Molecular Markers have been Employed for Taxonomic Identification and Tracking of Cryptic Hemipterans

Invasion of the B biotype *Bemisia tabaci* in the Americas and evidence for a species complex

Behaviourally distinct populations of *Bemisia tabaci* (Genn.) were first recognized by Julio Bird in Puerto Rico, having initially been referred to as host races (Bird, 1957) and later as 'biotypes' that could also be distinguished by differences in esterase patterns and biotic characters (Costa and Brown, 1991). Distinctive esterase patterns for the local 'A biotype' from the south-western USA, and the suspect invasive 'B' biotype, revealed the first genetic polymorphisms for *B. tabaci* (Fig. 6.1). Furthermore, the unique ability of the B biotype to cause silvering in *Cucurbita* spp.

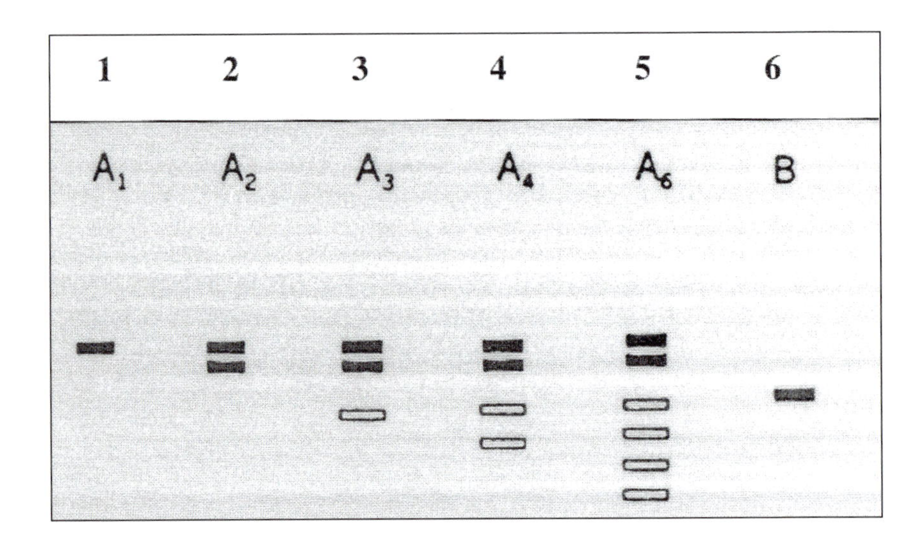

Fig. 6.1. First evidence for esterase polymorphisms in the Arizona 'A' biotype and the exotic 'B' biotype (Costa and Brown, 1991).

provided a diagnostic phenotype by which it could be tracked throughout the Americas and Caribbean region (Costa *et al.*, 1993), owing to the inabilty of New World *B. tabaci* to induce this phenotype in colonized cucurbit species. In addition, the B biotype was shown to have an extremely broad host range, to be highly fecund and to be capable of dispersing over long distances, compared with the A biotype or the 'Sida' or 'Jatropha biotype' from Puerto Rico (see Brown *et al.*, 1995), all of which were known to disperse primarily to nearby host plants. Thus, the 'A' and 'B' biotype nomenclature was coined to distinguish between these two biologically distinct *B. tabaci*, and their recognition added further credence to Bird's observation that *B. tabaci* exhibits host-range plasticity, as well as other behaviours that were subsequently discovered. Numerous esterase profiles were subsequently documented when *B. tabaci* was examined from different host plants and geographical localities worldwide, revealing a surprising degree of variability for the species (Fig. 6.2).

Biochemical markers (esterases and allozymes) (Costa *et al.*, 1993; Perring *et al*, 1993; Brown *et al.*, 2000) and several molecular markers, including the 18S ribosomal gene, for which a single nt polymorphism was identified for the A and B biotypes, two mitochondrial genes (for COI and 16S) (Frohlich *et al.*, 1999), the internal transcribed spacer (ITS1) region (De Barro *et al.*, 2000), which resolves polymorphisms to about the same degree as the mt 16S gene, and RAPDs (Gawel and Bartlett, 1993; Perring *et al.*, 1993), have collectively confirmed that a high degree of genetic polymorphism exists within this cryptic whitefly species (Rosell *et al.*, 1997) . Examination of worldwide populations of *B. tabaci* using morphological characters (Fig. 6.3) revealed no useful phenetic features that separated haplotypes distinguishable by biochemical and, subsequently, molecular markers.

Studies in which the mitochondrial cytochrome oxidase I and 16S genes were examined for a geographically representative suite of *B. tabaci* from different host plants revealed that the COI gene was more rapidly evolving than the 16S gene, and that it was also thus the most informative among molecular markers examined to date, with respect to phylogeographical relationships (Brown *et al.*, 1995; Frohlich *et al.*, 1999). When the mtCOI was used as a molecular marker, the A biotype clustered with other New World collections, whereas the B biotype grouped instead with its closest relatives from the Middle East/Mediterranean region, indicating the B biotype is an invasive, Old World species (Fig. 6.4). A subsequent study in Spain, using RAPD-PCR, revealed that the indigenous Spanish *B. tabaci* is represented by at least two haplotypes: one is host-specific (*Convolvulaceae*) and is most closely related to *B. tabaci* in western Africa (Brown, unpublished data), and the other that is most closely related to members of the longer clade that also contains the B biotype, the latter having been designated the 'Q' biotype. RAPD-PCR analysis of the Q biotype by Moya *et al.* (2001) indicated that the Q type is highly polymorphic compared with the B type, with which it was sympatric in southern Spain. This result corroborated the mtCOI data, which suggest that the biotype is non-indigenous to Spain, and confirmed the hypothesis that the B biotype was introduced there only recently. The B biotype is now known to occur throughout the Americas and has invaded and successfully established in most arid, irrigated agricultural areas, including those in Mexico, Guatemala, the Caribbean

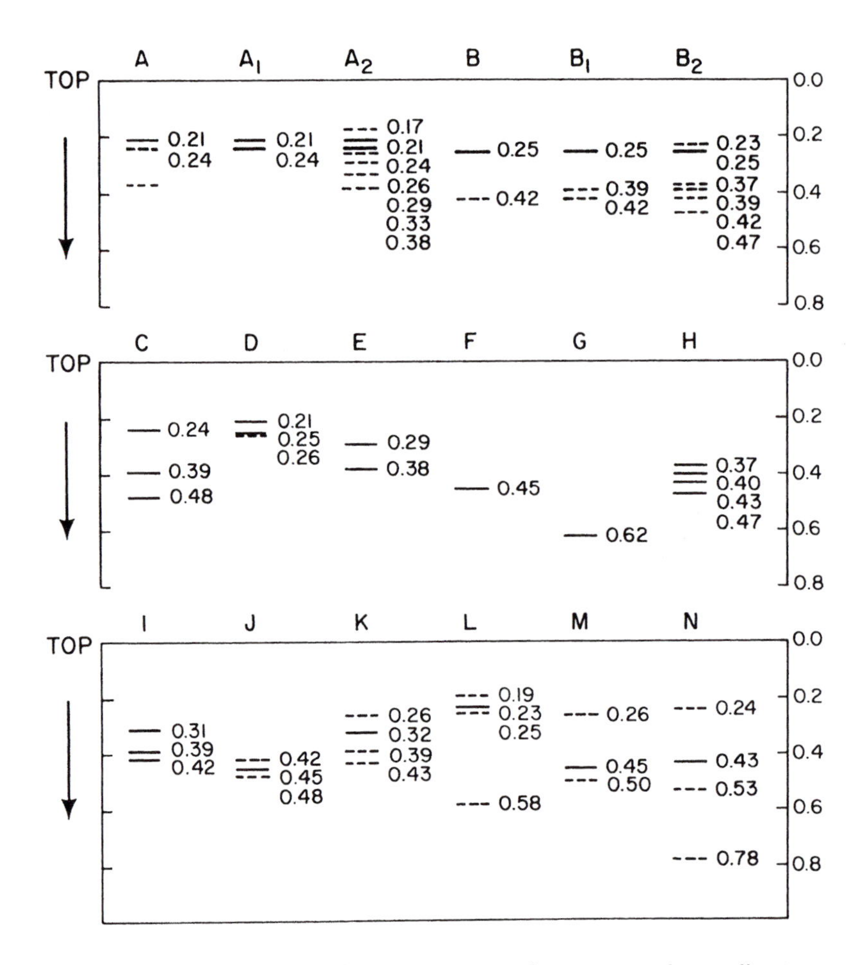

Fig. 6.2. Examples of differential esterase patterns for *Bemisia tabaci* collections from different hosts and geographical localities worldwide, which revealed even more widespread protein polymorphisms than expected within the species (from Brown *et al.*, 1995).

region and in certain South American countries (Brown, 2000; Kirk *et al.*, 2000; Viscarret *et al.*, 2003).

These observations, together with evidence that *B. tabaci* cannot be distinguished by morphological characters (Bedford *et al.*, 1994; Rosell *et al.*, 1997), have led to the hypothesis that *B. tabaci* is best described as a species complex (Brown *et al.*, 1995). In contrast, others have proposed that it is a complex of species (Perring *et al.*, 1993). Recent, more expanded, assessments of globally representative *B. tabaci* populations have revealed at least three major New World clades, and numerous distinctive clades comprising Old World *B. tabaci* (Brown, 2000; Viscarret *et al.*, 2003). It should be noted that a large database of mtCOI sequences is maintained at the University of Arizona for users to compare an

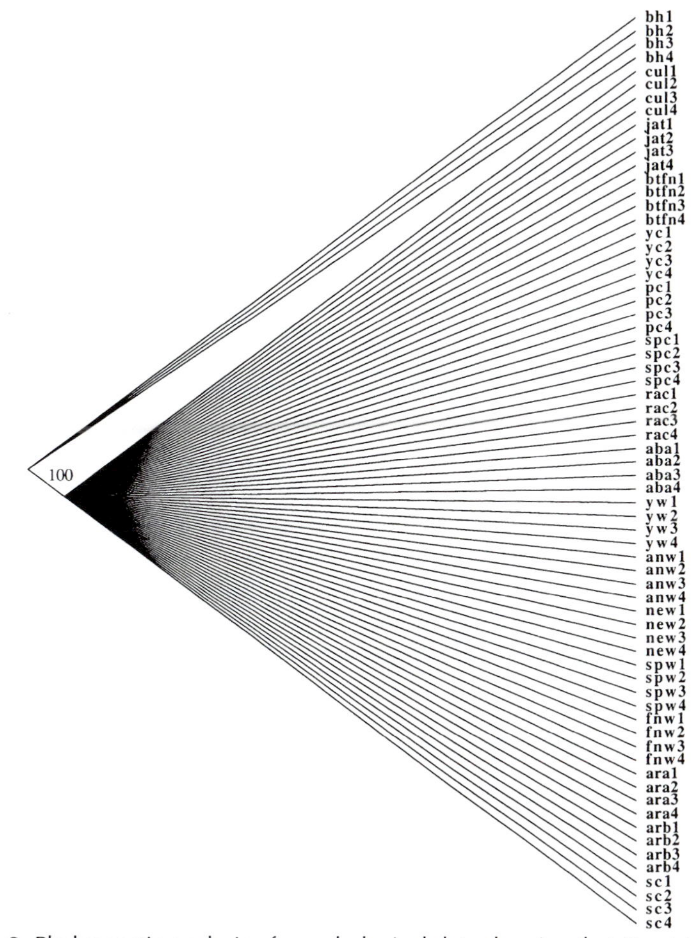

Fig. 6.3. Phylogenetic analysis of morphological data showing that *Bemisia tabaci* cannot be distinguished by phenetic characters (from Rosell *et al.*, 1997).

analogous COI fragment sequence, using the BLAST algorithm available at URL http://www.gemini.biosci.arizona.edu/whitefly. Sequences have also been deposited in the GenBank database.

Clearly, analysis of informative nuclear loci is now required to better elucidate the diversity of *B. tabaci*, leading to understanding of the evolution of this cryptic species complex.

Phylogeographical matching between *B. tabaci* mitochondrial COI haplotypes and natural enemies from the same geographical origin

In this study, natural enemy collections were differentiated using RAPD-PCR (D. Vacek, USDA-APHIS Mission Biocontrol Laboratory, Texas, personal communi-

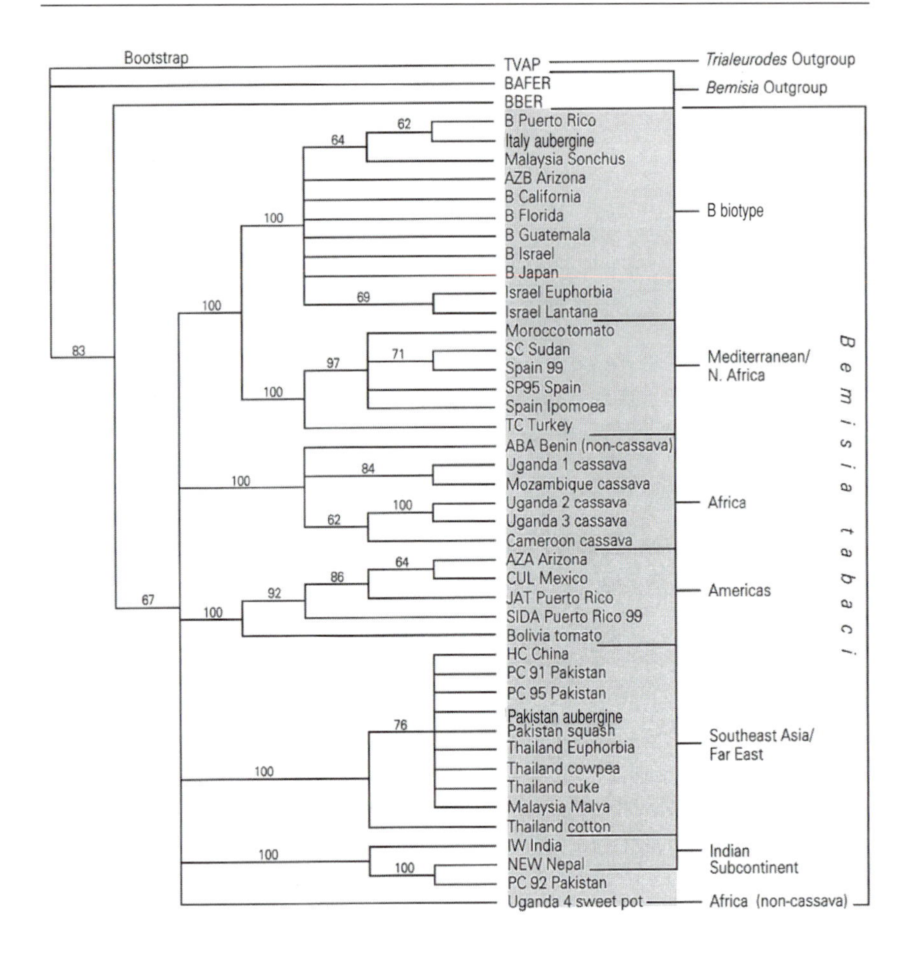

Fig. 6.4. Phylogenetic analysis of *Bemisia tabaci*, based on the mitochondrial cytochrome oxidase I gene sequence (800 bp) as a molecular marker. Collections from the western hemisphere (Americas) group as a single clade, while most from the eastern hemisphere form at least five geographical clades, including two from Africa, two from South-East Asia and the Far East (including the Indian subcontinent), and a large clade from the Mediterranean–North Africa–Middle East, which also includes the clade comprising the B biotype (from Brown, 2000).

cation), and mtCOI was used as a molecular marker to identify and determine the extent of variability in *B. tabaci* populations from which naturally occurring parasitoids were collected, in an attempt to genetically type discrete populations associated with the same host plant. The most abundant parasitoid from Spain was *Eretmocerus mundus* (Mercet), with apparent field parasitism of 39–44%. In Thailand, *Encarsia formosa* (Gahan), *Encarsia transvena* Timberlake, *Encarsia adrianae* Lopez-Avila, *Eretmocerus* sp. 1 and sp. 2 emerged, with apparent field parasitism of 1–65%. Based on mtCOI analysis, collections of *B. tabaci* from Thailand (and other South-East Asian collections) grouped separately from the invasive B biotype from Arizona and Florida and the target B type from Texas, USA, which grouped with its closest relatives from Spain (non-B) (now referred to as Q), Morocco, Turkey and the Middle East. Two other groups were resolved, which contained collections from India and all New World *B. tabaci*, respectively (Kirk *et al.*, 2000). Laboratory tests indicated that *E. mundus* from Spain parasitized more *B. tabaci* type B than did an *Eretmocerus* spp. that was native to Texas, or other exotic parasitoids evaluated. *E. mundus* from Spain also successfully parasitized *B. tabaci* biotype B when field released in a 0.94 million ha test area in Texas, and has significantly enhanced control of *B. tabaci* type B in California, USA. In contrast, parasitoids from Thailand failed to establish in the field in Texas, collectively suggesting a positive correlation between the centres of diversity of compatible parasitoid–host complexes (Goolsby *et al.*, 1998).

These preliminary results demonstrated that the mtCOI sequence, used as a molecular marker to ascertain the phylogeography for *B. tabaci*, together with RAPD-PCR identification of whitefly natural enemies, can assist greatly in achieving accurate estimates of relationships between biogeographic lineages of whiteflies and natural enemies (Kirk *et al.*, 2000). *E. mundus* M92014, collected in the arid region of southern Spain, the apparent zone of diversification for the B biotype lineage (Fig. 6.5), was shown to be largely responsible for the increase in parasitism and control of *B. tabaci* in broccoli in the Rio Grande Valley, Texas, during dry winter and spring seasons (Goolsby *et al.*, 1998), suggesting that climate/weather parameters were also involved.

This approach is expected to lead to the expedient selection of relevant natural enemies for other specific whitefly targets, and to allow for 'customized control' in amenable circumstances. Although natural enemies occur and parasitize, or are predators, in most locations in which *B. tabaci* occurs, the highly polyphagous nature of most *B. tabaci* haplotypes apparently precludes highly successful biological control of this insect as the sole method. None the less, both genealogical and climate-matching methods used in this study showed great promise, particularly for more 'specialist' invasive pests, a characteristic of many other whitefly species. This was the first example in which molecular methods were employed to identify and/or track both the host insect and its natural enemies, and a plausible correlation has been shown (Kirk *et al.*, 2000).

First demonstration that an invasive *B. tabaci* is associated with the spread of severe cassava mosaic disease in East Africa

The objective of this study was to investigate the hypothesis that an invasive biotype of *B. tabaci* was responsible for vectoring a more virulent recombinant

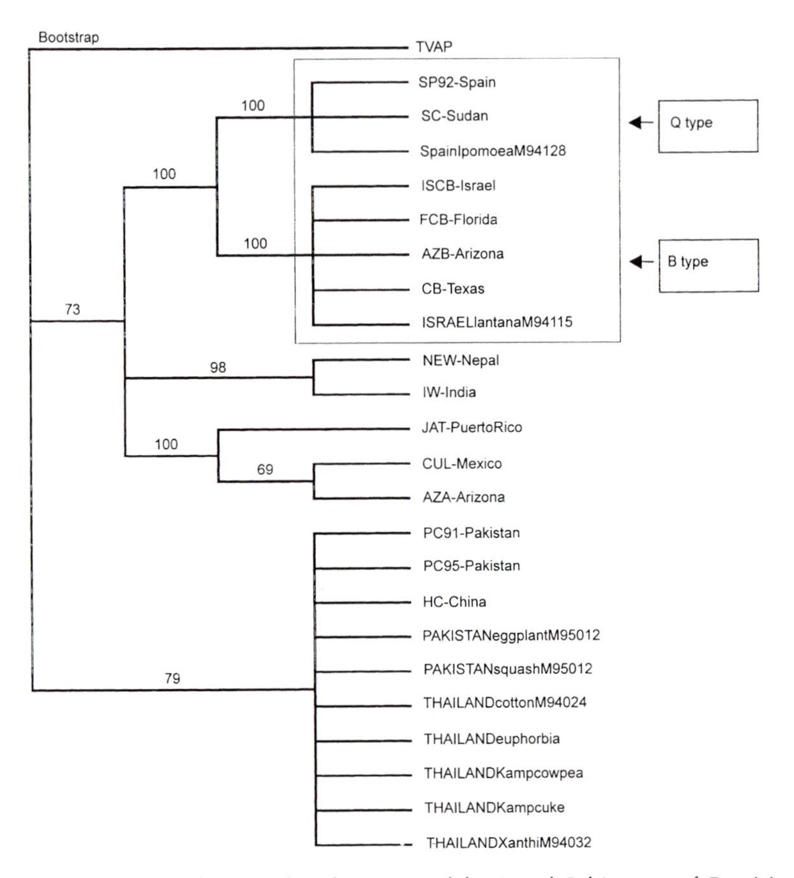

Fig. 6.5. Cladogram showing the placement of the B and Q biotypes of *Bemisia tabaci* in the Mediterranean–North Africa–Middle East, where natural enemies were identified for the control of the B biotype using augmentative releases. The bar graph (inset) shows the results of natural enemy screening using the B biotype under quarantine conditions in Texas (from Kirk *et al.*, 2000).

begomovirus (genus, *Begomovirus*; family, Geminiviridae) in cassava throughout East Africa, and hence for the 10-year epidemic of African cassava mosaic disease there. Whitefly samples were collected from symptomatic cassava plantings along east/west transects, according to the spread of the epidemic, which began in western Uganda and spread eastward toward and southward of Lake Victoria. This new, severe disease now is known to have spread throughout Uganda, and to Kenya and Tanzania (Legg *et al.*, 2002).

Assessment of the mtCOI gene (800 bp) of populations collected from 'Ahead of the disease front', 'At the front' and 'Behind the front' revealed at least two haplotypes of *B. tabaci*, designated 'local' or indigenous type, and the 'invader' or, putatively, non-indigenous type. In Uganda, the 'invader' was consistently associated with severe disease symptoms in cassava as the disease front moved south-westwardly, whereas, the 'local' *B. tabaci* was found 'At the front' and 'Ahead of the front', where typical, less severe symptoms occurred, as had been known prior to the 'severe' virus epidemic. These observations suggested that a previously unknown haplotype was associated with the spread, and possibly the introduction, of the severe, recombinant begomovirus. Further, certain evidence suggests that the 'invader' is more fecund than the 'local' type. The closest relatives of UG2 'invader' hailed from Cameroon (westward).

Comparison of mtCOI sequences for UG1 and UG2 and well-studied *B. tabaci* reference populations indicated that the two Ugandan populations exhibited about 8% divergence, suggesting that they represent distinct sub-Saharan African lineages. Neither Ugandan genotype cluster was identified as the widely distributed, polyphagous and highly fecund B biotype of Old World origin, with which they both diverged by about 8%. Within-genotype cluster divergence of UG1, at 0.61 ± 0.1%, was twice that of UG2 (at 0.35 ± 0.1%). Mismatch analysis (Slatkin and Hudson, 1991; Rogers and Harpending, 1992; Rogers, 1995) suggested that UG2 has undergone a recent population expansion and may be of non-Ugandan origin, whereas UG1 has diverged more slowly, and is likely to be an indigenous genotype (Legg *et al.*, 2002). Documenting divergence and population growth curves for extant populations within which an invasion is suspected or known may well provide valuable indicators of new or impending upsurges in insect populations and associated pathogens that they transmit.

Upsurge of *Myndus crudus*, the vector of the phytoplasma inducing lethal yellowing disease of palm, and rapid disease spread following Hurricane Mitch

Lethal yellowing (LY) is caused by a phytoplasma that affects over 35 palm species; it is the most devastating disease of coconuts in the Caribbean and Americas. Dying palms with LY-like symptoms have been reported from the Caribbean for over 100 years. Epidemics destroyed millions of coconut palms in Jamaica and southern Florida in the 1960s and 1970s, before LY spread from the Caribbean Islands to Yucatan, Mexico, in 1979 and more recently (1997) into Belize and Honduras. Phytoplasmas, *Mollicute* plant pathogens, are transmitted by leafhoppers, planthoppers and psyllids (Kirkpatrick, 1992). The planthopper *Myndus crudus* (van Duzee) (Hemiptera, Fulgoromorpha, Cixiidae) is the only recognized vector of LY, although few experimental transmissions have been documented (Howard and Wilson, 2001). Control of LY has been based on resistant varieties, especially the 'MayPan', a 'Malayan Dwarf' × 'Panama Tall' hybrid, which has been planted extensively in Jamaica and Florida since the 1970s. Beginning in 1995, 'MayPan' in Jamaica succumbed to LY for the first time, and the epidemic reached crisis levels, with damage far greater than that which occurred in the epidemics of the 1960s.

In common with other phytoplasmas, the LY pathogen cannot be cultured, but molecular studies have determined that it exists as a group of near-identical strains in the western Caribbean region, belonging to the 16SrIV group of phytoplasmas. Little is known about phytoplasma–vector specificity, or about prospective intraspecific diversity for the LY vector, *M. crudus*. Information regarding dispersal patterns of *M. crudus*, and flight capability locally or regionally, is also lacking. Although severe LY outbreaks and its rapid spread are well documented, the mechanisms underlying long-distance spread are unknown. Anecdotally, such outbreaks have been associated with the aftermath of severe weather events in the region, the last episode being associated with Hurricane Mitch in 1998.

The objective of this analysis was to explore the use of the mtCOI gene as a molecular marker to examine the biodiversity and distribution of *M. crudus* in the Caribbean region, in relation to the LY disease. It has been hypothesized that periodic outbreaks and (apparent) increased spread of lethal yellowing phytoplasma are due to disruption of the local and regional systems' ecology following severe tropical storms. It is not known if populations of *M. crudus* are dispersed between islands and landmasses in the Caribbean/Central America/Mexico region during such storms and thereby phytoplasma variants are likewise redistributed, or if dispersal behaviour of local populations is altered such that feeding behaviour related to increased transmission frequency occurs, and is thereby the explanation for periodic increased disease incidence.

Using PCR and degenerate primers, a preliminary examination of *M. crudus* samples from Florida, Mexico and Honduras was undertaken, and individuals from a colony that was established in Jamaica in 1973 (John Innes Centre, UK) and maintained thereafter were also examined. *Myndus adiopodoumeensis* Synave from Ghana (Brown *et al.*, 2002) and *Circulifer tennellus* Baker, the beet leafhopper from North America, were included as the outgroup species and genus, respectively. Phylogenetic analysis revealed possibly two to three groups in samples examined from Honduras and the colony originating in Florida (Fig. 6.6). It is hoped that the mtCOI gene will serve as an informative molecular marker with which to examine more extensively the intraspecific variation of *M. crudus* vector populations in the region; however, particularly low intraspecific divergence may require the examination of different or additional genetic markers. Clearly, it is far too soon to draw any sound conclusions, owing to an insufficient sample size. None the less, one hypothesis can be advanced in future analyses. Given the narrow host range of the insect vector, the necessity of virus–vector specificity and the 'multiple islands' inhabited by this vector species, low nt divergence may in fact prevail in Caribbean *M. crudus* populations. If so, one or several factors may contribute to the putative isolation and, prospectively, reduced, or absence of, gene flow (Templeton, 1998), which could feasibly be compounded by island effects (Gillespie and Roderick, 2002). Analysis of a larger sample size and from additional hosts (grasses and palms) is expected to shed new light on this vector–phytoplasma pathogen complex, which has, to date, not been examined using molecular markers to investigate the genetic diversity or gene flow in this unique system.

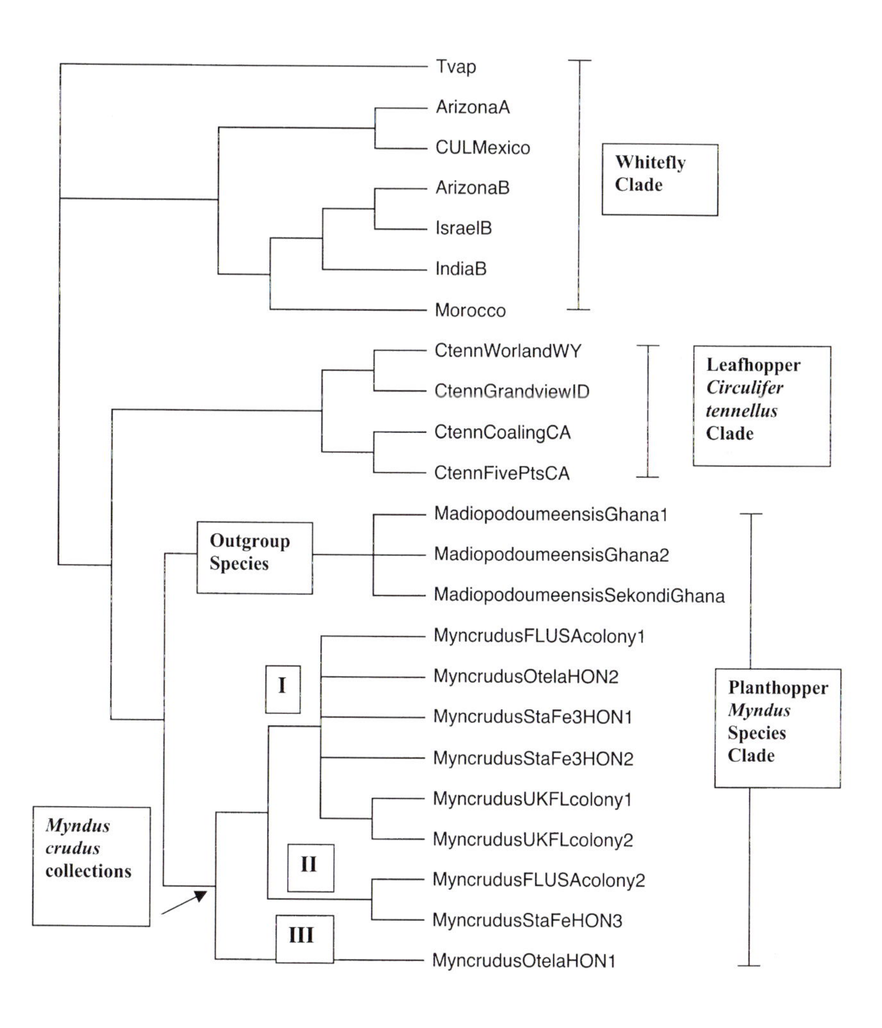

Fig. 6.6. Preliminary results of an analysis of the mitochondrial cytochrome oxidase I gene fragment for two *Myndus* species, revealing two to three groups of *Myndus crudus* in the Caribbean region. The outgroup is a *Myndus* species from Africa, and relationships to two whitefly species and the leafhopper *Circulifer tennellus* are also shown (Brown *et al*., 2002).

Divergence of aphids on non-cultivated hosts: determining haplotypes and plant hosts using a mitochondrial marker

The term 'biotype' has long been applied to describe biotic differences in aphids, and initially referred to biological phenotypic differences rather than morphological characters. Another view holds that aphid biotypes are infraspecific insect populations, with similar genetic composition for a particular biological attribute, and for which five categories have been defined (see references in Anstead *et al.*, 2002). Typically, aphid biotypes have been defined by their ability to succumb to, or overcome, host plant resistance, and/or as 'host races', which are based on an affinity for a particular repertoire of plant hosts. *Schizaphis graminum* (Rondani) (Hemiptera: Aphidae), the greenbug, is notable for its ability to induce toxic reactions on the hosts that it colonizes, and, at times, the reaction can be so severe that the host dies. This aphid also has a propensity to overcome insect resistance genes that have been introgressed in cereal grains, such as wheat; however, the mechanism by which this rapid resistance develops is not known. This study examined *S. graminum* field populations to determine how diverse greenbug populations are, and whether the greenbug actually occurs as a putatively diverse complex of host-adapted races, instead of a homogeneous genetic pool, thereby allowing selection for resistance to occur when aphids encounter resistant cultivars. If so, this could explain the persistent ability of the greenbug to rapidly overcome resistance in cultivated crops.

To test this hypothesis, the genetic variation for the mtCOI gene (1.0 kb) was examined for field collections of greenbug, 24 collected from non-cultivated hosts and 12 collected from cultivated hosts. Phylogenetic relatedness was predicted from the mtCOI sequence data set using maximum likelihood, distance (NJ) and maximum parsimony methods. Three clades were resolved by the analysis, irrespective of the algorithm, and variation within each clade was low (Anstead *et al.*, 2002), but intraclade differences were as high as 5%, a previously unprecedented level of nt diversity for this species. Gene flow was evident between one biotype in all three clades. There was no relationship between biotype and haplotype. Populations in the three clades were grouped by those that colonized: (i) johnsongrass (*Sorghum halepense*), (ii) *Agropyron* spp. and (iii) wheat plus four non-cultivated hosts other than *Agropyron* spp. Such partitioning by haplotype strongly suggested the presence of host-adapted races in *S. graminum*. Interestingly, past studies from which the authors concluded that host-adapted races did not occur, based on genetic evidence (minimal to no nt divergence), examined primarily aphids from cultivated crop species. Clearly, sufficient sample size and representative sources (cultivated and non-cultivated plant hosts) are important sampling considerations when attempting to examine the genetic structure and, subsequently, reconstructing the evolutionary history of insect populations.

Tracking Mediterranean fruit fly introductions in California using composite genotypes

The Mediterranean fruit fly, *Ceratitis capitata* (Wiedemann) (Diptera: Tephritidae), is a damaging insect pest of cultivated fruits, which is endemic to sub-Saharan Africa. It was introduced from Africa to the Mediterranean region about 200 years ago and became established in Central and South America during the past 100 years (Meixner *et al.*, 2002).

Many extant outbreaks are hypothesized to initiate from small founder populations, or possibly single females that hitchhike on imported fruit. There is neither information regarding genetic stability or heterogeneity in existing populations, nor are the origins of prospective introductions known. In an effort to better understand the source(s) of new outbreaks, microsatellite and mtDNA markers were used to examine three polymorphic restriction sites and two microsatellite loci. Collections were made following outbreaks in California during 1992–1994 and 1997–1999, in which 359 flies were trapped and analysed genetically.

Results of the study for which composite genotypes were obtained revealed that five separate and independent introductions had occurred between 1992 and 1998. The majority of collections were of the same genotype (AAA), indicating a single introduction into the Los Angeles (LA) area in 1992. The same genotype was identified in northern and southern California in the same year, but individuals shared microsatellite alleles absent in AAA and lacked others common to AAA, suggesting that an additional independent introduction had occurred from a different source. In 1993, a third introduction was documented, based on novel or microsatellite alleles. In that same year, the BBN mt haplotype, together with microsatellite alleles, consistent with an origin in Hawaii, was discovered. This same haplotype was trapped again in 1998 but in San Diego County (south of LA). Single flies that represented two additional introductions were trapped in Santa Clara County (1997) and El Monte (Los Angeles County) (AAC type) (1999), in total seven introductions detected in 5 years. It is thought that the AAA flies captured in many years might be part of a persistent medfly population in California, or that they have been reintroduced routinely from the same source. Evidence indicates that certain haplotypes may have come from outside locations that received fruit from international sources, including Spain, with which one profile was matched in 1993. Collectively, results indicate that a worldwide medfly colonization process is probably ongoing, due to the illegal or legal transport of infested produce to and from areas already colonized by this insect. Results also suggest that both point-source introductions and more widespread distribution events occur. This conclusion can be drawn because, given such diverse populations, not all can be explained as coming from either resident sources or point introductions within current quarantined areas. Thus, individual travellers are probably among the lowest of risks, whereas importers of fruits from exotic locales who receive shipments outside current quarantine restrictions probably represent the greater risk (Meixner *et al.*, 2002).

Conclusions

Both indirect and direct DNA-sequence-based technologies that permit the assessment of genetic variation (nuclear and mitochondrial genome markers), together with phenetically based identification and corroboration of phenotypes or biotic traits, if available, are ideally suited for tracing the origin, distribution and dispersal patterns of insect pests and vectors of viral and phytoplasma pathogens of plants. The further ability to distinguish genetic variants, ranging from 'biotypes' to cryptic species and species complexes, based on nt polymorphisms, provides a powerful tool for tracing potentially destructive exotic insect introductions or local upsurging populations resulting from altered selection pressures, often due to human activities. A mosaic of molecular markers for particularly problematic, cosmopolitan insects permits rapid and accurate interception of introductions, and allows for discrimination of resident populations and those originating from point introductions or more widespread sources. Biological control efforts to match origins of target insects with those of beneficial, co-evolved natural enemies holds much promise for augmentative biological control of insects, weedy plants and possibly other invasive species. Accordingly, employment of population-genetic tools and DNA sequence databases assembled for important cryptic insect pests and virus vectors will become an increasingly valuable resource for tracing their origin, in pursuit of elucidating the origin of particularly damaging phenotypes, and toward the identification of useful natural enemies. Such databases will constitute an invaluable resource to facilitate rapid recognition of introduced species, or of outbreaks by indigenous genetic variants that result from changing conditions and/or agricultural practices. Collectively, rapid, genetics-based identification systems for insect pest/vector host–natural enemy complexes will thus permit more expedient selection of relevant natural enemies for specific pest/vector targets, ultimately facilitating 'customized biological control', and will provide invaluable surveillance capabilities to rapidly recognize prospective exotic introductions or unanticipated local outbreaks, in relation to invasive species and insecticide failures.

Acknowledgements

I would like to express my gratitude to the organizers of the symposium and editors of this book for their hard work and dedication to this important and timely effort. I would also like to thank Drs A. Kirk and M.C. Bon for their expert contributions to this presentation during the symposium. Finally, I am grateful to numerous friends and colleagues in many different countries, who are unfortunately too numerous to name here, for their generosity and meticulous care in providing high-quality insect samples used in certain of the analyses described. Without their support, dedication and generosity of time and resources, many of these studies would not have been possible.

References

Agusti, N., Vicente, M.C. de and Gabarra, R. (2000) Developing SCAR markers to study predation on *Trialeurodes vaporariorum*. *Insect Molecular Biology* 9, 263–268.

Aikhionbare, F.O. and Mayo, Z.B. (2000) Mitochondrial DNA sequences of greenbug (Homoptera: Aphididae) biotypes. *Genetic Analysis: Biomolecular Engineering* 16, 199–205.

Anstead, J.A., Burd, J.D. and Shufran, K.A. (2002) Mitochondrial DNA sequence divergence among *Schizaphis graminum* (Hemiptera: Aphididae) clones from cultivated and non-cultivated hosts: haplotype and host associations. *Bulletin of Entomological Research* 92, 17–24.

Anthony, N., Brown, J.K., Markham, P.G. and ffrench-Constant, R.H. (1995) Molecular analysis of cyclodiene resistance-associated mutations among populations of the sweetpotato whitefly *Bemisia tabaci*. *Pesticide Biochemistry and Physiology* 51, 220–228.

Avise, J.C. (2000) *Phylogeography: the History and Formation of Species*. Harvard University Press, Cambridge, Massachusetts.

Avise, J.C., Arnold, J., Ball, R.M., Bermingham, E., Lamb, T.N., Joeph, E., Reeb, C.A. and Saunders, N.C. (1987) Intraspecific phylogeography: the mitochondrial DNA bridge between population genetics and systematics. *Annual Review of Ecology and Systematics* 18, 489–522.

Bedford, I.D., Markham, P.G., Brown, J.K. and Rosell, R.C. (1994) Geminivirus transmission and biological characterization of whitefly (*Bemisia tabaci*) biotypes from different world regions. *Annals of Applied Biology* 125, 311–325.

Bird, J. (1957) A whitefly-transmitted mosaic of *Jatropha gossypifolia*. *Technical Paper. Agricultural Experiment Station, Puerto Rico* 22, 1–35.

Black, W.C., DuTeau, N.M., Puterka, G.J., Nechols, J.R. and Pettorini, J.M. (1992) Use of the random amplified polymorphic DNA polymerase chain reaction (RAPD-PCR) to detect DNA polymorphisms in aphids (Homoptera: Aphididae). *Bulletin of Entomological Research* 82, 151–159.

Bonizzoni, M., Katsoyannos, B.I., Marguerie, R., Guglielmino, C.R., Gasperi, G., Malacrida, A. and Chapman, T. (2002) Microsatellite analysis reveals remating by wild Mediterranean fruit fly females, *Ceratitis capitata*. *Molecular Ecology* 11, 1915–1921.

Brown, J.K. (2000) Molecular markers for the identification and global tracking of whitefly vector–begomovirus complexes. *Virus Research* 71, 233–260.

Brown, J.K. (2001) The molecular epidemiology of begomoviruses. In: Khan, J.A. and Dykstra, J. (eds) *Trends in Plant Virology*. The Haworth Press, New York, pp. 279–316.

Brown, J.K., Frohlich, D.R. and Rosell, R. (1995) The sweetpotato/silverleaf whiteflies: biotypes of *Bemisia tabaci* (Genn.), or a species complex? *Annual Review of Entomology* 40, 511–534.

Brown, J.K., Perring, T.M., Cooper, A.D., Bedford, I.D. and Markham, P.G. (2000) Genetic analysis of *Bemisia* (Homoptera: Aleyrodidae) populations by isoelectric focusing electrophoresis. *Biochemical Genetics* 38, 13–25.

Brown, J.K., Dollet, M., Doyle, M.M., Harrison, N.A., Jones, P. and Philippe, R. (2002) Investigations into the phylogeography and ecology of *Myndus crudus* van Duzee (Hemiptera, Fulgoromorpha, Cixiidae), the leafhopper vector of coconut lethal yellowing. *The 11th International Auchenorrhyncha Congress*, Berlin–Potsdam, 4–9 August.

Buchner, P. (1965) *Endosymbiosis of Animals with Plant Microorganisms*. John Wiley & Sons, New York.

Caballero, R., Torres-Jerez, I. and Brown, J.K. (2001) Two distinct *Wolbachia* identified in *Bemisia tabaci* (Genn.) are associated with uni-directional cytoplasmic incompatibility in infected and uninfected biotypes. *Proceedings of the Sixth International Whitefly and Geminivirus Workshop*, Ragusa, Sicily, 27 February–2 March.

Campbell, B.C., Campbell-Steffen, J.D., Sorensen, J.T. and Gill, R.J. (1995) Paraphyly of Homoptera and Auchenorrhyncha inferred from 18S rDNA nucleotide sequences. *Systematic Entomology* 20, 175–194.

Coats, S.A., Brown, J.K. and Hendrix, D.L. (1994) Biochemical characterization of biotype-specific esterases in the whitefly *Bemisia tabaci* Genn. (Homoptera: Aleyrodidae). *Insect Biochemistry and Molecular Biology* 24, 723–728.

Costa, H.S. and Brown, J.K. (1991) Variation in biological characteristics and in esterase patterns among populations of *Bemisia tabaci* (Genn.) and the association of one population with silverleaf symptom development. *Entomologia Experimentalis et Applicata* 61, 211–219.

Costa, H.S., Brown, J.K., Sivasupramaniam, S. and Bird, J. (1993) Regional distribution, insecticide resistance, and reciprocal crosses between the 'A' and 'B' biotypes of *Bemisia tabaci*. *Insect Science and its Application* 14, 127–138.

Costa, H.S., Wescot, D.M., Ullman, D.E, Rosell, R., Brown, J.K. and Johnson, M.W. (1995) Morphological variation in *Bemisa* endosymbionts. *Protoplasma* 189, 194–202.

Craig, C.L. (2001) Evolution, theory of. In: Levin, S.A. (ed.) *Encyclopedia of Biodiversity*, Vol. 2. Academic Press, New York, pp. 671–681.

De Barro, P.J., Driver, F., Trueman, J.W.H. and Curran, J. (2000) Phylogenetic relationships of world populations of *Bemisia tabaci* (Gennadius) using ribosomal ITS1. *Molecular Phylogenetics and Evolution* 16, 29–36.

Ebert D., Haag, C., Kirkpatrick, M., Riek, M., Hottinger, J.W. and Pajunen, V.I. (2002) A selective advantage to immigrant genes in a *Daphnia* metapopulation. *Science* 295, 485– 488.

Estoup, A., Jarne, P. and Cornuet, J.-M. (2002) Homoplasy and mutation model at microsatellite loci and their consequences for population genetics analysis. *Molecular Ecology* 11, 1591–1604.

Figueroa, C.C., Simon, J.-C., Le Gallic, J.-F. and Niemeyer, H.M. (1999) Molecular markers to differentiate two morphologically-close species of the genus *Sitobion*. *Entomologia Experimentalis et Applicata* 92, 217–225.

Figueroa, C.C., Muro-Loayza R. and Niemeyer, H.M. (2002) Temporal variation of RAPD-PCR phenotype composition of the grain aphid *Sitobion avenae* (Hemiptera: Aphididae) on wheat: the role of hydroxamic acids. *Bulletin of Entomological Research* 92, 25–33.

Frohlich, D., Torres-Jerez, I., Bedford, I.D., Markham, P.G. and Brown, J.K. (1999) A phylogeographic analysis of the *Bemisia tabaci* species complex based on mitochondrial DNA markers. *Molecular Ecology* 8,1593–1602.

Gaunt, M.W. and Miles, M.A. (2002) An insect molecular clock dates the origin of the insects and accords with palaeontological and biogeographic landmarks. *Molecular Biology and Evolution* 19, 748–761.

Gawel, N.J. and Bartlett, A.C. (1993) Characterization of differences between whiteflies using RAPD-PCR. *Insect Molecular Biology* 2, 33–38.

Gillespie, R.G. and Roderick, G.K. (2002) Arthropods on islands: colonization, speciation, and conservation. *Annual Review of Entomology* 47, 595–632.

Goolsby, J.A., Ciomperlik, M., Legaspi, B.C., Legaspi, J.C. and Wendel, L.E. (1998) Laboratory and field evaluation of exotic parasitoids of the sweetpotato whitefly, *Bemisia tabaci* (Gennadius) (Biotype 'B') (Homoptera: Aleyrodidae) in the lower Rio Grande Valley of Texas. *Biological Control* 12, 127–135.

Howard, F.W. and Wilson, M.R. (2001) Hemiptera: Auchenorrhyncha. In: Howard, F.W., Moore, D., Giblin-Davis, R.M. and Abadj, R.G. (eds) *Insects on Palms*. CAB International, Wallingford, UK, pp. 128–161.

Hoy, M.A., Jeyaprakash, A., Morakote, R., Lo, P.K.C. and Nguyen, R. (2000) Genomic analyses of two populations of *Ageniaspis citricola* (Hymenoptera: Encyrtidae) suggest that a cryptic species may exist. *Biological Control* 17, 1–10.

Huelsenbeck, J.P., Rannala, B. and Masly, J.P. (2000) Accommodating phylogenetic uncertainty in evolutionary studies. *Science* 288, 2349–2350.

Kirk A.A., Lacey, L.A., Brown, J.K., Ciomperlik, M.A., Goolsby, J.A., Vacek, D.C., Wendel, L.E. and Napompeth, B. (2000) Variation within the *Bemisia tabaci* s.l. species complex (Hemiptera: Aleyrodidae) and its natural enemies leading to successful biological control of *Bemisia* biotype B in the USA. *Bulletin of Entomological Research* 90, 317–327.

Kirkpatrick, B.C. (1992) Mycoplasmalike organisms – plant and invertebrate pathogens. In: Balows, A., Truper, H.G., Dworkin, M., Harder, W. and Schleifer K.H. (eds) *The Prokaryotes*, Vol. 4. Springer Verlag, New York, pp. 4050–4067.

Kondo, R., Satta, Y., Matsuura, E.T., Ishiwa, H., Takahata, N. and Chigusa, S.I. (1990) Incomplete maternal transmission of mitochondrial DNA in *Drosophila*. *Genetics Society of America* 126, 657–663.

Kwon, O. and Ishikawa, H. (1992) Unique structure in the intergenic and 5′ external transcribed spacer of the ribosomal RNA gene from the pea aphid *Acyrthosiphon pisum*. *European Journal of Biochemistry* 206, 935–940.

Legg, J., French, R., Rogan, D., Okao-Okuja, G. and Brown, J.K. (2002) A distinct *Bemisia tabaci* (Gennadius) (Hemiptera: Sternorrhyncha: Aleyrodidae) genotype cluster is associated with the epidemic of severe cassava mosaic virus disease in Uganda. *Molecular Ecology* 11, 1219–1229.

Lunt, D.H., Zhang, D.X., Szymura, J.M. and Hewitt, G.M. (1996) The insect cytochrome oxidase I gene: evolutionary patterns and conserved primers for phylogenetic studies. *Insect Molecular Biology* 5, 153–165.

Manzari, S., Plaszek, A., Belshaw, R. and Quicke, D.L.J. (2002) Morphometric and molecular analysis of the *Encarsia inaron* species group (Hymenoptera: Aphelinidae), parasitoids of whiteflies (Hemiptera: Aleyrodidae). *Bulletin of Entomological Research* 92, 165–175.

Massonnet, B., Leterme, N., Simon, J.-C. and Weisser, W.W. (2001) Characterization of microsatellite loci in the aphid species *Macrosiphoniella tanacetaria* (Homoptera, Aphididae). *Molecular Ecology Notes* 1, 14–15.

Massonnet, B., Leterme, N., Simon, J.-C. and Weisser, W.W. (2002) Characterization of microsatellite loci in the aphid species *Metopeurum fuscoviride* (Homoptera, Aphididae). *Molecular Ecology Notes* 2, 127–129.

Matsuda, H. and Ishii, K. (2001) A synthetic theory of molecular evolution. *Genetic Systems* 76, 149–158.

Mayr, E. (1992) Speciation and macroevolution. *Evolution* 36, 1119–1132.

McCauley, D.E., Raveill, J. and Antonovics, J. (1995) Local founding events as determinants of genetic structure in a plant metapopulation. *Heredity* 75, 630–636.

Medina, M. and Walsh, P.J. (2000) Molecular systematics of the order *Anaspidea* based on mitochondrial DNA sequence (12S, 16S, and COI). *Molecular Phylogenetics and Evolution* 15, 41–58.

Meixner, M.D., McPheron, B.A., Silva, J.G., Gasparich, G.E. and Sheppard, W.S. (2002) The Mediterranean fruit fly in California: evidence for multiple introductions and persistent populations based on microsatellite and mitochondrial DNA variability. *Molecular Ecology* 11, 891–899.

Morin, S., Williamson, M.S., Goodson, S.J., Brown, J.K., Tabashnik, B.E. and Dennehy, T.J. (2002) Mutations in the *Bemisia tabaci* sodium channel gene associated with resistance to a pyrethroid plus organophosphate mixture. *Insect Biochemistry and Molecular Biology* 32, 1781–1791.

Moritz, C., Dowling, T.E. and Brown, W.M. (1987) Evolution of animal mitochondrial DNA: relevance for population biology and systematics. *Annual Review of Ecology and Systematics* 18, 269–292.

Moya, A., Guirao, P., Cifuentes, D., Beitia, F. and Cenis, J.L. (2001) Genetic diversity of Iberian populations of *Bemisia tabaci* (Hemiptera: Aleyrodidae) based on random amplified polymorphic DNA-polymerase chain reaction. *Molecular Ecology* 10, 891–897.

Mullis, K.R. and Faloona, F.A. (1987) Specific synthesis of DNA *in vitro* via a polymerase-catalyzed chain reaction. *Methods in Enzymology* 155, 335–350.

Mun, J.H., Song, Y.H., Heong, K.L. and Roderick, G.K. (1999) Genetic variation among Asian populations of rice planthoppers, *Nilaparvata lugens* and *Sogatella furcifera* (Hemiptera: Delphacidae): mitochondrial DNA sequences. *Bulletin of Entomological Research* 89, 245–253.

Nikolaidis, N. and Scouras, Z.G. (1996) The *Drosophila montium* subgroup species. Phylogenetic relationships based on mitochondrial DNA analysis. *Genome* 39, 874–883.

Perring, T.M., Cooper, A.D., Russell, R.J., Farrar, C.A. and Bellows, T.S. Jr (1993) Identification of a whitefly species by genomic and behavioral studies. *Science* 259, 74–77.

Pimentel, D. (2001) Agricultural invasions. In: Levin, S.A. (ed.) *Encyclopedia of Biodiversity*, Vol. 1. Academic Press, New York, pp. 71–83.

Pimentel, D., Lach, L., Zuniga, R. and Morrison, D. (2000) Environmental and economic costs associated with nonindigenous species in the United States. *Bioscience* 50, 53–65.

Roehrdanz, R.L., Degrugillier, M.E. and Black, W.C. (2002) Novel rearrangements of arthropod mitochondrial DNA detected with long-PCR: applications to arthropod phylogeny and evolution. *Molecular Biology and Evolution* 19, 841–849.

Rogers, A.R. (1995) Genetic evidence for a pleistocene population explosion. *Evolution* 49, 608–615.

Rogers, A.R. and Harpending, H. (1992) Population growth makes waves in the distribution of pairwise genetic differences. *Molecular Biology and Evolution* 9, 552–569.

Rosell, R.C., Bedford, I.D., Frohlich, D.R., Gill, R.J., Brown, J.K. and Markham, P.G. (1997) Analysis of morphological variation in distinct populations of *Bemisia tabaci* (Homoptera: Aleyrodidae). *Annals of the Entomological Society of America* 90, 575– 589.

Saccone, C., Gissi, C., Lanave, C., Larizza, A., Pesole, G. and Reyes, A. (2000) Evolution of the mitochondrial genetic system: an overview. *Gene* 261, 153–159.

Saiki, R.K., Scharf, S., Faloona, F., Mullis, K.B., Horn, G.T., Erlich, H.A. and Arnheim, N. (1985) Enzymatic amplification of β-globin genomic sequences and restriction site analysis for diagnosis of sickle cell anemia. *Science* 230, 1350–1354.

Simon, C., Frati F., Beckenbach, A., Crespi, B., Liu, H. and Flook, P. (1994) Evolution, weighting, and phylogenetic utility of mitochondrial gene sequences and a compilation of conserved polymerase chain reaction primers. *Annals of the Entomological Society of America* 87, 651–701.

Slatkin, M. and Hudson, R.R. (1991) Pairwise comparisons of mitochondrial DNA sequences in stable and exponentially growing population. *Genetics* 229, 555–562.

Sloane, M.A., Sunnucks, P., Wilson, A.C.C. and Hales, D.F. (2001) Microsatellite isolation, linkage group identification and determination of recombination frequency in the peach-potato aphid, *Myzus persicae* (Sulzer) (Hemiptera: Aphididae). *Genetic Research* 77, 251–260.

Stern, D.L., Aoki, S. and Kurosu, U. (1997) Determining aphid taxonomic affinities and life cycles with molecular data: a case study of the tribe Cerataphidini (Hemiptera: Aphidoidea: Hormaphididae). *Systematic Entomology* 22, 81–96.

Templeton, A.R. (1998) The role of molecular genetics in speciation studies. In: DeSalle, R. and Schierwater, B. (eds) *Molecular Approaches to Ecology and Evolution.* Birkhauser Verlag, Basel, pp. 131–149.

VillaBlanca, F.X., Roderick, G.K. and Palumb, S.R. (1998) Invasion genetics of the Mediterranean fruit fly: variation in multiple nuclear introns. *Molecular Ecology* 7, 547– 560.

Viscarret, M.M., Torres-Jerez, I., Agostini de Manero, E., López, S.N., Botto, E.E. and Brown, J.K. (2003) Mitochondrial DNA evidence for a distinct clade of New World *Bemisia tabaci* (Genn.) (Hemiptera: Aleyrodidae) from Argentina and Bolivia, and presence of the Old World B biotype in Argentina. *Annals of the Entomological Society of America* 96, 65–72.

Von Haeseler, A. (1999) Maximum likelihood tree reconstruction. *Zoology* 102, 101–110.

Zchori-Fein, E. and Brown, J.K. (2002) Diversity of prokaryotes associated with *Bemisia tabaci* (Genn.) (Hemiptera: Aleyrodidae). *Annals of the Entomological Society of America* 95, 711–718.

Zhang, D.X. and Hewitt, G.M. (1996) Assessment of the universality and utility of a set of conserved mitochondrial COI primers in insects. *Insect Molecular Biology* 6, 143–150.

7

Predicting Evolutionary Change in Invasive, Exotic Plants and its Consequences for Plant–Herbivore Interactions

H. Müller-Schärer and T. Steinger

Université de Fribourg, Département de Biologie, Unité Ecologie et Evolution, chemin du Musée 10, CH-1700 Fribourg, Switzerland

Introduction

Invasion ecology, the study of the distribution and spread of organisms in habitats to which they are not native, has received considerable attention during past decades (Groves and Burdon, 1986; Drake *et al.*, 1989; Vitousek *et al.*, 1996; Williamson, 1996; Lonsdale, 1999; Walker, 1999; Alpert *et al.*, 2000; Mack *et al.*, 2000). This is mainly a consequence of the increased awareness of the major threats posed by invasions to biodiversity, ecosystem integrity, agriculture and human health (Lonsdale, 1999; Mack *et al.*, 2000). Two questions have dominated most of the studies in this context: which species are most likely to become invasive, and which habitats are most susceptible to invasion (Alpert *et al.*, 2000; Kolar and Lodge, 2001). Surprisingly, the evolutionary genetics of invasive species remained relatively unexplored despite the profound effect of genetic characteristics of populations on their capacity for range expansion (Ellstrand and Schierenbeck, 2000; Tsutsui *et al.*, 2000) and on species interactions (Carroll *et al.*, 2001; Siemann and Rogers, 2001). In fact, in a recent review on this topic, Lee (2002) concluded that 'the invasion success of many species might depend more heavily on their ability to respond to natural selection than on broad physiological tolerance or plasticity'. Natural selection and genetic drift can alter the genetic structure of invading populations, and hence affect not only the process of adaptation to the new physical environment, but also the plentiful biotic interaction encountered in the new habitat.

The invasion process is generally divided into two phases: the initial introduction and establishment, and the spread into the new environment. Many of the species that become successful invaders do so only after a long lag time (Sakai *et al.*, 2001, and references therein). Some of the most frequently suggested explanations for this time lag are listed in Table 7.1. Several evolutionary explanations have been put forward, but few empirical data yet exist to test these hypotheses. Populations may be poorly adapted initially, but after a period of selection they may be able to expand. This assumption is confirmed by a recent review of published studies on the conditions that promote rapid adaptive evolution (Reznick and Ghalambor, 2001). Most of the studies cited show examples of species colonizing new habitats. In addition, a common feature of many studies given in this review is the combination of directional selection, the presence of genetic variation, and at least a short-term opportunity for population increase. The opportunity for population growth, together with availability of genetic variation, may be a key factor that promotes rapid evolution, since directional selection might otherwise be expected to cause population decline and lead to an extinction vortex (Silvertown and Charlesworth, 2001). Unfortunately, time series data on recently introduced plant populations do not exist (Bone and Farres, 2001).

Table 7.1. Explanations for a time lag between establishment and population increase during an invasion process.

Ecological explanations
density-dependent forces and Allee effects (propagule pressure, pollen limitation)
lag phase of an exponential growth curve
lag phase varies with the detection threshold
Evolutionary explanations
purging of genetic load responsible for inbreeding depression
accumulation of additive genetic variation
hybridization and polyploidization
recovery from loss of genetic diversity and fitness
adaptation to a new environment (including absence of antagonists)
Sociological explanation
public awareness follows an exponential growth curve (multiplication of news)

Adaptation to local conditions is a particularly important form of evolution in plant populations, as plants are sessile. This makes plants ideal organisms for the study of local adaptations, as we know the environment they are exposed to. Indeed, many of the best examples of rapid evolution involve invasive plant species (Reznick and Ghalambor, 2001). This may be due to the generation of genetic variation (on which selection can act), such as through hybridization (e.g. Ellstrand and Schierenbeck, 2000), and/or strong directional selection exerted by abiotic, but particularly also by biotic, factors. Indeed, interactions with competitors and antagonists may differ strongly between the native and introduced range (Thompson, 1998; Bone and Farres, 2001).

Biological invasions by higher plants have increased in importance, with increasing human activities affecting both dispersal (trade and travel) and habitat availability (environmental change). Exotic plants constitute one of the most serious threats to biodiversity (Lonsdale, 1999, and references therein) and it seems that there is now no nature reserve in the world outside Antarctica that is without introduced plant species (Usher, 1988). The classical approach of biological control, referring to the introduction of a specialist enemy to control an invasive exotic species, has been the most successful control strategy against exotic plant invaders (Julien and Griffiths, 1999), and will probably remain the curative control measure of choice against environmental weeds in the near future, due to its effectiveness, low cost and relative environmental safety (Cronk and Fuller, 1995; Crutwell-McFadyen, 1998; van Klinken and Edwards, 2002; but see Louda *et al.*, 1997; Callaway *et al.*, 1999; Pearson *et al.*, 2000).

Two contrasting hypotheses dominate the literature on plant invasions; one assumes that trait combinations pre-adapt species to become good invaders ('pre-adaptation hypothesis'), whereas the other view postulates successful invasion as the outcome of rapid evolutionary change in the new habitat ('post-invasion evolution hypothesis'). If evolutionary change occurs, knowledge of its pace and direction is important to predict the impact of subsequent biological control attempts. The fact that certain correlates of invasion success have been identified, and that invasions can sometimes be reversed by biological control, support the view of pre-adapted invaders and the simple release from their natural enemies, respectively. However, recent theories (e.g. Blossey and Nötzold, 1995) and empirical evidence (e.g. Ellstrand and Schierenbeck, 2000) indicate that invasiveness can evolve. In this chapter, we will explore potential changes in plant traits and how this might influence species interactions, specifically subsequent biological control attempts by introducing specialist insect herbivores.

Framework and scope of our study

In this review, we adopt a quantitative-genetic framework to explore microevolutionary processes in invasive plants. We chose this approach because most of the plant traits relevant to our study are known to be quantitative (polygenic), as opposed to traits such as herbicide and pathogen resistance, where often only a few major genes are involved (Burdon and Thompson, 1992). Adaptation can be understood by studying how the interplay between natural selection and genetic variability translates into evolutionary change. According to a basic model of quantitative-genetic theory, adaptive evolutionary change (R) in quantitative traits can be predicted from knowledge of the selection differential (S) and the heritability (h^2), through $R = S \times h^2$ (Falconer and Mackay, 1996).

We will start by examining the direction and magnitude of expected selection pressures, and explore what phenotypic traits may be favoured by selection acting on invasive plants (S in the above equation). In a second part, we will examine processes that affect the amount of additive genetic variation (h^2 in the above equation) in invading plant populations and, therefore, their ability to evolve, and discuss consequences for plant–herbivore interactions and biocontrol success. Thirdly, we will compare these predictions with observed evolutionary responses (R in the above equation) and, again, explore how this might affect interactions with herbivorous insects, such as potential biological control agents. Finally, new studies will be proposed that link biocontrol with invasion ecology, an approach that might result in synergies contributing to the advance of both disciplines.

Selection Pressure in Invaded Habitats and Expected Selection on Plant Traits

Traits favoured by selection can be studied in several ways, including: (i) looking at the invaded habitats and the selection regime they impose, and (ii) adopting a comparative approach that contrasts invasive with non-invasive plant species, to predict traits associated with invasion.

Properties of invaded habitats

It has proven easier to identify habitat types prone to invasion than to identify traits associated with invasiveness (Lonsdale, 1999). There is a general agreement that disturbance can strongly affect habitat invasibility, mainly through changes due to fire, grazing or creation of gaps. D'Antonio *et al.* (2000) recently proposed that not only does increased disturbance affect invasibility, but that there has been a general departure from natural disturbance regimes, which facilitates invasion. Furthermore, low levels of environmental stress have been identified as a factor promoting invasions by plants (Alpert *et al.*, 2000). Absence of environmental stress might shift the competitive balance between invasives and natives. For instance, low abiotic stress may favour invasive plants because they are better able than natives to take advantage of high resource availability (Alpert *et al.*, 2000; Keane and Crawley, 2002). It was hypothesized that these factors probably interact, and it is mainly the combination of altered disturbance with high resource availability that renders habitats invasible (Alpert *et al.*, 2000). Besides low levels of nutrients and water, high levels of environmental stress also includes factors that limit resource acquisition (Alpert *et al.*, 2000, and references therein), such as competition and herbivory. Thus, low levels of competition and reduced levels of herbivory in a novel habitat might favour its colonization, but will also be of importance in view of a subsequent introduction of biological control agents. These habitats should, therefore, initially strongly select for 'ruderals' with short generation time, high growth rate, high fecundity and good dispersal (Grime, 1977, 2001; Stearns, 1992).

General traits associated with plant invaders

Traits found to be associated with invaders, based on interspecific studies, can also be used to infer expected selection at the intraspecific level, occurring during the invasion process. Such traits that are relevant for our study are listed in Table 7.2. Besides native range (species that occur more widely), rapid dispersal was found to be a good predictor of the first phase of the invasion process, i.e. during pick-up and initial introduction (see, for example, reviews by Alpert *et al.*, 2000; Sakai *et al.*, 2001). However, these traits are not good predictors of a species' invasiveness. This may be because the second phase of invasion, spread into the habitat, is habitat specific (Alpert *et al.*, 2000). Table 7.2 lists traits of the second phase that might be of special importance for the habitat types described above. Flowering time is an important trait for colonization, as it directly affects fitness in different habitats (Neuffer and Hurka, 1999). Time at first reproduction is a further important life-history trait, with early reproduction being favoured when colonizing habitats that are characterized by low levels of environmental stress (Rejmanek and Richardson, 1996; Crawley, 1997). High relative growth rate (RGR) of seedlings was found to be the most significant life-history trait separating invasive from non-invasive *Pinus* species, and specific leaf area (SLA) was the main component responsible for differences in RGR (Grotkopp *et al.*, 2002). Finally, several hypotheses have been put forward with regard to species interactions that provide mechanisms for invasions by an exotic plant species (see below).

Table 7.2. Expected selection on plant traits at two stages of the invasion process into novel habitats (based on interspecific studies).

During establishment (including probability of initial introduction)
rapid dispersal (short generation time, long fruiting period, small seed size, prolonged seed viability, transport by wind)
Spread into the new habitat
flowering phenology (latitudinal and altitudinal gradient)
life cycle (annual/perennial; mono-/polycarpic)
relative growth rate (RGR) and its components (mainly specific leaf area, SLA)
response to resident competitors and antagonists

Traits related to competitors and antagonists

Human disturbance of native communities can create new habitats to which natives are not yet adapted (e.g. increased grazing regime, nutrient enrichment), decreasing their competitive ability towards potentially pre-adapted exotic species (see above). On the other hand, exotics might have a greater inherent competitive ability than native species, as a result of their different evolutionary history (e.g. faster uptake of limiting resources) (Lonsdale, 1999; Callaway and Aschehoug, 2000; Keane and Crawley, 2002).

Besides these plant–plant interactions, the release from antagonists in the new habitat has been proposed as one of the most important factors contributing to the success of exotics as invaders (Williamson, 1996). In this benign environment, resources normally lost to antagonists may be allocated to growth and/or reproduction by a plastic phenotypic response. The assumptions and predictions of the enemy release hypothesis, also referred to as herbivore escape or ecological release hypothesis, has recently been explored in considerable detail by Keane and Crawley (2002).

Based on optimal defence theory (reviews by Herms and Mattson, 1992; Zangerl and Bazzaz, 1992), an alternative and evolutionary mechanism has been proposed by Blossey and Nötzold (1995), by which well-defended exotic plants might benefit indirectly from a lack of specialist antagonists and become invasive. The EICA (evolution of increased competitive ability) hypothesis states that, because a plant has limited resources to partition to enemy defence and competitive ability, an exotic in an antagonist-free environment will, over time, evolve to invest less in defence. This evolutionary trade-off will allow exotics to invest more in fitness components other than defence, such as increase in size, biomass and/or reproductive effort (Story *et al.*, 2000). Thus, for a common environment, the EICA hypothesis predicts that, compared with genotypes from its native range, genotypes from a plant's introduced range will: (i) grow faster and/or produce more seeds; and (ii) be less well defended, and thus specialized herbivores (i.e. potential biological control agents) will show improved performance (Blossey and Nötzold, 1995).

As with most theories on the evolution of plant defences, these proposed changes in the evolutionary trajectory hinge on the presence of fitness costs of defence. Strauss *et al.* (2002), who recently updated the seminal review paper by Bergelson and Purrington (1996) on resistance costs, found significant fitness reductions associated with herbivore resistance in 82% of the studies in which genetic background was controlled. The magnitude of direct costs ranged from 6 to 15%, and 40% of the studies investigated ecological costs (= indirect costs expressed as altered species interactions), with costs ranging from –20 to 58%. These magnitudes are generally greater than those reported by Bergelson and Purrington (1996) and indicate a strong selective disadvantage to resistance in herbivore-free environments, given that the relative fitness of the resistant genotype in the absence of the selective agent is [1 – (cost/100)]. The same conclusion can be drawn by comparing selection gradients, a measure of the selection intensity, for resistance traits, given in the review by Strauss *et al.* (2002), with 393 selection gradient estimates from 19 plant studies published in a recent review on the strength of phenotypic selection in natural populations (Kingsolver *et al.*, 2001). Selection gradients of resistance traits generally lay in the uppermost quartile (Fig. 7.1), thus greatly influencing fitness. In their review, Bergelson and Purrington (1996) have found costs of pathogen resistance to be nearly twice as high as those for herbivore resistance, indicating even stronger overall selection

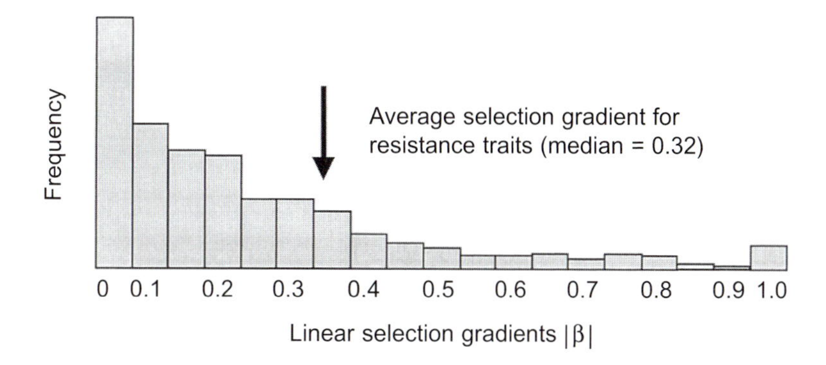

Fig. 7.1. Frequency distribution of the absolute values of the linear selection gradient estimates |β| (*N* = 393 from 19 plant studies; adapted from the database of Kingsolver *et al.*, 2001), in relation to the average selection gradient for resistance traits (trichomes, furanocoumarins and glucosinolates; *N* = 4) (from Strauss *et al.*, 2002).

towards reduced defence in the novel antagonist-free habitat. However, these general considerations have to be taken with some caution, as discussed in the following.

Caveats

Tolerance, the neglected defence strategy

Tolerance, i.e. the ability of plants to buffer the negative effects of antagonists on fitness through compensatory growth and reproduction after damage, constitutes an alternative to resistance as an evolutionary response to selection imposed by consumers. Heritable variation in tolerance has been widely documented, but with mixed support for fitness costs (reviewed by Strauss and Agrawal, 1999). Thus, in a novel, antagonist-free environment, evolution of increased competitive ability may also be expected at the expense of tolerance. If tolerance instead of resistance is involved in defence, then relaxed defence in the novel habitat (as hypothesized in the EICA hypothesis) cannot be detected by measuring insect performance, although invasive plants are assumed to be less able to tolerate herbivory. As tolerance does not impose selection on the consumers, the dynamics of plant–herbivore relationships are expected to be more stable (Roy and Kirchner, 2000; Tiffin, 2000).

Tolerance traits, such as compensatory growth, might also directly increase a plant's competitive ability. Lennartsson *et al.* (1997) recently compared life histories and morphologies of populations with consistently different histories of attack from antagonists. They showed that overcompensation occurred only in historically grazed or mown populations of field gentians (*Gentianella campestris*). Overcompensation was largely associated with changes in flowering phenology in these populations, as well as axillary branching. Most interestingly, such selection by herbivores may have resulted in a higher overall plant fitness in these environments, even in the absence of herbivory (Strauss and Agrawal, 1999). Thus, plants introduced from areas with a long selection history might be more competitive in the novel habitat as compared with plants introduced from regions with little herbivore pressure. If these plant traits are maintained in the novel habitat, subsequent biological control will be more difficult.

Defend and compete

The growth differentiation balance hypothesis (Herms and Mattson, 1992) and similar hypotheses predict that plants will evolve to 'grow *or* defend', by either investing resources in defence against herbivory and growing more slowly, or investing in growth to outcompete neighbours and relax defence against antagonists. Empirical evidence of the assumed underlying physiological trade-off between growth and differentiation is based on growth responses of single plant individuals. In a recent experimental study, Siemens *et al.* (2002) used *Brassica rapa* lines genetically divergent in constitutive levels of both glucosinolates and myrosinase (which breaks down glucosinolates into toxic compounds), which are known to affect herbivory on plants in the *Brassica* family. Defence costs (measured as growth rates) were significant in the absence of competition, but, in contrast to theoretical predictions (Bergelson and Purrington, 1996; Weis and Hochberg, 2000) and empirical evidence (Baldwin, 1998), costs of defence were not detectable in the more stressful competitive environment. This was explained by the fact that these secondary compounds also functioned as allelopathic agents against interspecific competitors, such that allelopathic benefits outweighed costs (Siemens *et al.*, 2002). Thus, 'grow and defend' might be possible because of the dual function of some secondary metabolites (Herms and Mattson, 1992; Inderjit and Moral, 1997).

The generalist–specialist dilemma

Up to now, we have assumed an antagonist-free situation in the novel, invaded habitat. However, generalist herbivores may always be present and continue to impose selection on plant defence traits. Surprisingly, little information is available on generalist enemy impact on introduced plants as compared with their native range, or as compared with co-occurring natives (Maron and Vila, 2001; Keane and Crawley, 2002). Several studies indicate a genetic basis for variation in secondary plant compounds, leading to variation in plant susceptibility to herbivores (Van der Meijden, 1996, and references therein). Van der Meijden (1996) specifically addressed the generalist–specialist dilemma for plant defence: high concentrations of the same plant chemical, such as glucosinolates or pyrrolizidine alkaloids, may deter generalist herbivores, while at the same time attracting specialist herbivores that use these chemicals as cues to find or identify their food plant. A theoretical model predicts that when attacked by specialist and generalist herbivores, plants should produce neither very low nor very high levels of defence substances, and that intraspecific variation in defence concentrations can be maintained by a shifting balance between selection pressure exerted by generalists and by specialists (Van der Meijden, 1996). Thus, a novel specialist-free environment may allow an invading plant species to defend optimally against generalists by increasing its concentrations of secondary plant metabolites. In addition, specific chemicals that only deter specialist herbivores (Shinoda *et al.*, 2002) will become redundant in these habitats. Both these mechanisms will lead to increased attractiveness and reduced resistance to specialist herbivores. Biocontrol impact is therefore expected to be higher on these introduced plant populations than on populations in the native habitat.

Genetic Variation in Invasive Plants

Quantitative-genetic theory predicts that the evolutionary response to directional selection is proportional to the magnitude of additive genetic variation in a population (Falconer and Mackay, 1996; Lynch and Walsh, 1998). An analysis of the ecological and genetic processes determining genetic diversity of invasive populations is therefore pivotal to an understanding of the role of adaptive changes for invasion success.

Reduced genetic variation in invasive plant populations?

The process of invasion into a new territory generally involves one or several successive episodes during which population size is highly reduced. Theory on the genetic consequences of small population size is well developed and predicts a loss of genetic diversity in proportion to the effective size of a population (Nei *et al.*, 1975; Brown and Marshall, 1981; Barrett and Husband, 1990). This is because founders of a colonizing population carry only a small subset of the genetic information in the source population (founder effect). If a population remains small for several generations after initial establishment, then genetic drift will become strong enough to lead to an additional loss of allelic diversity through the fixation of alleles at polymorphic loci. In addition, the frequency of heterozygotes will decrease in small populations as a result of increased mating among relatives (inbreeding).

A number of models have been proposed to describe the population-genetic consequences of colonization (Barrett and Husband, 1990). Among the simplest is the continent–island model, which assumes unidirectional dispersal of propagules from a large source population to small, isolated sink populations. Although simple, this model may nevertheless be adequate to describe the genetic effects of many examples of invasions in which a long-distance dispersal event has led to population establishment on a new continent or island. The continent–island model predicts a reduction in within-population diversity and increased differentiation from the source, depending on number of founding individuals, their total genetic diversity, and the rate of increase in population size after founding. The effects of genetic bottlenecks will be particularly strong when migration, and therefore gene flow, from the native range is restricted after initial establishment, e.g. by geographic barriers. In contrast, multiple introductions from different source populations will counteract the effect of genetic bottlenecks, especially when source populations are genetically highly structured.

To date there are still few empirical studies comparing the genetic structure of plant populations in the introduced and native range, although the situation may change rapidly as a result of the widespread availability of molecular genetic marker technology. Available studies have focused mainly on the static description of spatial patterns of genetic variation, with the principal aim of reconstructing the historical process of migration and expansion (but see Saltonstall, 2002). Although valuable, there is a lack of more integral studies examining directly key processes of invasions (number of introductions, size of founder populations, levels of gene flow after introduction, identity and genetic structure of source populations) over more than one generation. The paucity of data on these processes currently restricts direct tests of theoretical models dealing with the genetics of migration and colonization.

A number of empirical studies confirmed the theoretical prediction of reduced genetic variation in invasive plant populations (reviews by Brown and Marshall, 1981; Barrett and Richardson, 1986). For example, invasive populations of the weed *Rubus alceifolius* were characterized by a single different genotype in three Indian Ocean islands, whereas great genetic variability was found in the native range in South-East Asia (Amsellem *et al.*, 2000). The spread of these genotypes after introduction seems to have been favoured by the switch towards an apomictic breeding system (Amsellem *et al.*, 1991). Indeed, a reduction of genetic variation in the introduced range may only be common in species that reproduce primarily by selfing, apomixis or clonal spread, whereas similar levels of genetic variation are often observed in outcrossing species, or when multiple introductions of propagules from different source populations have occurred (Novak and Mack, 1993; Schierenbeck *et al.*, 1995; Novak and Welfley, 1997; Meekins *et al.*, 2001).

Low levels of genetic variation may influence plant–antagonist dynamics, in both the short and the long term. Data from several agricultural studies comparing disease dynamics in crop monocultures and in multiline mixtures have demonstrated that low levels of genetic variation can accelerate the development of epidemics within the field (Finckh and Wolfe, 1997; Garrett and Mundt, 1999; Zhu *et al.*, 2000). Similarly, in their review on biological control and the reproductive mode of weeds, Burdon and Marshall (1981) found that apomictic and other asexually reproducing plants were effectively controlled more often than sexually reproducing plants (but see Chaboudez and Sheppard, 1995). This was explained by the lack of genetic variation in populations of the clonal weeds. Thus, genetic uniformity in introduced plant populations is expected to increase biocontrol efficacy.

Genetic diversity in invasive plant populations may not only reduce spread of the antagonists, but may also allow selection by biological control agents for less-susceptible target populations, which is well illustrated by the *Chondrilla juncea* programme in Australia. Successful control of the most widespread 'narrow-leaved' form of the rosette-forming apomict *C. juncea* was achieved by the 1971 release of the rust pathogen *Puccinia chondrillina* collected from natural populations in Italy – a spectacular biocontrol success. However, the reduction of populations of this genotype resulted in the spread of the more-resistant 'intermediate-leaved' form, previously suppressed by the competitive dominance of the narrow-leaved plants (Cullen and Groves, 1977; Burdon *et al.*, 1981, 1984). Genetic diversity (only three genotypes in the case of *C. juncea* in Australia!) in resistance towards a biocontrol agent may therefore reduce biocontrol sustainability in the long term.

Small population size during initial colonization will result in increased levels of inbreeding, due to both mating between close relatives and increased self-fertilization. The direct (negative) effects of inbreeding on plant fitness have been widely demonstrated, but surprisingly little is known about indirect effects mediated by altered interaction with plant antagonists. Carr and Eubanks (2002) found that the detrimental effects of selfing on plant biomass and flower production in *Mimulus guttatus* increased by up to three times when plants were attacked by an insect herbivore as compared with unattacked control plants. In a study with *Silene alba*, Ouborg *et al.* (2000) reported altered levels of resistance to a fungal pathogen due to inbreeding. Effects ranged from inbreeding depression to inbreeding enhancement, depending on the population studied. The general finding of large between-population variation in inbreeding effects demonstrates that the consequences of inbreeding at the population level will be largely dependent on the genotypic composition of the founder population.

Caveats

Low correlation between molecular- and quantitative-genetic variation

Although the recent spread of molecular marker techniques may boost studies on the population-genetic consequences of plant invasions, caution must be applied when one tries to use molecular data to infer the adaptive potential of populations. Adaptive changes in invasive plants will mostly depend on genetic variation in quantitative traits of ecological importance (e.g. growth rate, phenology, reproduction, herbivore resistance). These traits show complex, polygenic inheritance, and their expression is generally highly dependent on environmental conditions. Based on the results of a meta-analysis of published data, Reed and Frankham (2001) concluded that the correlation between molecular- and quantitative-genetic estimates of within-population variation is generally low, questioning the value of molecular marker data to predict a population's capacity for adaptive evolutionary change (see also Steinger *et al.*, 2002).

Various factors may be responsible for this low correlation (Lynch, 1996; Reed and Frankham, 2001). For example, the effect of population bottlenecks on additive genetic variation of quantitative traits may deviate from a simple model of genetic drift when trait variation is influenced by non-additive effects of alleles (dominance, epistasis). Several theoretical studies have shown that epistasitic and/or dominance variance can be converted to additive genetic variance as populations pass through a bottleneck (Goodnight, 1987, 1988; Whitlock *et al.*, 1993; Cheverud and Routman, 1996). Thus, predicted reductions in additive genetic variance of bottlenecked, invading populations may be limited, or even reversed, allowing them to maintain their adaptive capacity during colonization. Several experimental studies have confirmed the predicted increase in additive genetic variation following one or several bottlenecks (Bryant *et al.*, 1986; Lopez-Fanjul and Villaverde, 1989; Ruano *et al.*, 1996), although the range of taxa studied is limited and plant studies are almost entirely missing (but see Waldmann, 2001).

An additional complication for the assessment of the evolutionary potential of invasive populations arises from the general observation that additive genetic variance in quantitative traits is not a fixed property of a population but may be strongly dependent on environmental conditions. Few generalizations seem currently possible that would allow prediction of the environmental influence on the expression of additive genetic variation. Hoffmann and Parsons (1996) proposed the hypothesis that environmental stress will tend to increase genetic variance and heritability, and there is some empirical support for this prediction (reviewed by Jenkins *et al.*, 1997), but other studies found the opposite (for plants see Sultan and Bazzaz, 1993; Bennington and McGraw, 1996).

Increased genetic variance through hybridization

Invasive populations can escape genetic impoverishment after bottlenecks not only through gene flow with native populations or co-introduced populations that were previously isolated, but also through introgression of genes from other plant species, both native and introduced. Interspecific hybridization is being increasingly recognized as an important mechanism stimulating the evolution of invasiveness. In their review, Ellstrand and Schierenbeck (2000) list 28 examples where invasiveness was preceded by hybridization. A number of hypotheses have been put forward to explain the superior competitive ability of hybrid taxa, which may contribute to invasion success. First, hybridization may generate genetic variation, that is, provide the raw material for rapid adaptation. This can involve the recombination of phenotypic traits from both parental taxa, or the generation of transgressive phenotypes that are extreme relative to either parent. Rieseberg *et al.* (1999) noted that transgressive segregation may be the rule rather than the exception, and is likely to contribute to the evolutionary success of hybrids, in particular when transgression involves tolerances to various biotic and abiotic factors. Secondly, hybrids may show heterosis, i.e. their fitness can surpass that of both parents. Although heterosis will be eroded quickly by sexual recombination, several mechanisms exist that maintain heterosis (vegetative propagation, apomixis, allopolyploidy, permanent translocation heterozygosity). Thirdly, hybridization may reduce the genetic load resulting from the accumulation of mildly deleterious alleles in small populations.

How will hybridization affect the interactions of plants with their antagonists? Fritz *et al.* (1999) recently reviewed published plant studies analysing the consequences of hybridization for herbivore and pathogen resistance. A surprising result was that, for cases where parental taxa differed in resistance levels to herbivores or pathogens, hybrids resembled the susceptible parent more often than the resistant parent. In quite a number of cases, hybrids exhibited transgressive phenotypes in the direction of increased susceptibility; that is, hybrids were less resistant than either parent. The generality of these results is not known, but because of the large implications for biological control programmes and the widespread occurrence of hybridization in invasive plants, a closer look at the genetics of resistance in intra- and interspecific hybrids would be valuable.

Observed Evolutionary Response

Far more predictions than experimental data are available on genetic changes in introduced plants, mainly as the genetics and evolution of invasive species have received far less attention than their ecology (Sakai *et al.*, 2001). Bone and Farres (2001) estimated rates of evolution in plants from previously published studies, and found widespread evidence for rapid evolutionary change in response to changing environmental conditions. Although estimated rates of evolution of invasive plants were, in general, relatively low, most interestingly, exceptionally high rates (>0.5 haldanes) were found for traits associated with loss of herbivore resistance, indicating the potential for rapid evolution associated with the presence or absence of the plant's natural enemies.

Using their compilation of estimated rates of evolution in plants, Bone and Farres (2001) explored specific trends with regard to differences among traits, life-history correlates and responses to environmental conditions. Physiological traits (e.g. resistance to heavy metals, herbicide or herbivores) were found to evolve faster than morphological trends (e.g. leaf length, biomass, achene volume). Annual plants may be expected to adjust more quickly in absolute time because of their short generation time, while long-lived plants may experience stronger selection pressure per generation (assuming constant selection pressure) and therefore evolve more quickly when generation time is standardized. Indeed, no overall difference in rates of evolution was found between annuals and perennials, when similar traits were compared.

Table 7.3 lists a selection of studies that addressed experimentally genetic changes in plants introduced into a novel habitat. Table 7.3a summarizes studies with *Lythrum salicaria* which compared plant performance and herbivore defence traits of plant populations from native (Europe) and introduced (USA, Australia) habitats. Although increased performance and reduced defence, as predicted by the EICA hypothesis, have been found repeatedly in introduced plants, as compared with their native conspecifics, the results here remain equivocal, as do those for the proposed cause of change (Table 7.3a). These *L. salicaria* studies, which probably constitute the best currently available dataset on this issue, indicate dependency of the findings on the test conditions and the herbivore species involved (Willis *et al.*, 2000; Thébaud and Simberloff, 2001). Future studies should involve reciprocal transplants (Gandon and Van Zandt, 1998), experimental reduction of potential maternal effects, the explicit testing of tolerance as an alternative to defence by resistance (Strauss and Agrawal, 1999; Willis *et al.*, 1999; DeJong and van der Meijden, 2000; Rogers and Siemann, 2002), and the control/monitoring of local antagonists colonizing the test plants.

Selected studies on other plant species are compiled in Table 7.3b. Thus, an increasing number of studies document evolutionary changes in a variety of plant traits during the invasion process. Possible outcomes for plant performance that might be expected when specialized herbivores are introduced as biological control agents are listed in Table 7.4, based on both theoretical predictions and empirical evidence (e.g. Crawley, 1983, 1997). The hypothesis of a change from a prevalent annual or biennial, monocarpic habit in the native temperate European habitat to a biennial or perennial, polycarpic habit associated with the invasion process into North American temperate regions, and the expected

Table 7.3. Genetic changes observed in introduced plant populations.

(a) Studies with *Lythrum salicaria.*

Type of experiment (pot/garden/field, type of transplant	Studied traits of plants and herbivores	Change in studied trait in novel habitat	Proposed cause of change	References
Pot experiment in the open field, transplant (1 site)	Biomass Height Leaf-feeder performance Root-feeder performance	Increased Increased No effect Increased	PIE	1
Garden plot, transplant (1 site)	Shoot biomass Colonization by leaf-feeder	Increased Preferred	PIE	2
Pot/garden plot, reciprocal transplant (2 sites)	Plant performance variates	Slightly increased or no effect	Multiple introduction SOR	3
Greenhouse, transplant (1 site)	Performance of specialist Performance of generalist Leaf phenolic content (defence)	No effect No effect Decreased	Little evidence in support of PIE	4

(b) Selected studies involving other plant species (see text for details)

Plant spp., type of experiment	Studied traits of plants and herbivores	Change in studied trait in novel habitat	Proposed cause of change	References
Four biennials, garden plot, transplant	Plant performance variates	No effect	No genetic response (plasticity)	5
Spartina alterniflora, greenhouse, transplant	Herbivore resistance	Decreased	SGI (+ PIE/drift)	6
	Aggregation by planthoppers	Increased		
Solidago spp., garden plot and field study, transplant	Flowering time	Earlier in the north	PIE	7
	Size at flowering	Reduced		
Capsella bursa pastoris, garden plot, transplant	Flowering time	Early in desert, late in coastal and snowy forest sites	SOR	8
Sapium sebiferum, garden plot, transplant	Plant size	Increased	PIE	9
	Probability of seed production	Increased		
	Foliar C/N ratio	Increased		
	Tannin concentrations	Reduced (defence)		
Senecio jacobaea, greenhouse/garden plot, reciprocal transplant	Plant size, reproductive	Increased	SOR and/or PIE	10
	Biomass, secondary plant compounds	Increased		
	Herbivore resistance	Reduced		
Centaurea maculosa, *Tripleurospermum perforatum*, *Verbascum thapus*, *Senecio jacobaea*	Life cycle	Perennial and polycarpic, flowering in the first year	SOR and PIE	11 12 13 10, 14

SGI, single genotype introductions, an extreme founder effect; SOR, sorting-out, differential establishment of genotypes; PIE, post-invasion evolution that includes mutation and recombination.

1, Blossey and Nötzold (1995); 2, Blossey and Kamil (1996); 3, Willis and Blossey (1999); 4, Willis *et al.* (1999); 5. Willis *et al.* (2000); 6, Daehler and Strong (1997); 7, Weber and Schmid (1998); 8, Neuffer and Hurka (1999); 9, Siemann and Rogers (2001); 10, J. Joshi, Zürich, Switzerland, 2002, personal communication; 11, Müller (1989a); 12, H. Hinz, Delémont, Switzerland, 2002, personal communication; 13, Juvik and Juvik (1992); 14, P. McEvoy, Corvallis, Oregon, 2002, personal communication.

Table 7.4. Expected consequence of observed genetic changes in introduced plant populations, and expected consequence for herbivore impact on individual plants.

Observed changes in	Expected herbivore impact on individual plants
Herbivore preference and and performance	Corrected for plant size, increased preference and performance (due to relaxed resistance) may result in increased impact
Plant biomass	Corrected for herbivore level, increased biomass may lead to reduced impact, but larger plants may attract more herbivores, equalizing this effect
Flowering phenology	Earlier and shorter flowering period may reduce overlap with specialized seed-feeding insects
Plant life cycle	Prevalence of a perennial and polycarpic life cycle may have evolved in the absence of root-feeders, but plants may then be more vulnerable to subsequent biological control agents
Relative growth rate and specific leaf area	Thinner and less-rough leaves may facilitate feeding by chewing and sucking insects

the invasion process into North American temperate regions, and the expected differential impact of root-herbivores appear to be especially worthwhile for explicit testing. In their native range, both between- and within-population variation in life cycle has been observed for a number of plant species, with partial evidence of a genetic basis (e.g. see the references in Table 7.3b). As an example, *Centaurea maculosa* ('*maculosa*' in North America is generally a misnomer for *Centaurea stoebe* ssp. *micranthos*; cf. Ochsmann, 2001) introduced into British Columbia (Canada) and Montana (USA) are nearly exclusively short-lived, perennial tetraploids (native to eastern Europe), while the most abundant and widely distributed *C. maculosa* in western Europe (= *C. stoebe* ssp. *stoebe*; Ochsmann, 2001) is a diploid biennial (Müller, 1989a,b). These diploid populations are also the main source of subsequent introductions of specialized root-feeders (Müller, 1989b; Müller *et al.*, 1989). As shown by a coarse population model derived from data from a common environment experiment, the potential for high seed production early in life, together with the perennial life cycle, may have favoured the spread of the tetraploid plants in North America (Müller, 1989a). Root-feeding herbivores are known preferentially to attack and kill larger plants (e.g. Müller, 1983, 1989b; Wesselingh *et al.*, 1993, 1997), and we might therefore speculate that a predominantly perennial, polycarpic life cycle might have been able to evolve (most likely through sorting-out or hybridization processes) and spread only in environments free of (specialized) root-herbivores, such as after transcontinental introductions. A similar conclusion can be derived from a modelling study by Klinkhamer *et al.* (1997), which found the optimal life-history solution to be semelparity (monocarpic life cycle) or mast years, if herbivores attracted by flowering plants mainly reduce plant survival. Interestingly, we would then expect an increased effect of root-feeding biological control agents on these introduced genotypes as compared with the conspecifics from their native range.

Once the impact on individual plants is known, the more crucial question of how this will translate into population-level interactions remains, as the success of biological weed control is measured in terms of reductions in population density and range of weed infestations (Crawley, 1989a). If plant recruitment is not seed limited, then insects that reduce seed production will have no impact on plant population dynamics. On the other hand, herbivore functional and numerical responses are seldom linear and will vary between species and environments (Crawley, 1997). Thus, owing to the many unknown aspects in this scaling-up process, generalizations on the impact of expected changes in plant traits during the invasion process on biological control success will remain problematic. Only detailed case-by-case studies, based on thorough experimental investigations, will allow us to make realistic predictions.

A pragmatic alternative approach is to look at plant attributes related to successes and failures in weed biocontrol projects. Plant attributes associated with good control were found to be a biennial life cycle (as opposed to annual), no seed bank, low power of regrowth, limited seed dispersal and genetic uniformity (see below) (Crawley, 1989b, 1997), which partly confirms our predictions.

Outlook

Recent developments in evolutionary theory, quantitative genetics and molecular tools offer a new integral approach to investigating genetic factors that affect invasion success and its consequences for classical biological control measures. Recently published reviews on genetics and invasion (Lee, 2002), resistance costs to herbivory (Strauss *et al.*, 2002) and the evolution of host specificity (van Klinken and Edwards, 2002), to list only a few, document the increasing scientific interest in combining ecological and evolutionary aspects related to biological invasions and weed biocontrol. It is being increasingly recognized that biological control offers an ideal opportunity to merge disciplines such as population dynamics with population genetics. The topic of this book is therefore well in line with these recent developments.

We are well aware that the scope of our study, i.e. predicting consequences of evolutionary change in invasive plants for plant–herbivore interactions, will involve unravelling innumerable processes, from changes in gene frequencies to plant fitness, population dynamics and community interactions, a clearly unrealistic objective. We hope, however, to have pointed to further interesting studies that address important processes at the various levels of integration. Six such emerging topics are briefly outlined below. In particular, studies in the context of classical biological weed control programmes offer a great opportunity for testing hypotheses that might result in findings contributing to the advance of both theory and practical application, in both biological control and invasion ecology.

Emerging research topics

1. Reciprocal transplant experiments could be used in combination with a phenotypic or genotypic selection analysis (Lande and Arnold, 1983; Mauricio and Mojonnier, 1997; Joshi *et al.*, 2001) to examine the direction and strength of selection in the native and the novel habitat. This would not only allow one to contrast current selection pressures acting on plant traits with evolved changes in these traits, but also to assess the adaptive value of trait variation in different environments.

2. Little is still known on evolutionary trajectories of defence traits when selection pressures are relaxed in the novel environment. It seems especially worthwhile to explore this trade-off for plants differing in their history of selection by herbivores, under different competitive environments, and in the presence and absence of native generalist herbivores. The introduction of biological control agents is expected to reverse the selection pressure on defence traits (see above). Where and when biological control agents were introduced are often well documented, allowing exploration of the pace of evolutionary change in reversed direction. Results of such studies would allow an extension of the EICA hypothesis by including selection history (in both the native and the introduced area) and the effects of generalist herbivores present in the invaded habitat.

3. Tolerance as a defence mechanism needs further attention. Mainly due to the fact that tolerance does not directly affect preference and performance of the antagonists and, thus, does not impose selection on the antagonists, plant–herbivore relationships and population dynamics might be expected to remain more stable (and biological control more sustainable) when tolerance is involved than when resistance is involved (Fineblum and Rausher, 1995; Strauss and Agrawal, 1999; Roy and Kirchner, 2000; Tiffin and Inouye, 2000).

It was even recently suggested that evolution of tolerance could promote an apparently mutualistic relationship between herbivores and their host plant. Selection from herbivores, resulting in increased branching, could have served to increase overall plant fitness even in the absence of herbivory (Järemo *et al.*, 1999; Strauss and Agrawal, 1999). This, in turn, could give these plants a competitive advantage over plants that have evolved under reduced levels of herbivory. If these plant traits are maintained in the new habitat through competitive interactions with native plant species, biocontrol success could be hampered. These hypotheses warrant rigorous testing.

4. Changes in the prevalent life cycle between the native and invaded habitat, and their effects on antagonists, could be studied in reciprocal transplant experiments under open-field conditions. A long-term study to follow the population dynamics of both the insect and the plant populations would be especially rewarding, with the plants grown in mono- and mixed populations of native and introduced genotypes, and herbivores being excluded or allowed to interfere.

5. The observed increased susceptibility in hybrids towards their natural enemies remains intriguing (Fritz *et al.*, 1999). The environmental dependency of hybrid resistance relative to the resistance levels of the parental taxa, especially, might-

deserve more careful studies. Assuming different defence chemicals to be involved in the parents, we might speculate joint occurrence in the hybrid, but at concentrations too low to influence insect performance negatively, resulting in increased susceptibility.

6. Increased synergism between biological control projects and basic research should be easy to achieve. As an example, host-specificity screening tests generally involve individuals of the weed species to be controlled in both the introduced (target) and the native (host) area. Careful examination of plant responses, as well as preference and performance of the candidate specialist herbivores, would allow the study of trade-offs between plant defence and fitness for native and introduced plant genotypes. Further, collection and storage of seed material from the target populations prior to release, together with a carefully elaborated release design, would allow the genetic composition of the plants during the invasion process to be followed in the presence and absence of specialist antagonists. Similarly, potential genetic changes in the control agents during their population increase and spread could be followed by comparing population samples in the release area at regular time intervals.

Acknowledgements

We thank Dieter Ebert, Jasmin Joshi, Peter McEvoy, Urs Schaffner, Klaas Vrieling, René Sforza and two anonymous referees for helpful comments. H.M.S. appreciated the hospitality of D. Matthies and the University of Marburg, Germany while writing parts of this review. This review was partly supported by the Swiss National Science Foundation through grant numbers 31–65356.01, to H.M.S., and 31–67044.01, to T.S., and through the National Centre of Competence in Research (NCCR) 'Plant Survival'.

References

Alpert, P., Bone, E. and Holzapfel, C. (2000) Invasiveness, invasibility and the role of environmental stress in the spread of non-native plants. *Perspectives in Plant Ecology, Evolution and Systematics* 3, 52–66.

Amsellem, L., Noyer, J.L., Le Bourgeois, T. and Hossaert-McKey, M. (2000) Comparison of genetic diversity of the invasive weed *Rubus alceifolius* Poir. (Rosaceae) in its native range and in areas of introduction, using amplified fragment length polymorphism (AFLP) markers. *Molecular Ecology* 9, 443–455.

Amsellem, Z., Sharon A. and Gressel, J. (1991) Abolition of selectivity of two mycoherbicidal organisms and enhanced virulence of avirulent fungi by an invert emulsion. *Phytopathology* 81, 985–988.

Baldwin, I.T. (1998) Jasmonate-induced responses are costly but benefit plants under attack in native populations. *Proceedings of the National Academy of Sciences USA* 95, 8113–8118.

Barrett, S.C.H. and Husband, B.C. (1990) The genetics of plant migration and colonization. In: Brown, A.H.D., Clegg, M.T., Kahler, A.L. and Weir, B.S. (eds) *Plant Population Genetics, Breeding, and Genetic Resources.* Sinauer, Sunderland, Massachusetts, pp. 254–277.

Barrett, S.C.H. and Richardson, B.J. (1986) Genetic attributes of invading species. In: Groves, R.H. and Burdon, J.J. (eds) *Ecology of Biological Invasions.* Cambridge University Press, Cambridge, pp. 21–33.

Bennington, C.C. and McGraw, J.B. (1996) Environment-dependence of quantitative genetic parameters in *Impatiens pallida. Evolution* 50, 1083–1097.

Bergelson, J. and Purrington, C.B. (1996) Surveying patterns in the cost of resistance in plants. *American Naturalist* 148, 536–558.

Blossey, B. and Kamil, J. (1996) What determines the increased competitive ability of non-indigenous plants? In: Moran, V.C. and Hoffman, J.H. (eds) *Proceedings of the IX International Symposium on Biological Control of Weeds,* Stellenbosch, South Africa, January, pp. 19–26.

Blossey, B. and Nötzold, R. (1995) Evolution of increased competitive ability in invasive nonindigenous plants: a hypothesis. *Journal of Ecology* 83, 887–889.

Bone, E. and Farres, A. (2001) Trends and rates of microevolution in plants. *Genetica* 112–113, 165–182.

Brown, A.H.D. and Marshall, D.R. (1981) Evolutionary changes accompanying colonization in plants. In: Scudder, G.C.E. and Reveal, J.L. (eds) *Evolution Today, Proceedings of the II International Congress on Systematics and Evolution in Biology,* pp. 351–363.

Bryant, E.H., McCommas, S.A. and Combs, L.M. (1986) The effect of an experimental bottleneck upon quantitative genetic-variation in the housefly. *Genetics* 114, 1191–1211.

Burdon, J.J. and Marshall, D.R. (1981) Biological control and the reproductive mode of weeds. *Journal of Applied Ecology* 18, 649–658.

Burdon, J.J. and Thompson, J.N. (1992) Gene-for-gene coevolution between plants and parasites. *Nature* 360, 121–125.

Burdon, J.J., Groves, R.H. and Cullen, J.M. (1981) The impact of biological control on the distribution and abundance of *Chondrilla juncea* in south-eastern Australia. *Journal of Applied Ecology* 18, 957–966.

Burdon J.J., Groves, R.H., Kaye, P.E. and Speer, S.S. (1984) Competition in mixtures of susceptible and resistant genotypes of *Chondrilla juncea* differentially infected with rust. *Oecologia* 64, 199–203.

Callaway, R.M. and Aschehoug, E.T. (2000) Invasive plants versus their new and old neighbors: a mechanism for exotic invasions. *Science* 290, 521–523.

Callaway, R.M., DeLuca, T.H. and Belliveau, W.M. (1999) Biological-control herbivores may increase competitive ability of the noxious weed *Centaurea maculosa. Ecology* 80, 1196–1201.

Carr, D.E. and Eubanks, M.D. (2002) Inbreeding alters resistance to insect herbivory and host plant quality in *Mimulus guttatus* (Scrophulariaceae). *Evolution* 56, 22–30.

Carroll, S.P., Dingle, H., Famula, T.R. and Fox, C.W. (2001) Genetic architecture of adaptive differentiation in evolving host races of the soapberry bug, *Jadera haematoloma. Genetica* 112–113, 257–272.

Chaboudez, P. and Sheppard, A.W. (1995) Are particular weeds more amenable to biological control? A reanalysis of mode of reproduction and life history. In: Delfosse E.S. and Scott R.R. (eds) *Proceedings of the Eighth International Symposium on Biological Control of Weeds.* CSIRO Publications, Melbourne, Australia, pp. 95–102.

Cheverud, J.M. and Routman, E.J. (1996) Epistasis as a source of increased additive genetic variance at population bottlenecks. *Evolution* 50, 1042–1051.

Crawley, M.J. (1983) *Herbivory – the Dynamics of Animal–Plant Interactions*. Blackwell, London.

Crawley, M.J. (1989a) Insect herbivores and plant population dynamics. *Annual Review of Entomology* 34, 531–564.

Crawley, M.J. (1989b) The successes and failures of weed biocontrol using insects. *Biocontrol News and Information* 10, 213–223.

Crawley, M.J. (1997) *Plant Ecology*. Blackwell Scientific Publications, Oxford.

Cronk, Q.C.B. and Fuller, J.L. (1995) *Plant Invaders: the Threat to Natural Ecosystems*. Chapman & Hall, London.

Crutwell-McFadyen, R.E.C. (1998) Biological control of weeds. *Annual Review of Entomology* 43, 369–393.

Cullen, J.M. and Groves, R.H. (1977) The population biology of *Chondrilla juncea* L. in Australia. *Ecological Society of Australia* 10, 121–134.

Daehler, C.C. and Strong, D.R. (1997) Reduced herbivore resistance in introduced smooth cordgrass (*Spartina alterniflora*) after a century of herbivore-free growth. *Oecologia* 110, 99–108.

D'Antonio, C.M., Dudley, T.L. and Mack, M. (2000) Disturbance and biological invasions: direct effects and feedbacks. In: Walker, L.R. (ed.) *Ecosystems of the World: Ecosystems of Disturbed Ground*. Elsevier Science, New York, pp. 429–468.

DeJong, T.J. and van der Meijden, E. (2000) On the correlation between allocation to defence and regrowth in plants. *Oikos* 88, 503–508.

Drake, J.A., Mooney, H.A., di Castri, F., Groves, R.M., Kruger, F.J., Rejmanek, M. and Williamson, M. (eds) (1989) *Biological Invasions: a Global Perspective*. John Wiley & Sons, Chichester.

Ellstrand, N.C. and Schierenbeck, K.A. (2000) Hybridization as a stimulus for the evolution of invasiveness in plants? *Proceedings of the National Academy of Sciences USA* 97, 7043–7050.

Falconer, D.S. and Mackay, T.F.C. (1996) *Introduction to Quantitative Genetics*. Longman, Harlow.

Finckh, M.R. and Wolfe, M.S. (1997) The use of biodiversity to restrict plant disease and some consequences for farmers and society. In: Jackson, L.E. (ed.) *Ecology in Agriculture*. Academic Press, New York, pp. 203–237.

Fineblum, W.L. and Rausher, M.D. (1995) Evidence for a tradeoff between resistance and tolerance to herbivore damage in a morning glory. *Nature* 377, 517–520.

Fritz, R.S., Moulia, C. and Newcombe, G. (1999) Resistance of hybrid plants and animals to herbivores, pathogens, and parasites. *Annual Review of Ecology and Systematics* 30, 565–591.

Gandon, S. and Van Zandt, P.A. (1998) Local adaptation and host–parasite interactions. *Trends in Ecology and Evolution* 13, 214–216.

Garrett, K.A. and Mundt, C.C. (1999) Epidemiology in mixed host populations. *Phytopathology* 89, 984–990.

Goodnight, C.J. (1987) On the effect of founder events on epistatic genetic variance. *Evolution* 41, 80–91.

Goodnight, C.J. (1988) Epistasis and the effect of founder events on the additive genetic variance. *Evolution* 42, 441–454.

Grime, J.P. (1977) Evidence for the existence of three primary strategies in plants and its relevance to ecological and evolutionary theory. *American Naturalist* 111, 1169–1194.

Grime, J.P. (2001) *Plant Strategies, Vegetation Processes and Ecosystem Properties.* John Wiley & Sons, Chichester, UK.

Grotkopp, E., Rejmanek, M. and Rost, T.L. (2002) Toward a causal explanation of plant invasiveness: seedling growth and life-history strategies of 29 pine (*Pinus*) species. *American Naturalist* 159, 396–419.

Groves, R.H. and Burdon, J.J. (1986) *Ecology of Biological Invasions.* Cambridge University Press, Cambridge.

Herms, D.A. and Mattson, W.J. (1992) The dilemma of plants: to grow or defend. *Quarterly Review of Biology* 67, 283–335.

Hoffmann, A.A. and Parsons, P.A. (1996) *Evolutionary Genetics and Environmental Stress.* Oxford University Press, Oxford.

Inderjit, P. and Moral, D. (1997) Is separating resource competition from allelopathy realistic? *Botanical Review* 63, 221–230.

Järemo, J., Tuomi, J., Nilsson, P. and Lennartsson, T. (1999) Plant adaptation to herbivory: mutualistic versus antagonistic coevolution. *Oikos* 84, 313–320.

Jenkins, N.L., Sgro, C.M. and Hoffmann, A.A. (1997) Environmental stress and the expression of genetic variation. In: Bijlsma, R. and Loeschcke, V. (eds) *Environmental Stress, Adaptation and Evolution* (*Experientia Supplementum*), Birkhäusen Verlag Basel, pp. 79–96.

Joshi, J., Schmid, B., Caldeira, M.C., Dimitrakopoulos, P.G., Good, J., Harris, R., Hector, A., Huss-Danell, K., Jumpponen, A., Minns, A., Mulder, C.P.H., Pereira, J. S., Prinz, A., Scherer-Lorenzen, M., Siamantziouras, A.-S.D., Terry, A.C., Troumbis, A.Y. and Lawton, J.H. (2001) Local adaptation enhances performance of common plant species. *Ecology Letters* 4, 536–544.

Julien, M.H. and Griffiths, M.W. (1999) *Biological Control of Weeds. A World Catalogue of Agents and Their Target Weeds.* CAB International, Wallingford, UK.

Juvik, J.O. and Juvik, S.P. (1992) Mullein (*Verbascum thaspsus*): the spread and adaptation of a temperate weed in the montane tropics. In: Stone, C.P., Smith, C.W. and Tunison, J.T. (eds) *Alien Plant Invasions in Native Ecosystems of Hawaii: Management and Research.* University of Hawaii Cooperative National Park Resources Study Unit, Honolulu, pp. 254–270.

Keane, R.M. and Crawley, M.J. (2002) Exotic plant invasions and the enemy release hypothesis. *Trends in Ecology and Evolution* 17, 164–170.

Kingsolver, J.G., Hoekstra, H.E., Hoekstra, J.M., Berrigan, D., Vignieri, S.N., Hill, C.E., Hoang, A., Gibert, P. and Beerli, P. (2001) The strength of phenotypic selection in natural populations. *American Naturalist* 157, 245–261.

Klinkhamer, P.G.L., Kubo, T. and Iwasa, Y. (1997) Herbivores and the evolution of semelparous perennial life-history of plants. *Journal of Evolutionary Biology* 10, 529–550.

Kolar, C.S. and Lodge, D.M. (2001) Progress in invasion biology: predicting invaders. *Trends in Ecology and Evolution* 16, 199–204.

Lande, R. and Arnold, S.J. (1983) The measurement of selection on correlated characters. *Evolution* 37, 1210–1226.

Lee, C.E. (2002) Evolutionary genetics of invasive species. *Trends in Ecology and Evolution* 17, 386–391.

Lennartsson, T., Tuomi, J. and Nilsson, P. (1997) Evidence for an evolutionary history of overcompensation in the grassland biennial *Gentianella campestris* (Gentianaceae). *American Naturalist* 149, 1147–1155.

Lonsdale, W.M. (1999) Global patterns of plant invasions and the concept of invasibility. *Ecology* 80, 1522–1536.

Lopez-Fanjul, C. and Villaverde, A. (1989) Inbreeding increases genetic variance for viability in *Drosophila melanogaster*. *Evolution* 43, 1800–1804.

Louda, S.M., Kendall, D., Connor, J. and Simberloff, D. (1997) Ecological effects of an insect introduced for the biological control of weeds. *Science* 277, 1088–1090.

Lynch, M. (1996) A quantitative-genetic perspective on conservation issues. In: Avise, J.C. and Hamrick, J.L. (eds) *Conservation Genetics: Case Histories From Nature.* Chapman & Hall, New York, pp. 471–501.

Lynch, M. and Walsh, B. (1998) *Genetics and Analysis of Quantitative Traits.* Sinauer Associates, Sunderland, Massachusetts.

Mack, R.N., Simberloff, D., Lonsdale, W.M., Evans, H., Clout, M. and Bazzaz, F.A. (2000) Biotic invasions: causes, epidemiology, global consequences, and control. *Ecological Applications* 10, 689–710.

Maron, J.L. and Vila, M. (2001) When do herbivores affect plant invasions? Evidence for the natural enemies and biotic resistance hypothesis. *Oikos* 95, 361–373.

Mauricio, R. and Mojonnier, L.E. (1997) Reducing bias in the measurement of selection. *Trends in Ecology and Evolution* 12, 433–436.

Meekins, J.F., Ballard, H.E. and McCarthy, B.C. (2001) Genetic variation and molecular biogeography of a North American invasive plant species (*Alliaria petiolata*, Brassicaceae). *International Journal of Plant Science* 162, 161–169.

Müller, H. (1983) Untersuchungen zur Eignung von *Stenodes straminea* Haw. (Lep Cochylidae) für die biologische Bekämpfung von *Centaurea maculosa* Lam. (gefleckte Flockenblume) (Compositae) in Kanada. *Mitteilungen der Schweizerischen Entomologischen Gesellschaft* 56, 329–342.

Müller, H. (1989a) Growth pattern of diploid and tetraploid spotted knapweed, *Centaurea maculosa* Lam. (Compositae) and effects of the root-mining moth *Agapeta zoegana* (L.) (Lep.: Cochylidae). *Weed Research* 29, 103–111.

Müller, H. (1989b) Structural analysis of the phytophagous insect guilds associated with the roots of *Centaurea maculosa* Lam., *C. diffusa* Lam., and *C. vallesiaca* Jordan in Europe: 1. Field observations. *Oecologia* 78, 41–52.

Müller, H., Stinson, C.A.S., Marquardt, K. and Schröder, D. (1989) The entomofaunas of roots of *Centaurea maculosa* Lam., *C. diffusa* Lam. and *C. vallesiaca* Jordan in Europe: niche separation in space and time. *Journal of Applied Entomology* 107, 83–95.

Nei, M., Maruyama, T. and Chakraborty, R. (1975) Bottleneck effect and genetic-variability in populations. *Evolution* 29, 1–10.

Neuffer, B. and Hurka, H. (1999) Colonization history and introduction dynamics of *Capsella bursa-pastoris* (*Brassicaceae*) in North America: isozymes and quantitative traits. *Molecular Ecology* 8, 1667–1681.

Novak, S.J. and Mack, R.N. (1993) Genetic-variation in *Bromus-tectorum* (Poaceae) – comparison between native and introduced populations. *Heredity* 71, 167–176.

Novak, S.J. and Welfley, A.Y. (1997) Genetic diversity in the introduced clonal grass *Poa bulbosa* (Bulbous bluegrass). *Northwest Science* 71, 271–280.

Ochsmann, J. (2001) On the taxonomy of spotted knapweed (*Centaurea stoebe* L.). In: Smith, L. (ed.) *Proceedings of the First International Knapweed Symposium of the Twenty-First Century.* Coeur d'Alene, Idaho, pp. 33–41.

Ouborg, N.J., Biere, A. and Mudde, C.L. (2000) Inbreeding effects on resistance and transmission-related traits in the *Silene–Microbotryum* pathosystem. *Ecology* 81, 520–531.

Pearson, D.E., McKelvey, K.S. and Ruggiero, L.F. (2000) Non-target effects of an intro-

duced biological control agent on deer mouse ecology. *Oecologia* 122, 121–128.
Reed, D.H. and Frankham, R. (2001) How closely correlated are molecular and quantitative measures of genetic variation? A meta-analysis. *Evolution* 55, 1095–1103.
Rejmanek, M. and Richardson, D.M. (1996) What attributes make some plant species more invasive? *Ecology* 77, 1655–1661.
Reznick, D.N. and Ghalambor, C.K. (2001) The population ecology of contemporary adaptations: what empirical studies reveal about the conditions that promote adaptive evolution. *Genetica* 112–113, 183–198.
Rieseberg, L.H., Archer, M.A. and Wayne, R.K. (1999) Transgressive segregation, adaptation and speciation. *Heredity* 83, 363–372.
Rogers, W.E. and Siemann, E. (2002) Effects of simulated herbivory and resource availability on native and exotic tree seedlings. *Basic and Applied Ecology* 3, 297–307.
Roy, B.A. and Kirchner, J.W. (2000) Evolutionary dynamics of pathogen resistance and tolerance. *Evolution* 54, 51–63.
Ruano, R.G., Silvela, L.S., Lopez-Fanjul, C. and Toro, M.A. (1996) Changes in the additive variance of a fitness-related trait with inbreeding in *Tribolium castaneum*. *Journal of Animal Breeding and Genetics-Zeitschrift Fur Tierzuchtung Und Zuchtungsbiologie* 113, 93–97.
Sakai, A.K., Allendorf, F.W., Holt, J.S., Lodge, D.M., Molofsky, J., With, K.A., Baughman, S., Cabin, R.J., Cohen, J.E., Ellstrand, N.C., McCauley, D.E., O'Neil, P., Parker, I.M., Thompson, J.N. and Weller, S.G. (2001) The population biology of invasive species. *Annual Review of Ecology and Systematics* 32, 305–332.
Saltonstall, C. (2002) Cryptic invasion by a non-native genotype of the common reed, *Phragmites australis*, into North America. *Proceedings of the National Academy of Sciences USA* 99, 2445–2449.
Schierenbeck, K.A., Hamrick, J.L. and Mack, R.N. (1995) Comparison of allozyme variability in a native and an introduced species of *Lonicera*. *Heredity* 75, 1–9.
Shinoda, T., Nagao, T., Nakayama, M., Serizawa, N., Koshioka, M., Okabe, H. and Kawai, A. (2002) Identification of a triterpenoid saponin from a crucifer, *Barbarea vulgaris*, as a feeding deterrent to the diamonback moth, *Plutella xylostella*. *Journal of Chemical Ecology* 28, 587–599.
Siemann, E. and Rogers, W.E. (2001) Genetic differences in growth of an invasive tree species. *Ecology Letters* 4, 514–518.
Siemens, D.H., Garner, S.H., Mitchell-Olds, T. and Callaway, R.M. (2002) Cost of defense in the context of plant competition: *Brassica rapa* may grow and defend. *Ecology* 83, 505–517.
Silvertown, J.W. and Charlesworth, D. (2001) *Introduction to Plant Population Ecology*, 4th edn. Blackwell Scientific Publications, Oxford.
Stearns, S.C. (1992) *The Evolution of Life Histories*. Oxford University Press, Oxford.
Steinger, T., Haldimann, P., Leiss, K. and Müller-Schärer, H. (2002) Does natural selection promote population divergence? A comparative analysis of population structure using amplified fragment length polymorphism markers and quantitative genetic traits. *Molecular Ecology* 11, 2583–2590.
Story, J.M., Good, W.R., White, L.J. and Smith, L. (2000) Effects of interaction of the biocontrol agent *Agapeta zoegana* L. (Lepidoptera: Cochylidae) and grass competition on spotted knapweed. *Biological Control* 17, 182–190.
Strauss, S.Y. and Agrawal, A.A. (1999) The ecology and evolution of plant tolerance to herbivory. *Trends in Ecology and Evolution* 14, 179–185.

Strauss, S.Y., Rudgers, J.A., Lau, J.A. and Irwin, R.E. (2002) Direct and ecological costs of resistance to herbivory. *Trends in Ecology and Evolution* 17, 278–285.

Sultan, S.E. and Bazzaz, F.A. (1993) Phenotypic plasticity in *Polygonum persicaria*. III. The evolution of ecological breadth for nutrient environment. *Evolution* 47, 1050–1071.

Thébaud, C. and Simberloff, D. (2001) Are plants really larger in their introduced ranges? *American Naturalist* 157, 231–236.

Thompson, J.N. (1998) Rapid evolution as an ecological process. *Trends in Ecology and Evolution* 13, 329–332.

Tiffin, P. (2000) Are tolerance, avoidance, and antibiosis evolutionarily and ecologically equivalent responses of plants to herbivores? *American Naturalist* 155, 128–138.

Tiffin, P. and Inouye, B.D. (2000) Measuring tolerance to herbivory: accuracy and precision of estimates made using natural versus imposed damage. *Evolution* 54, 1024–1029.

Tsutsui, N.D., Suarez, A.V., Holway, D.A. and Case, T.J. (2000) Reduced genetic variation and the success of an invasive species. *Proceedings of the National Academy of Sciences USA* 97, 5948–5953.

Usher, M.B. (1988) Biological invasions of nature reserves: a search for generalizations. *Biological Conservation* 44, 119–135.

van der Meijden, E. (1996) Plant defence, an evolutionary dilemma: contrasting effects of (specialist and generalist) herbivores and natural enemies. *Entomologia Experimentalis et Applicata* 80, 307–310.

van Klinken, R.D. and Edwards, O.R. (2002) Is host-specificity of weed biological control agents likely to evolve rapidly following establishment? *Ecology Letters* 5, 590–596.

Vitousek, P.M., D'Antonio, C.M., Loppe, L.L. and Westbrooks, R. (1996) Biological invasions as global environmental change. *American Scientist* 84, 468–478.

Waldmann, P. (2001) Additive and non-additive genetic architecture of two different-sized populations of *Scabiosa canescens*. *Heredity* 86, 648–657.

Walker, L.R. (1999) *Ecosystems of Disturbed Ground*. Elsevier, Amsterdam.

Weber, E. and Schmid B (1998) Latitudinal population differentiation in two species of *Solidago* (*Asteraceae*) introduced into Europe. *American Journal of Botany* 85, 1110–1121.

Weis, A.E. and Hochberg, M.E. (2000) The diverse effects of intraspecific competition on the selective advantage to resistance: a model and its predictions. *American Naturalist* 156, 276–292.

Wesselingh, R.A., de Jong, T.J., Klinkhammer, P.G.L., Vandijk, M.J. and Schlatmann, E.G.M. (1993) Geographical variation in threshold size for flowering in *Cynoglossum officinale*. *Acta Botanica Neerlandica* 42, 81–91.

Wesselingh, R.A., Klinkhammer, P.G.L., de Jong, T.J. and Boorman, L.A. (1997) Threshold size for flowering in different habitats: effect of size-dependent growth and survival. *Ecology* 78, 2118–2132.

Whitlock, M.C., Phillips, P.C. and Wade, M.J. (1993) Gene interaction affects the additive genetic variance in subdivided populations with migration and extinction. *Evolution* 47, 1758–1769.

Williamson, M. (1996) *Biological Invasions*. Chapman & Hall, London.

Willis, A.J. and Blossey, B. (1999) Benign environments do not explain the increased vigour of non-indigenous plants: a cross-continental transplant experiment. *Biocontrol Science and Technology* 9, 567–577.

Willis, A.J., Thomas, M.B. and Lawton, J.H. (1999) Is the increased vigour of invasive weeds explained by a trade-off between growth and herbivore resistance? *Oecologia* 120, 632–640.

Willis, A.J., Memmott, J. and Forrester, R.I. (2000) Is there evidence for the post-invasion evolution of increased size among invasive plant species? *Ecology Letters* 3, 275–283.

Zangerl, A.R. and Bazzaz, F.A. (1992) Theory and pattern of plant defense allocation. In: Fritz, R.S. and Simms, E.L. (eds) *Plant Resistance to Herbivores and Pathogens: Ecology, Evolution and Genetics.* University of Chicago Press, Chicago, pp. 363–391.

Zhu, Y., Chen, H., Fan, J., Wang, Y., Li, Y., Chen, J., Yang, S., Hu, L., Leung, H., Mew, T.W., Teng, P.S., Wang, Z. and Mundt, C.C. (2000) Genetic diversity and disease control in rice. *Nature* 406, 718–722.

8

Experimental Evolution in Host–Parasitoid Interactions

A.R. Kraaijeveld

NERC Centre for Population Biology and Department of Biological Sciences, Imperial College at Silwood Park, Ascot, Berkshire SL5 7PY, UK

Introduction

For many insects, parasitoids are important natural enemies, and resistance against parasitoids, be it physiological, immunological or behavioural, has evolved in a large number of insects (Godfray, 1994). Parasitoids are among the most commonly used biological control agents. Given how widespread some form of resistance against parasitoids is in natural populations, it is surprising that the evolution of resistance in pests appears to be very rare when parasitoids, rather than pesticides, are used for control (Henter and Via, 1995; Holt and Hochberg, 1997).

A number of non-mutually exclusive reasons have been proposed to explain this rarity of the evolution of resistance against parasitoids in biological control projects (Holt and Hochberg, 1997; Lapchin, 2002). Among these are: (i) lack of genetic variation for resistance in the pest population; (ii) resistance against parasitoids is costly for the pest; (iii) resistance depends on the interaction between pest and parasitoid genotype; (iv) the parasitoid, unlike a pesticide, can coevolve with the pest; and (v) selection pressures exerted by parasitoids on the pest fluctuate in time and space.

Experimental evolution has been used very successfully to gain a better understanding of the ecological and evolutionary dynamics of enemy–victim interactions in several microbial model systems. Lenski and co-workers, in a series of papers, used the bacterium *Escherichia coli* and a bacteriophage (in a way, the microbial equivalent of a parasitoid) to address questions related to the costs of resistance, the evolution of virulence and coevolutionary dynamics (Chao *et al.*, 1977; Levin and Lenski, 1983; Lenski, 1988; Bohannan and Lenski, 2000). Among other things, their work showed that both resistance in the host (the bacterium) and counter-resistance in the parasite (the phage) have a cost in terms of

 Genetics, Evolution and Biological Control (eds L.E. Ehler, R. Sforza and T. Mateille)

competitive ability. Buckling and Rainey (2002) used *Pseudomonas fluorescens* and a bacteriophage and showed that both hosts and parasites evolved resistance and counter-resistance. In addition, they found evidence for genotype × genotype interactions, with bacterial populations becoming more resistant to their own than to other phage populations.

It is important to realize that there is a subtle, but crucial, difference between experimental evolution and selection experiments. In the latter, there is a very specific trait or combination of traits that is selected for. Therefore, given genetic variation and a strong-enough selection pressure, evolution will usually proceed as expected. Experimental evolution, on the other hand, brings an organism into a certain situation and then forces it to find an evolutionary solution. This evolutionary solution may or may not involve expected traits. Therefore, the outcome of experimental evolution is less predictable, and less repeatable, than that of a typical selection experiment (Velicer and Lenski, 1999; Matos *et al.*, 2002).

In this chapter, I want to explore whether experimental evolution using host–parasitoid interactions can shed light on why resistance to parasitoids rarely, if ever, evolves under biological control. First I will bring together some observations from biological control projects which are suggestive of evolution taking place, either in the pest or in the parasitoid. Subsequently, I will focus in more detail on two host–parasitoid model systems which have been subject to experimental evolution in the laboratory.

Field 'Experiments'

Probably the most often cited example of the possible evolution of resistance against an introduced natural enemy is that of the larch sawfly (*Pristiphora erichsonii*). The ichneumonid parasitoid *Mesoleius tenthredinis* was released in British Columbia, where the larch sawfly had become a problem, in the 1930s. About a decade later, the parasitoid appeared to be largely unable to complete development in several areas of Canada, and it was thought that the sawfly host, under selection pressure from the parasitoid, had evolved resistance (Muldrew, 1953). However, more recent work suggests that the decrease in effectiveness of the parasitoid was not due to the host evolving resistance, but may have been caused by a European strain of the sawfly (which has a much higher level of resistance against the parasitoid) being released together with the parasitoid and outcompeting the Canadian strain (Wong, 1974; Godfray, 1994).

A stronger, though not conclusive, case for evolution after release of a biocontrol agent is that of the ichneumonid parasitoid *Bathyplectes curculionis*, which was introduced in California to control the lucerne weevil *Hypera postica*. Fifteen years after release, the level of encapsulation of the parasitoid had dropped from 40% to 5%, suggesting evolution of increased levels of counter-defence in the parasitoid (van den Bosch, 1964; Salt and van den Bosch, 1967). Interestingly, the ability to prevent encapsulation in a related weevil species, *Hypera brunneipennis*, had decreased, which suggests a possible trade-off between counter-defences against the two host species. As the study was not replicated and no proper controls were available, it is impossible to be sure that evolution of counter-defence took place in *B. curculionis*, but the scenario is certainly a plausible one.

Besides traits connected to defence and counter-defence, evolutionary change can also occur at a behavioural level. The braconid parasitoid *Cotesia glomerata* was introduced into the USA at the end of the 19th century to control caterpillars of *Pieris rapae*. Comparison of European and American parasitoids shows that they differ in aspects of their searching behaviour, suggesting evolutionary change after introduction (Le Masurier and Waage, 1993; Vos, 2001). European parasitoids prefer to attack the gregarious *Pieris brassicae* and, with the absence of this species in America, adaptation to the more solitary *P. rapae* can be expected in the USA. However, as in the previous example, lack of replication and proper controls make it hard to prove that this is really what happened. Van Nouhuys and Via (1997) found evidence for genetic differentiation in behavioural traits between *C. glomerata* populations from wild and cultivated habitats, suggesting that the difference in host distribution between the two habitats imposes different selection pressures on the parasitoid populations.

Besides possible evolutionary changes in field populations of pests or parasitoids after introduction, anecdotal evidence exists to suggest that evolutionary changes can also occur in mass-rearings of parasitoids (Henter *et al.*, 1996; McGregor *et al.*, 1998; Rojas *et al.*, 1999). As before, mass-rearings typically lack replication and proper controls, and therefore observed changes, whether in physiological or behavioural traits, can not definitely be said to be due to adaptation to a different host.

Experimental Evolution in Host–Parasitoid Systems

In this section I want to consider in more detail two host–parasitoid systems which have been subject, under laboratory circumstances, to experimental evolution: house flies (*Musca domestica*) and their pupal parasitoid *Nasonia vitripennis*, and fruit flies (*Drosophila melanogaster*) and their larval parasitoids.

House flies

N. vitripennis is a chalcidoid parasitoid which attacks pupae from a range of cyclorraphous Diptera. Pimentel and co-workers (Pimentel *et al.*, 1963, 1978; Pimentel, 1968; Zareh *et al.*, 1980) used *N. vitripennis* and house fly hosts to set up population cages in which the hosts were subject to parasitism, together with control cages in which parasitoids were absent. The principal aim of the experiments was to test the idea that coevolutionary interactions between host and parasitoid lead to more stable population dynamics, but here I will focus on the evolution of resistance and counter-resistance in this system.

Host evolution and cost of resistance

Several traits of the fly populations exposed to parasitism changed compared with the control populations. Host pupae became heavier and the length of the pupal period was reduced. Both changes can be understood in terms of the evolution of resistance against parasitoids. The thickness of the puparial wall was not measured, but it is likely that the heavier pupae had thicker puparial walls. Parasitoids are less likely to be able to drill through a thicker puparial wall, or take more time and are therefore more likely to give up (Morris and Fellowes, 2002). Shortening of the pupal period reduces the time window in which the host is susceptible to parasitism. Adult fly size was reduced in the populations exposed to parasitism. This may be the result of the reduction in pupal period and, therefore, can be seen as a cost of resistance.

Parasitoid evolution

As the experiments progressed, parasitoids appeared to evolve a reduced fecundity. This may be due to evolutionary change (and in that case a potential cost of counter-resistance), but it is also possible that parasitoids developing on smaller, faster-developing hosts are smaller themselves and therefore less fecund. Such a phenotypic effect has also been found in a system of gall insects and their parasitoids (Weis *et al.*, 1989), which shows an apparent coevolutionary arms race between host and parasitoid without actual evolutionary change in the parasitoid. Unfortunately, Pimentel and co-workers did not perform the crucial experiment to show that the parasitoids had indeed changed genetically: raising the parasitoids on control hosts. Also, the trait(s) that might have enabled the parasitoids to overcome the defence mechanisms of the host (thicker puparia and shorter vulnerable period) has not been identified.

Drosophila

In terms of experimental evolution, the most extensively studied host–parasitoid system is that of *D. melanogaster* and its larval parasitoids. Larvae of *D. melanogaster* are parasitized by a range of braconid and eucoilid endoparasitoids. In Europe, the most common species are the braconid *Asobara tabida* and the eucoilids *Leptopilina heterotoma* and *Leptopilina boulardi* (Carton *et al.*, 1986). Larvae can defend themselves against parasitism by an immunological response, called encapsulation (Nappi, 1975, 1981; Rizki and Rizki, 1984; Strand and Pech, 1995). After a parasitoid female oviposits an egg into the host's haemocoel, lamellocytes (a special type of blood cell) aggregate around the egg until it is covered by several layers of these blood cells. Subsequently, crystal cells, another type of blood cell, release enzymes into the haemocoel, which triggers a cascade of biochemical reactions (the phenoloxidase cascade). The end result of this cascade is the production of melanin, which forms a capsule around the egg. When successful, encapsulation leads to the death of the parasitoid egg and the survival of the adult fly.

Variation in host resistance

The ability of *D. melanogaster* larvae to encapsulate parasitoid eggs shows considerable variation in the field, both between and within populations. On a Europe-wide scale, populations with the highest encapsulation ability against *A. tabida* are found in the central Mediterranean (Kraaijeveld and van Alphen, 1995a; Fig. 8.1a), whereas low encapsulation ability is found in more northern parts of Europe and the extreme south-west and south-east. This variation across populations can be explained, at least partly, by the presence or absence of another host species for the parasitoid, *Drosophila subobscura*. Larvae of this species do not have the ability to encapsulate parasitoid eggs (Kraaijeveld and van der Wel, 1994). *D. subobscura* is common in northern parts of Europe but much rarer in the Mediterranean. Therefore, its absence as a common alternative host for the parasitoid in the Mediterranean leads to a stronger selection pressure on *D. melanogaster* for a well-developed immune system in these areas compared with more northern areas (Kraaijeveld and van Alphen, 1995a). In addition to variation between populations, there is also a high degree of within-population variation. Using isofemale lines, Carton and Boulétreau (1985) showed that the ability within one population of *D. melanogaster* to encapsulate eggs of *L. boulardi* varied from less than 10% to over 90% (Fig. 8.1b). Further indication for the existence of within-population genetic variation in encapsulation ability comes from early selection experiments using *D. melanogaster* and *L. heterotoma* (Schlegel-Oprecht, 1953; Hadorn and Walker, 1960; Walker, 1961), which resulted in an increase in the concentration of melanized haemocytes.

Cost of resistance

One explanation for the existence of within-population variation in resistance in *D. melanogaster* is that resistance is costly. A trade-off between resistance and some other fitness parameter(s), combined with variation in probability of parasitism, can maintain genetic variation in resistance within a population. The actual process of encapsulation uses up resources: larvae which have successfully encapsulated parasitoid eggs turn out to be smaller as adults, with lowered fecundity and reduced resistance to desiccation and starvation (Carton and David, 1983; Fellowes *et al.*, 1999a; Hoang, 2001).

However, costs of actual defence are costs that always have to be paid, as the alternative for the host is death. From an evolutionary perspective, the costs of *having the ability to defend* are more important than the costs of *actual defence*. Starting from a large field collection, Kraaijeveld and Godfray (1997) and Fellowes *et al.* (1998a) exposed replicated lines of *D. melanogaster* larvae to parasitism and only bred the next generation from flies that had been attacked and had survived. Their results show, first of all, that high resistance can evolve rapidly (in five generations from 5% to 55% against *A. tabida* and from 0.5% to 45% against *L. boulardi*). A trade-off was identified by comparison of the lines which had evolved high resistance with the control lines (with low resistance), in a range of life-history parameters. The only fitness parameter in which the evolved, high-resistance lines were inferior to the low-resistance control lines was larval competitive

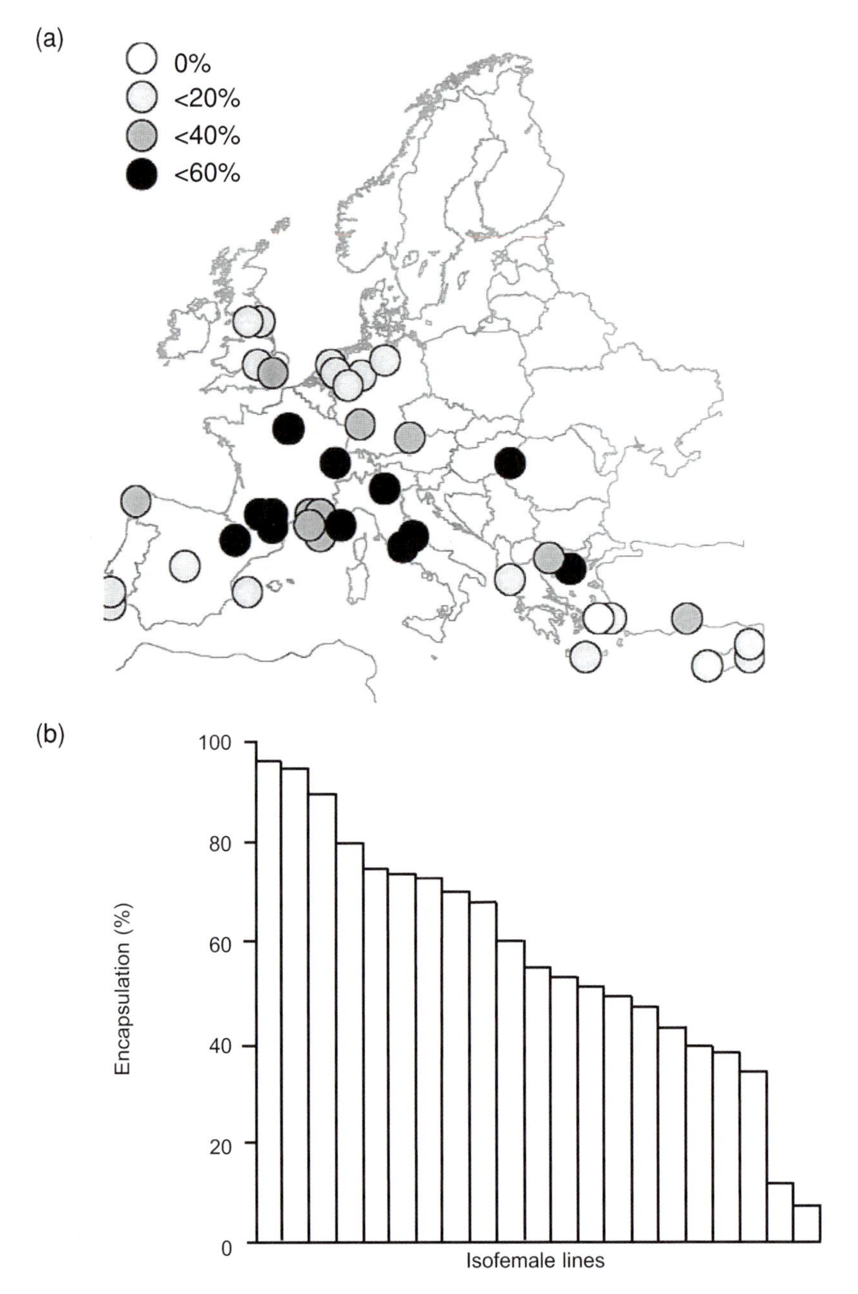

Fig. 8.1. (a) Variation in encapsulation rate of *Asobara tabida* eggs among field populations of *Drosophila melanogaster* collected across Europe (data taken from Kraaijeveld and van Alphen, 1995a). (b) Variation in encapsulation rate of *Leptopilina boulardi* eggs among isofemale lines from a single field population of *D. melanogaster* (data taken from Carton and Boulétreau, 1985).

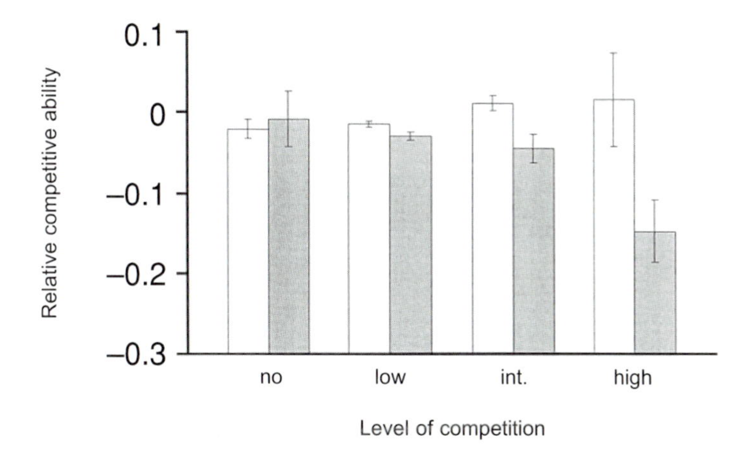

Fig. 8.2. Relative competitive ability (as measured against a reference strain) of lines of *Drosophila melanogaster* which had evolved resistance against *Asobara tabida* (grey bars) and control lines (white bars) at different levels of competition; mean of four replicate lines ± SE; competitive ability is calculated as $\ln(e/(r + 1))$, where e is the number of surviving experimental (evolved or control) flies and r the number of surviving reference flies (for experimental details, see Kraaijeveld and Godfray, 1997).

ability, but, interestingly, this trade-off only becomes apparent at low resource levels (Fig. 8.2). The reduction in competitive ability of the evolved lines is caused by a decrease in larval feeding rate (Fellowes *et al.*, 1998b), which explains why high encapsulation ability is essentially a neutral trait at high resource levels.

Cross-resistance and genotype specificity

The parasitoid species attacking *D. melanogaster* use different mechanisms to prevent encapsulation of their eggs. The eggs of *A. tabida* have a chorion with proteinaceous filaments, which cause the eggs to become embedded in host tissue, away from circulating haemocytes (Kraaijeveld and van Alphen, 1994; Eslin *et al.*, 1996). In contrast, the *Leptopilina* species inject virus-like particles into the host, which actively suppress the host's immune system by entering the haemocytes and causing them to apoptose (Rizki and Rizki, 1990; Dupas *et al.*, 1996). Given that different parasitoids use different counter-defence mechanisms, a trade-off could exist by which increased resistance against one parasitoid leads to decreased resistance against another. However, exposing the lines mentioned above (that had evolved resistance to one parasitoid species) to other parasitoid species showed that evolution of resistance against one parasitoid species leads either to no change or to an increase in resistance against another parasitoid species (Fellowes *et al.*, 1999b; Fig. 8.3a). In other words, there is no evidence for the existence of a trade-off of this kind.

Similarly, there is no evidence that resistance of *D. melanogaster* is determined by an interaction between parasitoid and host genotype: the lines that had evolved resistance against the parasitoid strain they were exposed to were also more resistant than the controls when they were parasitized by other parasitoid strains (Kraaijeveld and Godfray, 1999; Fig. 8.3b).

Variation in parasitoid counter-resistance and cost of counter-resistance

Like *D. melanogaster*, *A. tabida* also shows considerable variation between populations in the level of counter-resistance. Populations from southern Europe are much better able to prevent encapsulation of their eggs than populations from central and northern Europe. Correlated with this variation in counter-resistance is the degree to which eggs become embedded in host tissue after parasitism (Kraaijeveld and van Alphen, 1994). Kraaijeveld *et al.* (2001) set up replicated lines in which parasitoids from a large field collection were reared on a highly resistant strain of *D. melanogaster*; control lines were reared on *D. subobscura* (which has no immune reaction against parasitoids). After 17 generations, the level of counter-resistance had increased to 37% (compared with 11% in the control lines) and the evolved parasitoid lines showed an increase in the degree of egg embedding. Comparison of evolved and control lines in a range of life-history parameters showed no differences, except in one trait: the eggs from the evolved lines hatched about 2.5 hours later than those of the control lines. This small delay in hatching may be crucially important when host larvae are superparasitized (which is known to occur frequently in the field). As only one parasitoid larva can complete development in a host, two larvae present in the same host actively fight for possession of the host. In several *Drosophila* parasitoids, the oldest parasitoid larva has a higher probability of winning the competition for the host (van Strien-van Liempt, 1983; Visser *et al.*, 1992). So a likely cost of increased counter-resistance in *A. tabida* is a reduction in competitive ability.

Dupas and Boscaro (1999) reared a *L. boulardi* strain, which was genetically variable in counter-resistance to *Drosophila yakuba* (a close relative of *D. melanogaster*), on a susceptible strain of *D. melanogaster* in a replicated population cage set-up. After 10–15 parasitoid generations, counter-resistance against *D. yakuba* had significantly decreased, suggesting the existence of a cost. Unfortunately, because of the lack of a proper control, this experiment does not provide firm evidence of a cost of counter-resistance in *L. boulardi.*

Experimental coevolution

The experiments described above, in which either hosts evolved resistance or parasitoids evolved counter-resistance, were designed such that only one of the two could actually evolve. The logical next step is to bring host and parasitoid populations together in an experimental set-up that allows for more natural population dynamics and for coevolution. Boulétreau (1986) set up two experimental populations of *D. melanogaster* and reared them with *L. boulardi.* Comparison of the level of resistance of the host and counter-resistance of the parasitoid after about 50 host (= 25 parasitoid) generations suggested that both had increased.

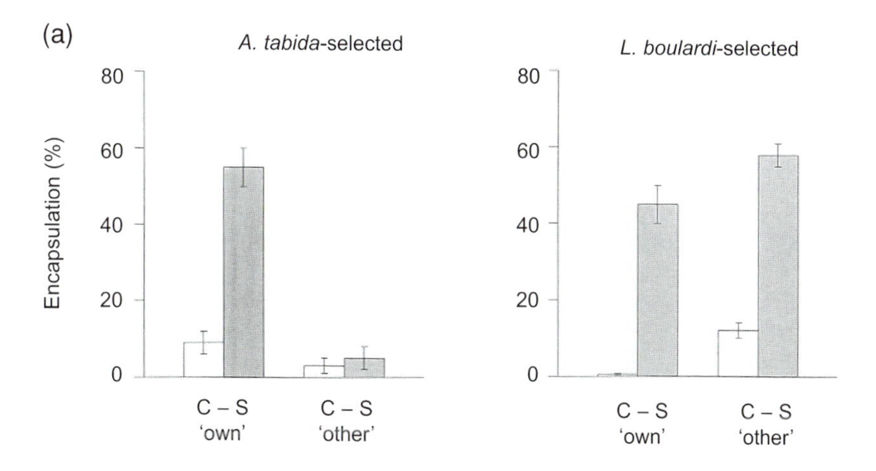

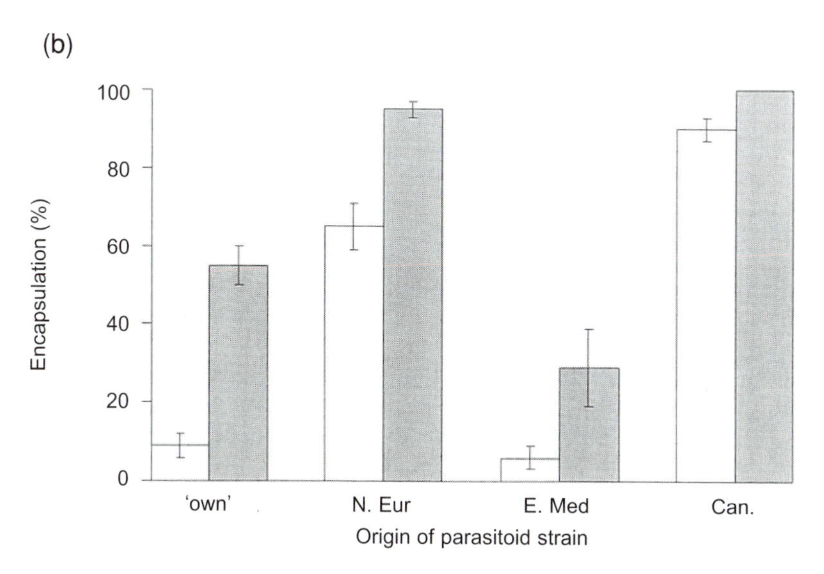

Fig. 8.3. (a) Cross-resistance by lines of *Drosophila melanogaster* which had evolved resistance (S, grey bars) against either *Asobara tabida* (left panel) or *Leptopilina boulardi* (right panel) and control lines (C, white bars). Within each panel, the left pair of bars ('own') shows the encapsulation rate of the parasitoid species used as a selection pressure and the right pair of bars ('other') shows encapsulation rate of the other parasitoid species. Mean of four replicate lines ± SE. (For experimental details, see Fellowes *et al.*, 1999b.) (b) Encapsulation rate of lines of *D. melanogaster* which had evolved resistance against *A. tabida* (grey bars), and control lines (white bars), when parasitized by the parasitoid strain they had evolved resistance against ('own') or by parasitoid strains from northern Europe (N. Eur.), the eastern Mediterranean (E. Med.) or Canada (Can.). Mean of four replicate lines ± SE. (Data taken from Kraaijeveld and Godfray, 1999.)

However, firm conclusions cannot be drawn from this experiment because there was no proper replicated control. Hughes and Sokolowski (1996) maintained three replicate *D. melanogaster* populations with *A. tabida* and found that, after 19 generations, larvae from the experimental populations had a resistance level of 39% compared with 23% for the control populations. Using a similar set-up, and the same host base population as used by Kraaijeveld and Godfray (1997), Green *et al.* (2000) found a level of resistance of 30% in the populations exposed to parasitoids, compared with 18% in the control populations; this increase was found after about 10 host generations. Green *et al.* (2000) also measured the level of counter-resistance in the parasitoid populations, but could find no evidence that counter-resistance had evolved in the approximately five parasitoid generations for which the experiment lasted.

Evolution of behavioural traits

In a parasitoid–host interaction, besides traits related to defence and counter-defence, other traits are potentially subject to evolutionary change. *D. melanogaster* shows a polymorphism in larval foraging behaviour (Sokolowski, 1980; Sokolowski *et al.*, 1986): 'rovers' move around while foraging, while 'sitters' stay in one place. Because *A. tabida* finds its host by reacting to the vibrations they cause, 'rovers' are more likely to be attacked by this parasitoid than 'sitters' (Kraaijeveld and van Alphen, 1995b). Therefore, it can be expected that in a population of *D. melanogaster* which is under attack by *A. tabida* the proportion of 'sitters' increases over time. However, in neither of the two population cage experiments mentioned above (Hughes and Sokolowski, 1996; Green *et al.*, 2000) was there a change in the relative proportions of 'rovers' and 'sitters' in the populations exposed to parasitoid attack, compared with the control populations.

An important behavioural trait of the parasitoid is host choice. Mollema (1991) showed that variation in host selection behaviour in *A. tabida* has a heritable component. Kraaijeveld *et al.* (1995) found that females from European *A. tabida* populations vary in their acceptance of *D. melanogaster* larvae for oviposition when offered together with *D. subobscura* larvae. Parasitoid females from northern and central European populations reject *D. melanogaster* to a much higher degree than parasitoids from southern European populations. This difference in host selection behaviour among populations is correlated with the ability of their eggs to prevent encapsulation (see above). Rolff and Kraaijeveld (2001) compared the host selection behaviour of the lines that had evolved higher levels of counter-defence (see above) with their controls. They found that females from the evolved lines accepted *D. melanogaster* larvae more readily than females from the control lines, showing that host choice behaviour had also evolved in these lines. The most likely explanation for this apparent link between a physiological and a behavioural trait, both in the field populations and in the lines that were reared on *D. melanogaster*, is parallel simultaneous evolution of counter-defence and host choice, rather than a direct genetic link between the two traits (Rolff and Kraaijeveld, 2001).

Discussion

The evolution of resistance in pest insects against biological control agents appears to be very rare, whereas evolution of resistance against pesticides is commonly observed. A number of non-mutually exclusive explanations have been proposed to account for this difference (Holt and Hochberg, 1997; Lapchin, 2002). The evolution of host resistance against parasitoids has been studied in a small number of model systems (house flies and *Drosophila*), using an experimental evolution approach. With the understanding obtained from these experiments, we can try and shed more light on why evolution of resistance against parasitoids is so much rarer than against pesticides.

Lack of genetic variation for resistance

Lack of genetic variation will obviously prevent any evolution of resistance in pest populations. Both the house fly and the *Drosophila* populations used in the various experiments responded to selection for increased survival when under attack from parasitoids. In *D. melanogaster* larvae, this led to an increase in the ability to encapsulate parasitoid eggs, the major immunological defence reaction that larvae have against parasitoids. In the case of the house fly pupae, keeping in mind that insect pupae do not have a cellular immune response, selection exerted by parasitoids led to an increase in the thickness of the puparial wall, the main barrier parasitoids have to overcome to oviposit in a pupa. The clear, and in the case of *D. melanogaster* often fast, response to selection shows that genetic variation for resistance is present in these populations and that, given a strong enough selection pressure, resistance can evolve quickly.

Of course, the presence of genetic variation for resistance in these two model host populations is no proof for the existence of comparable levels of genetic variation for resistance in populations of pest insects. It has become increasingly clear that at least a certain amount of genetic variation for resistance is common in natural populations of insects in the field. On the other hand, insect pest populations may be established, more often than not, by a limited number of individuals, leading to a potential reduction in the amount of genetic variation (for resistance and other traits). However, evolution of resistance against pesticides is far from rare, showing that genetic variation for this type of resistance is often present. In addition, Rosenheim *et al.* (1996) found no difference between introduced and native insect species in evolution of pesticide resistance. This suggests that bottlenecks in the establishment of pest populations are not sufficiently severe or common to reduce genetic variation sufficiently for evolution to become impossible. Although this could surely happen in individual cases, it seems unlikely that lack of genetic variation for resistance to parasitoids is a general explanation for the lack of evolution of this type of resistance under a biological control regime. Whether, and how fast, resistance can evolve will depend, of course, on the amount of genetic variation, the strength of the selection pressure exerted and the genetic architecture of the resistance traits.

Resistance is costly

If resistance is costly, these costs may prevent or slow down the evolution of increased resistance. Whether increased resistance evolves will then depend on the relative strengths of the opposing selection pressures of parasitoids and costs. Both the evolutionary experiments with house flies and those with *Drosophila* show that resistance against parasitoids is costly. The exact nature of the cost differs between the two systems, which is not surprising as the two host species defend themselves against parasitoid attack in very different ways. Despite these very different defence mechanisms, experimental evolution reveals a trade-off between resistance against parasitoids and another important fitness parameter. Research over the past decade on a variety of insect–parasite systems has shown that costs of resistance against parasites and parasitoids are widespread in insects (Kraaijeveld *et al.*, 2002) and are therefore likely to be a key part of the explanation as to why evolution of resistance is rare in pest insects.

Genotype × genotype interactions

If the level of resistance of an individual insect host depends on the interaction between its own genotype and that of the parasitoid attacking it, the resulting coevolutionary arms race will be cyclical, with different host and parasitoid genotypes becoming more and less common over time, rather than showing a general increase in host resistance and parasitoid counter-resistance. *D. melanogaster* populations that evolved resistance against one particular strain of parasitoid also showed a higher level of resistance than the controls when attacked by other parasitoid strains from across the parasitoid's geographic range. In addition, Green *et al.* (2000) found no evidence for evolution of genotype-specific resistance to parasitoid lines set up from one genetically variable parasitoid population, and a further attempt by Kraaijeveld and Godfray (2001) failed to find evidence for genotypic matching between *D. melanogaster* resistance and *A. tabida* counter-resistance at a Europe-wide scale. In fact, no convincing example of a genotype-matching aspect of resistance in host–parasitoid interactions has been reported so far (Kaltz and Shykoff, 1998). Whereas it cannot be ruled out that this might be found in a parasitoid–host interaction in the future, genotypic matching of resistance and counter-resistance seems very unlikely to be a general explanation for the lack of evolution of resistance to parasitoids in pest insects under biological control.

Parasitoids can coevolve

Unlike pesticides, parasitoid populations can potentially coevolve. Therefore an evolutionary increase in host resistance may be matched by a similar evolutionary increase in parasitoid counter-resistance, leading to the lack of a net increase in observed resistance levels in the field. In the house fly experiments, whether *N. vitripennis* evolved a counter-resistance mechanism (either physiological or behavioural) in response to the thickening of the puparial wall and the reduction in pupal period of its host is ambiguous. However, when *A. tabida* is reared on a highly resistant strain of *D. melanogaster*, it evolves higher levels of counter-resistance. This shows, first of all, that genetic variation in counter-resistance is present in this parasitoid and, secondly, that parasitoid counter-resistance can evolve in a relatively short amount of time. However, compared with the rate of evolution of resistance in the host, counter-resistance evolved only slowly. Evolution of counter-resistance was detected after 17 generations in one experiment, whereas in the population cage experiment by Green *et al.* (2000), using the same parasitoid base population, no change was detected after five generations. In addition to counter-resistance, behavioural traits related to host choice can also evolve in a comparable amount of time.

The *Drosophila–Asobara* system is the only host–parasitoid system in which rates of evolution of both resistance in the host and counter-resistance in the parasitoid have been determined, so until this has been studied in more host–parasitoid interactions the jury is still out on whether a parasitoid's ability to coevolve (as opposed to a biologically inert pesticide) is an important explanation for the lack of evolution of resistance in pest insects. It is important to keep in mind, though, that the selection pressures in the interaction between a host and a parasitoid population are inherently asymmetrical, in that every parasitoid needs to attack a host in order to produce offspring, but not every host has to defend itself against a parasitoid, as it may very well not be attacked. Therefore, by and large, the selection pressure acting on parasitoid counter-resistance is likely to be stronger than that on host resistance.

Selection pressures fluctuate in space and time

In both natural and pest insect populations, the selection pressure exerted by parasitoids will vary in time and space, which is likely to have an influence on the rate of evolution of resistance. As Kraaijeveld and Godfray (1997) and Green *et al.* (2000) used the same base population of flies, a comparison between the results of their experiments can give some insight into the effect of variation in selection pressure on the evolution of resistance. In the first set of experiments, only flies which had been parasitized and had encapsulated the parasitoid egg were used for the next generation, effectively ensuring a constant 100% attack rate. In the second set of experiments, attack rates were a function of the relative densities of hosts and parasitoids in the cages, and will therefore have fluctuated considerably over time. In addition, as a fixed amount of food was provided in these cages, independent of fly numbers, fluctuations in host densities will have led to fluctuations in the level of competition for food. Given that costs of resistance are density-dependent in *D. melanogaster* (see above), the expectation is that the rate of evolution of resistance is lower in the semi-natural population cage experiments than in the more artificial experiments by Kraaijeveld and Godfray (1997). Indeed, Green *et al.* (2000) found almost a doubling in resistance in about ten generations, whereas Kraaijeveld and Godfray (1997) saw resistance increase about tenfold in only five generations. In other words, and not unexpectedly, variation in the selection pressures acting on resistance will have an effect on its rate of evolution.

Concluding Thoughts

Why is the evolution of resistance to parasitoids in pests under biological control regimes so rare? Obviously, there has to be both genetic variation and a selection pressure in order for any evolution to take place, but these two are not necessarily sufficient. An interesting example for this comes from the study of a field population of the pea aphid (*Acyrtosiphon pisum*) and the parasitoid *Aphidius ervi.* Henter (1995) and Henter and Via (1995) showed that, despite within-population genetic variation for resistance in the host and substantial attack rates by the parasitoid, there was no detectable change in resistance in the population during a season. An obvious explanation for this lack of evolution would be the existence of considerable costs of resistance, but on-going work by Ferrari *et al.* (2001) has so far failed to find such a cost.

At least a pivotal part of the explanation is the existence of costs of resistance in the host/pest combined with variation in selection pressure exerted by the parasitoid. Owing to population dynamical processes, parasitoid numbers are more likely to fluctuate than a pesticide concentration. There is growing evidence that resistance mechanisms that rely on a quantitative change in the host, such as an increase in haemocyte numbers, are more likely to bear substantial costs than those that involve a qualitative change, such as a modification of a recognition protein (Cousteau *et al.*, 2000; Rigby *et al.*, 2002). As resistance to parasitoids will usually involve the deployment of an immune system, this suggests that, on average, costs of resistance to parasitoids are likely to be higher than against pesticides. However, further empirical work will be needed to confirm this. Interestingly, Bergelson and Purrington (1996), in a survey of costs of resistance in plants, found that costs of resistance against herbicides have been found more often than costs of resistance against herbivores.

Another potentially important aspect of costs of resistance is the relative magnitude of costs of resistance in the host and costs of counter-resistance in the parasitoid. Sasaki and Godfray (1999) constructed a model to explore this issue. The model included a genetically variable host and parasitoid population, non-specified costs of resistance and counter-resistance, and allowed the host and parasitoid population to show population dynamics. One of their main results was that when costs of resistance and counter-resistance are roughly comparable, evolutionary cycles appeared in which hosts alternately experienced selection for and against high resistance. However, when costs of resistance in the host were relatively high compared with costs of counter-resistance in the parasitoid, the majority of hosts did not invest in resistance but essentially traded off the cost of resistance against the risk of being attacked. Such a scenario may explain, for instance, why *D. subobscura* does not appear to have an immune response to parasitoid attack, despite suffering from high levels of parasitism at times.

Although this model result of hosts not evolving resistance, despite being under considerable selection pressure to do so, is reminiscent of the situation in many biocontrol projects using parasitoids, it remains to be seen whether this particular aspect of the cost of resistance also plays a role in explaining the rarity of the evolution of resistance in pest insects. However, experimental evolution strongly suggests that costs of resistance play a pivotal role in preventing pest insects from evolving resistance against their biocontrol agents.

References

Bergelson, J. and Purrington, C.B. (1996) Surveying patterns in the cost of resistance in plants. *American Naturalist* 148, 536–558.

Bohannan, B.J.M. and Lenski, R.E. (2000) Linking genetic change to community evolution: insights from studies of bacteria and bacteriophage. *Ecology Letters* 3, 362–377.

Boulétreau, M. (1986) The genetic and evolutionary interactions between parasitoids and their hosts. In: Waage, J.K. and Greathead, D. (eds) *Insect Parasitoids*. Academic Press, London, pp. 169–200.

Buckling, A. and Rainey, P.B. (2002) Antagonistic coevolution between a bacterium and a bacteriophage. *Proceedings of the Royal Society of London Series B* 269, 931–936.

Carton, Y. and Boulétreau, M. (1985) Encapsulation ability of *Drosophila melanogaster*: a genetic analysis. *Developmental and Comparative Immunology* 9, 211–219.

Carton, Y. and David, J.R. (1983) Reduction of fitness in *Drosophila* adults surviving parasitization by a cynipid wasp. *Experientia* 39, 231–233.

Carton, Y., Boulétreau, M., van Alphen, J.J.M. and van Lenteren, J.C. (1986) The *Drosophila* parasitic wasps. In: Ashburner, M. (ed.) *The Genetics and Biology of* Drosophila. Vol. 3e. Academic Press, London, pp. 347–394.

Chao, L., Levin, B.R. and Stewart, F.M. (1977) A complex community in a simple habitat: an experimental study with bacteria and phage. *Ecology* 58, 369–378.

Cousteau, C., Chevillon, C. and ffrench-Constant, R. (2000) Resistance to xenobiotics and parasities: can we count the cost? *Trends in Ecology and Evolution* 15, 378–383.

Dupas, S.M. and Boscaro, M. (1999) Geographic variation and evolution of immunosuppressive genes in a *Drosophila* parasitoid. *Ecography* 22, 284–291.

Dupas, S., Brehelin, M., Frey, F. and Carton, Y. (1996) Immune suppressive virus-like particles in a *Drosophila* parasitoid: significance of their intraspecific morphological variations. *Parasitology* 113, 207–212.

Eslin, P., Giordanengo, P., Fourdrain, Y. and Prevost, G. (1996) Avoidance of encapsulation in the absence of VLP by a braconid parasitoid of *Drosophila* larvae: an ultrastructural study. *Canadian Journal of Zoology* 74, 2193–2198.

Fellowes, M.D.E., Kraaijeveld, A.R. and Godfray, H.C.J. (1998a) Trade-off associated with selection for increased ability to resist parasitoid attack in *Drosophila melanogaster*. *Proceedings of the Royal Society of London Series B* 265, 1553–1558.

Fellowes, M.D.E., Kraaijeveld, A.R. and Godfray, H.C.J. (1998b) Association between feeding rate and defence against parasitoids in *Drosophila melanogaster. Evolution* 53, 1302–1305.

Fellowes, M.D.E., Kraaijeveld, A.R. and Godfray, H.C.J. (1999a) The relative fitness of *Drosophila melanogaster* (Diptera, Drosophilidae) that have successfully defended themselves against the parasitoid *Asobara tabida* (Hymenoptera, Braconidae). *Journal of Evolutionary Biology* 12, 123–128.

Fellowes, M.D.E., Kraaijeveld, A.R. and Godfray, H.C.J. (1999b) Cross-resistance following artificial selection for increased defence against parasitoids in *Drosophila melanogaster. Evolution* 53, 966–972.

Ferrari, J., Müller, C.B., Kraaijeveld, A.R. and Godfray, H.C.J. (2001) Clonal variation and covariation in aphid resistance to parasitoids and a pathogen. *Evolution* 55, 1805–1814.

Godfray, H.C.J. (1994) *Parasitoids – Behavioral and Evolutionary Ecology*. Princeton University Press, Princeton, New Jersey.

Green, D.M., Kraaijeveld, A.R. and Godfray, H.C.J. (2000) Evolutionary interactions between *Drosophila melanogaster* and its parasitoid *Asobara tabida. Heredity* 85, 450–458.

Hadorn, E. and Walker, I. (1960) Drosophila und Pseudeucoila. I. Selektionsversuch zur Steigerung der Abwehrreaktion des Wirtes gegen den Parasiten. *Revue Suisse de Zoologie* 67, 216–225.

Henter, H.J. (1995) The potential for coevolution in a host–parasitoid system. II. Genetic variation within a population of wasps in the ability to parasitize an aphid host. *Evolution* 49, 439–445.

Henter, H.J. and Via, S. (1995) The potential for coevolution in a host–parasitoid system. I. Genetic variation within an aphid population in susceptibility to a parasitic wasp. *Evolution* 49, 427–438.

Henter, H.J., Brasch, K. and van Lenteren, J.C. (1996) Variation between laboratory populations of *Encarsia formosa* in their parasitization behavior on the host *Bemisia tabaci. Entomologia Experimentalis et Applicata* 80, 435–441.

Hoang, A. (2001) Immune responses to parasitism reduces resistance of *Drosophila melanogaster* to desiccation and starvation. *Evolution* 55, 2353–2358.

Holt, R.D. and Hochberg, M.E. (1997) When is biological control evolutionarily stable (or is it)? *Ecology* 78, 1673–1683.

Hughes, K. and Sokolowski, M.B. (1996) Natural selection in the laboratory for a change in resistance by *Drosophila melanogaster* to the parasitoid wasp *Asobara tabida. Journal of Insect Behavior* 9, 477–491.

Kaltz, O. and Shykoff, J.A. (1998) Local adaptation in host–parasite systems. *Heredity* 81, 361–370.

Kraaijeveld, A.R. and Godfray, H.C.J. (1997) Trade-off between parasitoid resistance and larval competitive ability in *Drosophila melanogaster. Nature* 389, 278–280.

Kraaijeveld, A.R. and Godfray, H.C.J. (1999) Geographic patterns in the evolution of resistance and virulence in *Drosophila* and its parasitoids. *American Naturalist* 153, S61–S74.

Kraaijeveld, A.R. and Godfray, H.C.J. (2001) Is there local adaptation in *Drosophila*–parasitoid interactions? *Evolutionary Ecology Research* 3, 107–116.

Kraaijeveld, A.R. and van Alphen, J.J.M. (1994) Geographical variation in resistance of the parasitoid *Asobara tabida* against encapsulation by *Drosophila melanogaster* larvae: the mechanism explored. *Physiological Entomology* 19, 9–14.

Kraaijeveld, A.R. and van Alphen, J.J.M. (1995a) Geographical variation in encapsulation ability of *Drosophila melanogaster* larvae and evidence for parasitoid-specific components. *Evolutionary Ecology* 9, 10–17.

Kraaijeveld, A.R. and van Alphen, J.J.M. (1995b) Foraging behaviour and encapsulation ability of *Drosophila melanogaster* larvae: correlated polymorphisms? *Journal of Insect Behavior* 8, 305–319.

Kraaijeveld, A.R. and van der Wel, N.N. (1994) Geographic variation in reproductive success of the parasitoid *Asobara tabida* in larvae of several *Drosophila* species. *Ecological Entomology* 19, 221–229.

Kraaijeveld, A.R., Nowee, B. and Najem, R.W. (1995). Adaptive variation in host selection behaviour of *Asobara tabida*, a parasitoid of *Drosophila* larvae. *Functional Ecology* 9, 113–118.

Kraaijeveld, A.R., Hutcheson, K.A., Limentani, E.C. and Godfray, H.C.J. (2001) Costs of counterdefenses to host resistance in a parasitoid of *Drosophila*. *Evolution* 55, 1815–1821.

Kraaijeveld, A.R., Ferrari, J. and Godfray, H.C.J. (2002) Costs of resistance in insect–parasite and insect–parasitoid interactions. *Parasitology* 125, S71–S82.

Lapchin, L. (2002) Host–parasitoid association and diffuse coevolution: when to be a generalist? *American Naturalist* 160, 245–254.

Le Masurier, A.D. and Waage, J.K. (1993) A comparison of attack rates in a native and an introduced population of the parasitoid *Cotesia glomerata*. *Biocontrol Science and Technology* 3, 467–474.

Lenski, R.E. (1988) Experimental studies of pleiotropy and epistasis in *Escherichia coli*: I. Variation in competitive fitness among mutants resistant to virus T4. *Evolution* 42, 425–432.

Levin, B.R. and Lenski, R.E. (1983) Coevolution in bacteria and their viruses and plasmids. In: Futuyma, D.J. and Slatkin, M. (eds) *Coevolution*. Sinauer, Sunderland, Massachusetts, pp. 99–127.

Matos, M., Avelar, T. and Rose, M.R. (2002) Variation in the rate of convergent evolution: adaptation to a laboratory environment in *Drosophila subobscura*. *Journal of Evolutionary Biology* 15, 673–682.

McGregor, R., Hueppelsheuser, T., Luczynski, A. and Henderson, D. (1998) Collection and evaluation of *Trichogramma* species (Hymenoptera: Trichogrammatidae) as biological control of the oblique-banded leafroller *Choristoneura rosacea* (Harris) (Lepidoptera: Tortricidae) in raspberries and blueberries. *Biological Control* 11, 38–42.

Mollema, C. (1991) Heritability estimates of host selection behaviour by the *Drosophila* parasitoid *Asobara tabida*. *Netherlands Journal of Zoology* 41, 174–183.

Morris, R.J. and Fellowes, M.D.E. (2002) Learning and natal host influence host preference, handling time and sex allocation behaviour in a pupal parasitoid. *Behavioural Ecology and Sociobiology* 51, 386–393.

Muldrew, J.A. (1953) The natural immunity of the larch sawfly (*Pristiphora erichsonii* Htg.) to the introduced parasite *Mesoleius tenthredinis* Morley in Manitoba and Saskatchewan. *Canadian Journal of Zoology* 31, 313–332.

Nappi, A.J. (1975) Parasite encapsulation in insects. In: Maramorosch, K. and Shope, R. (eds) *Invertebrate Immunity*. Academic Press, New York, pp. 293–326.

Nappi, A.J. (1981) Cellular immune response of *Drosophila melanogaster* against *Asobara tabida*. *Parasitology* 83, 319–324.

Pimentel, D. (1968) Population regulation and genetic feedback. *Science* 159, 1432–1437.

Pimentel, D., Nagel, M.P. and Madden, J.L. (1963) Space–time structure of the environment and the survival of parasite–host systems. *American Naturalist* 97, 141–167.

Pimentel, D., Levin, S.A. and Olson, D.A. (1978) Coevolution and the stability of exploiter–victim systems. *American Naturalist* 112, 119–125.

Rigby, M.C., Hechinger, R.F. and Stevens, L. (2002) Why should parasite resistance be costly? *Trends in Parasitology* 18, 116–120.

Rizki, T.M. and Rizki, R.M. (1984) The cellular defense system of *Drosophila melanogaster*. In: King, R.C. and Akai, H. (eds) *Insect Ultrastructure*. Plenum Publishing, New York, pp. 579–603.

Rizki, R.M. and Rizki, T.M. (1990) Parasitoid virus-like particles destroy *Drosophila* cellular immunity. *Proceedings of the National Academy of Sciences USA* 87, 8388–8392.

Rojas, M.G., Morales-Ramos, J.A. and King, E.G. (1999) Response of *Catolaccus grandis* (Hymenoptera: Pteromalidae) to its natural host after ten generations of rearing on a factitious host, *Callosobruchus maculatus* (Coleoptera: Bruchidae). *Environmental Entomology* 28, 137–141.

Rolff, J. and Kraaijeveld, A.R. (2001) Host preference and survival in selected lines of a *Drosophila* parasitoid, *Asobara tabida*. *Journal of Evolutionary Biology* 14, 742–745.

Rosenheim, J.A., Johnson, M.W., Mau, R.F.L., Welter, S.C. and Tabashnik, B.E. (1996) Biochemical preadaptations, founder events, and the evolution of resistance in arthropods. *Journal of Economic Entomology* 89, 263–273.

Salt, G. and van den Bosch, R. (1967) The defence reactions of three species of *Hypera* (Coleoptera: Curculionidae) to an ichneumon wasp. *Journal of Invertebrate Pathology* 9, 164–177.

Sasaki, A. and Godfray, H.C.J. (1999) A model for the coevolution of resistance and virulence in coupled host–parasitoid interactions. *Proceedings of the Royal Society of London Series B* 266, 455–463.

Schlegel-Oprecht, E. (1953) Versuche zur Auslösung von Mutationen bei der zoophagen Cynipide *Pseudeucoila bochei* Weld und Befunde über die stammspezifische Abwehrreaktion des Wirtes *Drosophila melanogaster*. *Zeitschrift für Induktive Abstammungs – und Vererbungslehre* 85, 245–281.

Sokolowski, M.B. (1980) Foraging strategies of *Drosophia melanogaster*: a chromosomal analysis. *Behavioral Genetics* 10, 291–302.

Sokolowski, M.B., Bauer, S.J., Waiping, V., Rodriguez, L., Wong, J.L. and Kent, C. (1986) Ecological genetics and behaviour of *Drosophila melanogaster* in nature. *Animal Behaviour* 34, 403–408.

Strand, M.R. and Pech, L.L. (1995) Immunological basis for compatibility in parasitoid–host relationships. *Annual Review of Entomology* 40, 31–56.

van den Bosch, R. (1964) Encapsulation of the eggs of *Bathyplectes curculionis* (Thompson) (Hymenoptera: Ichneumonidae) in larvae of *Hypera brunneipennis* (Boheman) and *Hypera postica* (Gyllenhal) (Coleoptera: Curculionidae). *Journal of Invertebrate Pathology* 6, 343–367.

Van Nouhuys, S. and Via, S. (1997) Natural selection and genetic differentiation of behaviour between parasitoids from wild and cultivated habitats. *Heredity* 83, 127–137.

van Strien-van Liempt, W.T.F.H. (1983) The competition between *Asobara tabida* Nees von Esenbeck, 1834 and *Leptopilina heterotoma* (Thomson, 1862) in multiparasitized hosts. I. The course of competition. *Netherlands Journal of Zoology* 33, 125–163.

Velicer, G.J. and Lenski, R.E. (1999) Evolutionary trade-offs under conditions of resource abundance and scarcity: experiments with bacteria. *Ecology* 80, 1168–1179.

Visser, M.E., Luyckx, B., Nell, H.W. and Boskamp, G.J.F. (1992) Adaptive superparasitism in solitary parasitoids: marking of parasitized hosts in relation to the pay-off from superparasitism. *Ecological Entomology* 17, 76–82.

Vos, M. (2001) Foraging under incomplete information: parasitoid behaviour and community dynamics. PhD thesis, University of Wageningen, Wageningen, The Netherlands.

Walker, I. (1961) Drosophila und Pseudeucoila II. Schwierigkeiten beim Nachweis eines Selektionserfolges. *Revue Suisse de Zoologie* 68, 252–263.

Weis, A.E., McCrea, K.D. and Abrahamson, W.G. (1989) Can there be an escalating arms race without coevolution? Implications from a host–parasitoid simulation. *Evolutionary Ecology* 3, 361–370.

Wong, H.R. (1974) The identification and origin of the strains of the larch sawfly, *Pristiphora erichsonii* (Hymenoptera: Tenthredinidae), in North America. *Canadian Entomologist* 106, 1121–1131.

Zareh, N., Westoby, M. and Pimentel, D. (1980) Evolution in a laboratory host–parasitoid system and its effect on population kinetics. *Canadian Entomologist* 112, 1049–1060.

9

Interactions Between Natural Enemies and Transgenic Insecticidal Crops

J.J. Obrycki[1]*, J.R. Ruberson[2] and J.E. Losey[3]

[1]Department of Entomology, Iowa State University, Ames, IA 50011, USA; [2]Department of Entomology, University of Georgia, Tifton, GA 31793, USA; [3]Department of Entomology, Cornell University, Ithaca, NY 14853, USA

Introduction

Bioengineered transgenic insecticidal crops are among the most significant technological developments in insect pest management since the advent of synthetic insecticides (Persley, 1996; Cannon, 2000; Perlak *et al.*, 2001). Transgenic cotton and maize that produce insecticidal toxins derived from genes transferred from the bacterium *Bacillus thuringiensis kurstaki* (Btk), for protection against lepidopteran pests, have been planted widely in selected countries (Cannon, 2000; Fitt, 2000; Pray *et al.*, 2002; Shelton *et al.*, 2002). The planting of transgenic Bt maize is one of the most commonly used methods for suppression of the European corn borer, *Ostrinia nubilalis*, in maize-growing regions of the USA. Since 1996, the area of Bt cotton planted in the USA has increased nearly 2.5 times, to approximately 1.8 million ha (Perlak *et al.*, 2001; Shelton *et al.*, 2002). Transgenic *Bacillus thuringiensis tenebrionis* (Btt) potatoes created to resist the Colorado potato beetle, *Leptinotarsa decemlineata*, were grown on relatively small areas in the USA until 2001, when transgenic Bt potatoes were no longer sold (Shelton *et al.*, 2002). Additional transgenic insecticidal crops have been bioengineered for insect resistance using digestive inhibitors, e.g. snowdrop lectin (*Galanthus nivalis* agglutinin, GNA) and several protease inhibitors (Gatehouse *et al.*,1996; Michaud, 2000).

The widespread planting of transgenic crops with tissues containing high levels of insecticidal toxins has raised concerns about gene flow, selection for tolerant pest populations and ecological disruptions of food webs (Rissler and Mellon, 1996; Snow and Palma, 1997; Gould, 1998; Beringer, 2000; Cannon, 2000; Hails, 2000; Poppy, 2000; Watkinson *et al.*, 2000; Wolfenbarger and Phifer,

*See Contributors list for new address.

2000; Marvier, 2001; Obrycki *et al.*, 2001; Schmidt and Hilbeck, 2001; d'Oultremont and Gutierrez, 2002; Letourneau and Burrows, 2002; Manachini and Lozzia, 2002). The role of transgenic crops within the concept of integrated pest management has been discussed extensively (Persley, 1996; Waage, 1997; Hoy *et al.*, 1998; Way and van Emden, 2000; Hilbeck, 2001; Obrycki *et al.*, 2001). For example, Hoy *et al.* (1998) proposed that the conservation of natural enemies be considered in the design of transgenic crops because of their potential role in reducing rates of evolution of tolerant pest populations. Way and van Emden (2000) presented a thorough discussion of the potential to select for pest populations that will overcome Bt transgenic toxins and the likely role of biotic mortality factors in this process (Gould *et al.*, 1991). They also emphasized the importance of using appropriate comparisons when evaluating the relative environmental risks and benefits of transgenic crops (Way and van Emden, 2000). Hilbeck (2001) argued that the current use of transgenic insecticidal crops in a high toxin expression strategy is not compatible with the ecological principles of integrated pest management.

The potential benefits of transgenic insecticidal crops for natural enemies have been presented by several authors (Gould, 1998; Hoy *et al.*, 1998; Way and van Emden, 2000) and include: (i) reduction in insecticide use, thus limiting a major source of natural enemy mortality; (ii) increased prey/host availability for generalist natural enemies via increased densities of secondary insect pest species that may have been previously suppressed due to the use of insecticides for the primary target pest species; and (iii) altered behaviours, physiology or retarded development of insect herbivores that may increase their vulnerability to natural enemies. Balanced against these potential benefits are possible negative aspects of this transgenic technology on natural enemies, paralleling those common to the use of many insecticides: (i) significant reductions of populations of the target pest in transgenic fields; (ii) direct effects of transgenic-plant-produced toxins on natural enemies; and (iii) negative effects that are mediated through insect herbivore hosts, e.g. premature mortality, altered host suitability for growth and development of natural enemies, and altered host–parasitoid behavioural interactions.

The effects of transgenic insecticidal toxins on selected natural enemies have been considered (Schuler *et al.*, 1999a; Cannon, 2000; Groot and Dicke, 2002). Obrycki *et al.* (2001) specifically discussed the interactions between transgenic Btk maize and natural enemies. Schuler *et al.* (1999a) discussed the effects of transgenic insecticidal toxins on the behaviour, life history and ecology of natural enemies, including consideration of population dynamics of natural enemies and hosts within transgenic cropping systems. In a review of the influence of plant resistance on tri-trophic-level interactions, Hare (2002) presents ten examples of transgenic Bt crop–natural enemy interactions (eight involved predatory species and two focused on parasitoids). Seven of the ten studies showed positive interactions (additive or synergistic effects), one study showed no effect, and two negative interactions involved the predator, *Chrysoperla carnea*. By comparison, Hare (2002) also summarized 33 studies that evaluated the interactions between parasitoids and selectively bred resistant crop varieties: 27% were antagonistic, 53%

were additive, 10% were synergistic and 10% were disruptive. Seventeen interactions of predators with resistant crops were listed: 59% were additive, 23% were synergistic and 18% were antagonistic (Hare, 2002). In this review, similar percentages of positive interactions of predators and parasitoids were noted with bioengineered resistance factors and with traits derived from selective breeding techniques.

In this chapter, we emphasize recent laboratory and field studies of the interactions of natural enemies with transgenic insecticidal plants, focusing on their influence on biological control. After a decade examining the interactions between natural enemies and transgenic insecticidal toxins, are there emerging patterns that can be used as the basis for future investigations? As background to the interactions between transgenic crops and natural enemies, we briefly discuss the bases of studies that have documented negative interactions between microbial insecticide Bt formulations and natural enemies.

Bacillus thuringiensis Versus Predators and Parasitoids

Relatively few effects of Bt sprays on natural enemies have been documented, due to the selective nature of Bt toxins, modes of exposure and short exposure times (Glare and O'Callaghan, 2000). However, negative effects of microbial insecticide formulations of Bt on selected natural enemy species have been reported (Salama *et al.*, 1982; Flexner *et al.*, 1986; Croft, 1990; Laird *et al.*, 1990; Blumberg *et al.*, 1997). For example, when Bt causes premature death of a host, parasitoid species developing within these hosts also die (Blumberg *et al.*, 1997). Adverse effects of Bt infections on 13 parasitoid species (eight Braconidae and five Ichneumonidae), due to premature host death, were summarized by Brooks (1993). However, studies of the interactions between natural enemies and microbial sprays of Bt are not necessarily reflective of the possible Bt exposure and Bt expressions in transgenic crops, due to differences in the length of exposure, concentration and form of Bt toxins experienced by hosts and natural enemies. Emergence of two parasitoid species, *Cotesia marginiventris* and *Microplitis croceipes*, from larval *Heliothis virescens* hosts, continuously fed Bt toxins, was inversely related to *B. thuringiensis* concentration in *H. virescens* diets (Atwood *et al.*, 1997, 1999). This type of study is analogous to the potential exposure of larval parasitoids developing in or on hosts on transgenic crops that continuously produce relatively high concentrations of Bt toxins.

Similarly, few studies of the interactions of Bt sprays and predatory species have documented effects on predators, due to the selective activity of Bt toxins and the relatively short exposure times (Salama *et al.*, 1982; Flexner *et al.*, 1986; Croft, 1990). High concentrations of a microbial insecticide formulation of *B. thuringiensis tenebrionis* (Btt) for *L. decemlineata* reduced predation rates by adult *Coleomegilla maculata*, and prolonged larval development when applied to pollen fed to *C. maculata* larvae (Giroux *et al.*, 1994). These effects were observed when Bt concentrations were much higher than the recommended field application rates (Giroux *et al.*, 1994), but they do indicate the potential for adverse effects of high concentrations of Bt toxins on predatory species.

Insect predators

Consumption of transgenic plant tissues or Bt toxins in diet

Within the maize agroecosystem, several species of insect predators that prey on *O. nubilalis* also feed on maize pollen, thus the potential effects of maize pollen on these species have been examined. Consumption of transgenic maize pollen by immature stages of three predatory species commonly found in maize fields in the USA (*Coleomegilla maculata*, *Chrysoperla carnea* and *Orius insidiosus*) did not affect immature development or survival (Pilcher *et al.*, 1997). Transgenic pollen from Event 176, which expresses a relatively high level of toxin, was used in these studies. The most widely planted transgenic maize hybrids in the USA (events MON 810 and Bt 11) express much lower levels of Bt toxin in their pollen.

Development of predatory *Hippodamia convergens* larvae was not affected by Btk toxin (analogous to the insecticidal toxin produced by transgenic Bt cotton) incorporated in an artificial diet (Sims, 1995). Similarly, survival of *C. carnea* larvae was not affected by feeding on *Sitotroga cerealella* eggs coated with an aqueous solution of purified Btk toxin (Sims, 1995). As noted by Hilbeck and Bigler (2001), this method of exposure minimizes contact with *C. carnea* larvae, due to their mode of larval feeding. In contrast, increased mortality of green lacewing (*C. carnea*) larvae was observed when second- and third-instar *C. carnea* fed on an artificial diet containing Bt toxin (Hilbeck *et al.*, 1998a).

Predaceous Heteroptera may feed directly on plants, particularly during early nymphal stages, supplementing nutrition from prey species (Alomar and Wiedenmann, 1996). The longevity of field-collected nymphs and adults of selected species of predatory Heteroptera (*Geocoris punctipes*, *Geocoris pallens*, *Orius tristicolor*, *Nabis* spp. and *Lygus hesperus* (a facultative predator)) feeding solely on Btt-expressing potato tissues was compared with that on non-transgenic tissues by Armer *et al.* (2000). Longevity of individuals exposed to transgenic or non-transgenic potato tissue was similar, and no elevated levels of Btt toxin, based on ELISA, were detected in predators exposed to transgenic tissues (Armer *et al.*, 2000).

Consumption of prey exposed to transgenic insecticidal toxins

Preying upon *O. nubilalis* (highly susceptible to Btk toxin) or *Spodoptera littoralis* (less susceptible to Btk toxin) larvae that had fed on transgenic maize or a diet containing Bt toxins resulted in higher mortality of *C. carnea* larvae (Hilbeck *et al.*, 1998b, 1999). As summarized by Hilbeck and Bigler (2001), a series of laboratory studies (Hilbeck *et al.*, 1998a,b, 1999) show: (i) the susceptibility of *C. carnea* larvae to Bt toxins; (ii) that prey-mediated mortality of *C. carnea* larvae was higher than expected when prey fed on transgenic Bt plants; and (iii) that *C. carnea* larval feeding on a non-target lepidopteran species (*Spodoptera littoralis*), which shows slightly higher mortality and delayed development in response to Bt maize (event Bt 11) (Dutton *et al.*, 2002), may have detrimental effects on predator survival.

Previously, Lozzia *et al.* (1998) observed no effects on the development and survival of *C. carnea* preying on aphids (*Rhopalosiphum padi*) feeding on event 176 Bt maize. Recently, Dutton *et al.* (2002) quantified the effects on *C. carnea* larvae preying on three herbivorous species that had fed on (event 176) Bt maize: *R. padi*, *Tetranychus urticae* and *Spodoptera littoralis*. Higher mortality and delayed development of *C. carnea* larvae were observed when predators fed on *S. littoralis* larvae fed Bt maize than those fed non-Bt maize. *C. carnea* larvae were not adversely affected when fed *T. urticae* prey that had fed on Bt maize, even though *T. urticae* contained relatively higher concentrations of Bt toxin. Dutton *et al.* (2002) discuss possible mechanisms that may explain differences in the Bt maize–herbivore–*C. carnea* interactions, and point out that all published field studies have shown no effects of Bt maize or Bt cotton on densities of *C. carnea*.

Similar developmental times and survival rates were observed when the anthocorid predator, *Orius majusculus*, was fed a thrips species (*Anaphothrips obscurus*) that had been reared on either Bt or non-Bt maize (Zwahlen *et al.*, 2000). The developmental time and survival rate of *Coleomegilla maculata* larvae preying on *L. decemlineata* larvae exposed to Btt potato tissue for 24 h were similar to values observed for prey fed on non-Btt potatoes (Riddick and Barbosa, 1998). The effects of Btk rice lines on a non-target pest species, the brown leafhopper (*Nilaparvata lugens*) and its mirid predator (*Cyrtorhinus lividipennis*), were examined by Bernal *et al.* (2002). Altered feeding behaviours and greater volumes of acidic honeydew were produced by *N. lugens* feeding on several Bt rice lines; however, the Bt rice lines had no effect on several life-history parameters of *N. lugens* or *C. lividipennis* (Bernal *et al.*, 2002). Similarly, no detrimental effects on aphid consumption, development, survival or reproduction were reported for *Hippodamia convergens* larvae and adults feeding on *Myzus persicae* reared on transgenic Btt potatoes resistant to *L. decemlineata* (Dogan *et al.*, 1996). The basis for the lack of effects on predatory species is probably related to the failure of phloem-feeding aphid species to acquire Bt toxins from transgenic plants (Head *et al.*, 2001; Raps *et al.*, 2001; Dutton *et al.*, 2002).

In contrast to Bt crops, transgenic crops bioengineered for aphid resistance have been observed to cause tri-trophic-level effects. Adults of the ladybird predator (*Adalia bipunctata*) preying on *M. persicae* reared on transgenic potatoes expressing snowdrop lectin GNA showed reduced fecundity, egg viability and up to a 51% reduction in female longevity (Birch *et al.*, 1999). No acute toxicity of *A. bipunctata* larvae feeding on *M. persicae* containing snowdrop lectin GNA were observed; however, developmental delays were reported and individuals were smaller at pupation, but larvae consumed more aphids to reach pupation (Down *et al.*, 2000). The authors proposed that GNA tri-trophic-level effects were due to reductions in prey suitability of *M. persicae*. They also hypothesize that under field conditions, aphid prey will probably not be a limiting factor; thus, no effects (direct or indirect) would be expected for *A. bipunctata* larvae feeding on transgenic potato plants expressing GNA (Down *et al.*, 2000). If sub-lethal effects on female reproduction (Birch *et al.*, 1999) are considered in conjunction with the developmental delays and reduced size of immature stages, combined with the effects of GNA transgenic potatoes on *M. persicae* populations, possible negative field effects on *A. bipunctata* populations may be predicted.

Responses of predatory *Perillus bioculatus* females feeding on *Leptinotarsa decemlineata* injected with the proteinase inhibitor oryzacystatin I (OCI) were examined, to simulate potential effects of transgenic potatoes producing (OCI) on this predator (Ashouri *et al.*, 1998). Female survival was not affected; but fertility was reduced by 50%; the pre-oviposition period was prolonged; egg-mass size was reduced; and the rate of egg eclosion was reduced. These results demonstrate the sensitivity of *P. bioculatus* to OSI, and possible cascading effects of proteinase inhibitors produced by transgenic crops on insect predators (Malone and Burgess, 2000). Previously, Jorgensen and Lovei (1999) reported that the carabid beetle, *Harpalas affinis*, consumed fewer *Helicoverpa armigera* larvae fed a diet containing the proteinase inhibitor, bovine pancreatic trypsin inhibitor, compared with consumption of control larvae. The authors speculated that reduced suitability of prey larvae feeding on the proteinase inhibitor, or consumption of undigested proteinase inhibitor in the digestive system of the prey, may be involved.

Parasitoids

The effects of transgenic insecticidal toxins on parasitoid–host interactions may vary considerably, depending on the biology and physiological interactions between the parasitoid species and their hosts. For example, one would expect different responses from ecto- and endo-parasitoids, host-specific species and parasitoids with broader host ranges, and between those species that are synovigenic (females may feed on hosts) compared with proovigenic species. Idiobionts, i.e. parasitoids that kill hosts at the time of parasitism, preventing further development, and typically having broader host ranges, may be less affected by transgenic insecticidal toxins compared with koinobionts, parasitoids that tend to be endoparasitic and alter host developmental physiology to complete their life cycles. Adults of many parasitic species are known to consume nectar, pollen and honeydew (Thompson and Hagen, 1999), which may be another avenue of exposure to transgenic toxins for parasitoid species. Survival of adults of three species of parasitic wasps fed sucrose solutions (artificial honeydew) with snowdrop lectin GNA was significantly reduced (Romeis *et al.*, 2003).

Parasitism of hosts that have consumed transgenic insecticidal plant tissues

The influence of transgenic Bt oilseed rape (expressing Cry1Ac toxin) bioengineered for resistance to the diamondback moth, *Plutella xylostella*, on adult behaviour of the solitary endoparasitic braconid, *Cotesia plutellae*, was quantified by Schuler *et al.* (1999b). Parasitoid larvae developing in susceptible *P. xylostella* hosts fed Bt tissues died due to premature host death; 54% of parasitoids developing in Bt-tolerant hosts fed Bt tissue completed development. Schuler *et al.* (1999b) propose that detrimental effects due to transgenic Bt oilseed rape will be mitigated in the field due to the parasitoid's survival in Bt-tolerant *P. xylostella* hosts and similar host-finding behaviours of *C. plutella* females in response to damaged transgenic Bt and non-transgenic host plants (Schuler *et al.*, 1999b). Alternatively, the initial use of these transgenic crops may result in significant reductions of susceptible *P. xylostella* populations and, due to the inability of *C. plutellae* females to distinguish between Bt and non-Bt plants, reductions in local population densities of *C. plutellae* may be observed.

Using relatively large experimental cages (*c.* 4 m^3), Schuler *et al.* (2001) examined the effects of two transgenic oilseed rape (*Brassica napus*) cultivars, one producing Btk (Cry1Ac) and the second producing the proteinase inhibitor oryzacystatin I (OCI), on the braconid parasitoid *Diaeretiella rapae*, which parasitizes *M. persicae.* These cage experiments have advantages over field studies and provide more reality compared with laboratory studies; however, the number of cages available for these types of studies may limit treatments and replications. No detrimental effects of either transgenic line were observed on percentage host parasitism, percentage emergence from aphid mummies or the reduction of *M. persicae* densities by *D. rapae* in the cages (Schuler *et al.*, 2001).

Pre-imaginal survival decreased, and developmental times increased, when the gregarious, idiobiont species *Parallorhogas pyralophagus* developed on the stalkborer, *Eoreuma loftini*, fed Bt maize (event CBH 351; Cry 9C) for 48 h (Bernal *et al.*, 2002). Additionally, the positive relationship between host size and parasitoid brood size, observed when the host *E. loftini* was reared on non-transgenic maize, was not observed when hosts were fed Bt maize for 48 h (Bernal *et al.*, 2002).

In a study of several resistance mechanisms in potatoes, bioengineered or selectively bred for defence against *L. decemlineata*, Ashouri *et al.* (2001a) observed reductions in larval survival and adult size of the aphid parasitoid, *Aphidius nigripes*, developing in the aphid *Macrosiphum euphorbiae* reared on Btt potatoes. In contrast, the size and fecundity of adult *A. nigripes* increased when immatures developed in *M. euphorbiae* reared on a transgenic potato line expressing the protease inhibitor, rice cystatin I (OCI) (Ashouri *et al.*, 2001a). These results were explained by the influence of the resistant mechanisms on the size of the aphid host, which is indicative of host quality for *A. nigripes* (Ashouri *et al.*, 2001b). The effects of exposure to snowdrop lectin GNA via immature parasitoid development and adult host feeding behaviours of a second aphid parasitoid species (*Aphelinus abdominalis*), parasitizing *M. euphorbiae*, was quantified by Couty *et al.* (2001) and Couty and Poppy (2001). No direct detrimental effects of GNA on *A. abdominalis* were observed; however, Couty *et al.* (2001) documented indirect effects, mediated through decreased aphid host size, that reduced the size of adults and proportion of female parasitoids that had developed in GNA-fed *M. euphorbiae.*

The effects of snowdrop lectin GNA on the eulophid parasitoid *Eulophus pennicornis*, a gregarious ectoparasitoid of *Lacanobia olereacea*, were determined by Bell *et al.* (1999). Even though GNA decreased larval size and retarded development of *L. oleracea*, no adverse effects on size, longevity, egg load or fecundity of *E. pennicornis* females was reported. In a greenhouse study, Bell *et al.* (2001) documented a complementary interaction between transgenic GNA potatoes and parasitism by *E. pennicornis* that reduced feeding damage by *L. oleracea*. Mean size and longevity of female *E. pennicornis* was reduced in *L. oleracea* hosts feeding on GNA potatoes, but these effects did not influence the production of F_2 progeny (Bell *et al.*, 2001).

Transgenic sugarcane expressing snowdrop lectin GNA developed for the primary stalkboring pest of sugarcane in the southern USA, the Mexican rice borer (*E. loftini*), has no measurable negative effects on a second stalkboring insect pest, the sugarcane borer (*Diatraea saccharalis*) (Setamou *et al.*, 2002a). The braconid parasitoid *Cotesia flavipes* plays a major role in the biological control of *D. saccharalis*, thus Setamou *et al.* (2002b) examined the potential host-mediated interactions between this parasitoid and transgenic GNA sugarcane. Hosts feeding on GNA sugarcane did not affect host acceptance, total parasitoid development time or egg load of *C. flavipes* females; however, slight reductions in the following parameters were reported: host suitability, brood size, percentage adult eclosion, proportion of female offspring and female longevity (Setamou *et al.*, 2002b).

Field Surveys of Natural Enemies in Transgenic Insecticidal Fields

Predatory arthropods

Sampling from small-scale field plots and relatively large fields of transgenic and non-transgenic Btk maize has detected no consistent differences in abundance of the predatory insects typically found within the canopy of maize fields in the Midwestern USA (Orr and Landis, 1997; Pilcher *et al.*,1997; Pilcher, 1999; Wold *et al.*, 2001). The predatory species include several predacious Coleoptera in the families Coccinellidae (e.g. *Coleomegilla maculata*, *Cycloneda munda*, *Hippodamia convergens*, *Hippodamia tredecempunctata*, *Coccinella septempunctata* and *Harmonia axyridis*), Neuroptera (*Chrysoperla carnea* and *Chrysopa oculata*, and Hemerobiidae), Heteroptera (*Orius insidiosus*, *Nabis* spp.) and Syrphidae. In Italian maize fields, Lozzia (1999) and Manachini (2000) observed no differences in the abundance of ground beetles (Carabidae) and canopy-dwelling arthropod predators in Bt and non-Bt maize fields.

Following a 2-year field study in Oregon, USA, Reed *et al.* (2001) concluded that the use of transgenic Btt potatoes or weekly sprays of Bt for suppression of *Leptinotarsa decemlineata* did not affect the abundance of arthropod predators in potato fields. No host-specific natural enemies of *L. decemlineata* were observed by Reed *et al.* (2001). Based upon the lack of effects on the predatory fauna in potato fields, compared with bi-weekly sprays of pyrethroid insecticides,

Reed *et al.* (2001) concluded that the use of Btt potatoes was compatible with naturally occurring arthropod predators in Oregon potato agroecosystems. Previously, Riddick *et al.* (1998) observed reductions in the abundance of the carabid beetle, *Lebia grandis*, whose larvae develop as ectoparasites of *L. decemlineata* pupae, in Btt potato fields in Maryland. In contrast, the abundance of the predatory ladybird, *C. maculata*, was similar in Btt and non-Btt potato fields (Riddick *et al.*, 1998).

Surveys of transgenic Bt cotton fields in Arizona showed no effects on densities of adults of several predatory species (*Chrysoperla carnea*, *Hippodamia convergens*, *Geocoris punctipes*, *Orius tristicolor*, *Collops vittatus* and *Nabis* spp.) compared with levels in non-Bt cotton fields (Flint *et al.*, 1995). Similarly, an earlier study by Wilson *et al.* (1992) reported no effects of Bt cotton on abundance of *Chrysoperla carnea*, *Hippodamia convergens*, *Orius tristicolor*, *Collops vittatus* and *Nabis* spp.

In this chapter, we present initial results from an ongoing study quantifying the abundance and diversity of insect predators in Bt and non-Bt cotton fields in three commercial field pairings in Georgia, USA (Table 9.1) (a detailed description of the study will be published following its conclusion). In addition, estimated rates of predation of sentinel eggs of *Helicoverpa zea* placed in Bt and non-Bt fields during three field seasons are presented. Based upon weekly beat-cloth (16–40 samples from 2 m sections of rows per field per date) and whole-plant samples (40 plants examined per field per date), few significant differences in arthropod predator abundance were observed between Bt and non-Bt fields at the three locations in 2000 (paired *t*-test of weekly abundance averaged over the season, using sampling locations within fields as replicates). Differences were due to two species of ladybird predators, *Hippodamia convergens* and *Harmonia axyridis*, which varied in numbers of larvae and adults (Table 9.2). The similarities in abundance of arthropod predators in Bt and non-Bt fields in 2000 were probably due to the lack of insecticide use in any of the fields during the sampling periods. In contrast, 31 differences in predatory species were recorded in 2001, which were evenly distributed among the three locations (Table 9.3). No patterns in the differences were observed, and no one taxon differed significantly between treatments at the three locations. Greater insecticide use in 2001 most probably accounted for many of the differences observed (see Table 9.5 for description of insecticide use), although not all of these differences were in accordance with reductions expected from insecticide applications, i.e. higher numbers were occasionally observed in the treated portion of the field, which may have been due to a resurgence in aphid numbers following insecticide treatments.

On selected dates in 2000, 2001 and 2002, eggs of *H. zea* were placed in fields for 24 h to assess rates of egg predation. Egg survival in 2000 did not differ between Bt and conventional fields, or between locations on the plant where the eggs were placed (Table 9.4), consistent with the predator abundance pattern noted above. Survival of eggs tended to increase later in the season in Bt and non-Bt fields. Similar patterns were observed in 2001 and 2002 (Tables 9.5, 9.6), although survival during 2001, particularly later in the season, was generally higher than observed in 2000, consistent with the higher insecticide usage in 2001.

Table 9.1. Predatory taxa sampled in Georgia, USA, cotton fields during 2000, 2001 and 2002.

Taxon	Family	Comments
Geocoris punctipes	Geocoridae	Predominant species encountered; *G. uliginosus* observed occasionally
Orius insidiosus	Anthocoridae	
Green lacewings (eggs)	Chrysopidae	Species uncertain; all eggs laid singly
Green lacewings (larvae)	Chrysopidae	>90% *Chrysoperla rufilabris*
Green lacewings (pupae)	Chrysopidae	>95% *Chrysoperla rufilabris*
Brown lacewings (larvae)	Hemerobiidae	Species uncertain; all in genus *Micromus*
Coleomegilla maculata	Coccinellidae	
Hippodamia convergens	Coccinellidae	
Coccinella septempunctata	Coccinellidae	
Harmonia axyridis	Coccinellidae	
Scymnus sp.	Coccinellidae	Most of these (>70%) were *Scymnus loewii*, but other species were present
Diomus sp.	Coccinellidae	Species not distinguished
Damsel bugs	Nabidae	Nearly all were *Nabis* species, both *N. roseipennis* and *N. americoferus*; some *Nabicula* spp. were observed
Assassin bugs	Reduviidae	Almost all were *Sinea* species, probably *S. diadema*, but some *Zelus* spp. were observed
Podisus maculiventris	Pentatomidae	
Notoxus monodon	Anthicidae	
Solenopsis invicta	Formicidae	
Spiders	Various	Diverse group, represented primarily by members of the families Oxyopidae and Salticidae; the dominant species was the green lynx spider, *Peucetia viridans*
Flower flies	Syrphidae	Species not distinguished
Ground beetles	Carabidae	Various species and subfamilies; species not distinguished
Rove beetles	Staphylinidae	Species not distinguished, but at least two were present

Although differences were observed in densities of selected predatory species in Bt and non-Bt cotton fields, there were no patterns suggesting specific mechanisms. Predator activity was comparable between treatments, as documented by estimates of *H. zea* egg survival, despite some differences in pesticide use in 2001. No differences in predator abundance or activity could be directly attributable to

Table 9.2. Significant differences based upon paired *t*-tests observed in predatory arthropods sampled in 2000 in Bt and conventional (non-Bt) cotton fields.

		Shake samples		Whole-plant samples	
Taxon	Field	Bt	Conv.	Bt	Conv.
Hippodamia convergens (larvae)	Chula	0.08	0.67	NS	NS
H. convergens (adults)	Chula	0.52	1.65	NS	NS
Harmonia axyridis (adults)	Chula	NS	NS	0.17	0.02
H. axyridis (larvae)	Lindsey	NS	NS	0.33	0

Shake samples are mean number per sample date per 2m section of row (n = 16 such samples per field); whole-plant samples are mean number per date per five plants (n = 4 such samples per field); NS = no significant differences between means, $P > 0.05$.

Bt-transgenic cotton compared with non-Bt cotton, although a difference in sentinel egg loss was observed in 2002, this was most probably attributable to insecticide usage.

Parasitoids

In a 3-year field study at four locations in Iowa, USA, Pilcher (1999) used yellow sticky cards to monitor the seasonal abundance of adults of the polyembryonic parasitoid *Macrocentrus cingulum* (formerly *Macrocentrus grandii*), which parasitizes *O. nubilalis* larvae. The number of *M. cingulum* adults collected was highly correlated with *O. nubilalis* egg-mass densities in Bt and non-Bt fields. However, *M. cingulum* adult densities were 30–60% lower in Bt maize fields compared with non-Bt fields. In a second field study, Siegfried *et al.* (2001) collected diapausing *O. nubilalis* larvae in autumn from Bt and non-Bt maize fields in several locations in the Midwestern USA. Collections were made from transgenic (event 176) Bt maize, which has relatively high levels of Btk toxin expression in maize tissues early in season, but levels decrease later in the season. As a result, the development of a second generation of *O. nubilalis* larvae may be observed in fields of this transgenic Bt maize (Siegfried *et al.*, 2001). Two larval parasitoids of *O. nubilalis* were reared from these collections, *M. cingulum* (Braconidae) and *Eriborus terebrans* (Ichneumonidae); the relative levels of parasitism for each parasitoid species were not reported. Siegfried *et al.* (2001) observed an approximately 50% reduction in larval parasitism from 14% (non-Bt) to 7% (Bt). These lower rates of parasitism are probably due to the suppression of *O. nubilalis* populations early in the season by event 176 Bt maize (Siegfried *et al.*, 2001).

The reductions in densities of adult *M. cingulum* and larval parasitism due to *M. cingulum* and *E. terebrans* were expected, due to the significant reductions in

Table 9.3. Significant differences based upon paired *t*-tests observed in predatory arthropods sampled in 2001 in Bt and conventional (non-Bt) cotton fields.

		Shake samples		Whole-plant samples	
Taxon	Field	Bt	Conv.	Bt	Conv.
Geocoris punctipes (nymphs and adults)	Marchant	2.47	1.18	0.28	0.08
Orius insidiosus (nymphs and adults)	Marchant	NS	NS	0.90	0.27
Green lacewings (eggs)	Marchant	NA	NA	0.48	1.68
Brown lacewings (larvae)	Marchant	0.60	0.18	NS	NS
Hippodamia convergens (larvae)	Marchant	0.20	1.15	0.07	0.22
Scymnus sp. (adults)	Marchant	2.33	1.17	NS	NS
Notoxus monodon	Marchant	NS	NS	0.07	0
Diomus sp.	Marchant	0.08	0.45	NS	NS
Cotton aphids	Marchant	NA	NA	48.7	97.6
Green lacewings (eggs)	Chula	NA	NA	1.03	0.54
Brown lacewings (larvae)	Chula	0.13	0.43	NS	NS
Coccinella septempunctata (adults)	Chula	NS	NS	0.13	0.04
Harmonia axyridis (larvae)	Chula	NS	NS	0.16	0.01
Flower flies (larvae)	Chula	NS	NS	0.01	0.09
Orius insidiosus (nymphs and adults)	Lindsey	NS	NS	1.41	0.82
Damsel bugs (adults)	Lindsey	0.16	0.60	NS	NS
Notoxus monodon	Lindsey	1.99	0.90	NS	NS
Flower flies (larvae)	Lindsey	NS	NS	0.	0.06

Shake samples are mean number per sample date per 2m section of row (n = 20–40 such samples per field); whole-plant samples are mean number per date per five plants (n = 5–10 such samples per field); NS = no significant differences between means, $P > 0.05$.

larval *O. nubilalis* hosts in transgenic Bt maize. A previous field-plot study reported no effects of transgenic Bt maize on *E. terebrans* parasitism (Orr and Landis, 1997). However, in that study, relatively small non-transgenic plots were planted within transgenic plots; rates of parasitism were recorded from larval hosts on these non-transgenic plants.

The effects of transgenic Bt maize on parasitoid species needs to be considered in the context of overall levels of biotic and abiotic mortality factors influencing *O. nubilalis* populations (Phoofolo, 1997; Phoofolo *et al.*, 2001; Kuhar *et al.*, 2002). In a 3-year field study of biotic mortality of *O. nubilalis* in Iowa, USA, Phoofolo *et al.* (2001) observed that *M. cingulum* accounted for >95% of larval parasitism; *E. terebrans* was rarely found. During 1 year, parasitism levels peaked at 31%; however, the effect of *M. cingulum* on *O. nubilalis* densities was relatively low in the following 2 years. Similarly, low levels of parasitism by *M. cingulum* have

Table 9.4. Survival (after 24 h) of *Helicoverpa zea* eggs placed in Bt cotton and non-Bt cotton fields during 2000 (200 eggs placed in each field on each date).

Location	Collection date	Treatment	% eggs surviving after 24 h in upper and middle third of plant	
			Upper third	Middle third
Marchant[a]	18 July	Bt	9.2	2.5
		Non-Bt	26.7	22.5
	16 August	Bt	30.8	38.8
		Non-Bt	55.8	57.5
Chula[a]	23 June	Bt	11.7	16.3
		Non-Bt	3.3	6.3
	21 July	Bt	4.2	1.3
		Non-Bt	16.7	8.8
	10 August	Bt	12.5	17.5
		Non-Bt	25.8	25.0
Lindsey[a]	23 June	Bt	0	1.3
		Non-Bt	0	0
	14 July	Bt	16.7	16.3
		Non-Bt	29.2	28.8
	11August	Bt	30.0	20.0
		Non-Bt	15.0	8.8

[a]None of the fields received insecticides during the season.

been observed in other Midwestern USA locations (Siegel *et al.*, 1987a; Clark *et al.*, 1997). While results from studies conducted in Illinois, USA, showed that larval parasitism by *M. cingulum* had no effect on *O. nubilalis* densities (Siegel *et al.*, 1987b; Onstad *et al.*, 1991), the significant reductions of this parasitoid species in transgenic Bt fields warrant additional quantitative studies.

Insect pathogens

Two insect pathogen species infecting *O. nubilalis* in North America are *Nosema pyrausta*, a microsporidium, and *Beauveria bassiana*, an entomopathogenic fungus (Lewis and Cossentine, 1986; Lewis *et al.*, 2001). *N. pyrausta* appears to be specific to *O. nubilalis*; therefore, we would expect declines in host densities, due to the use of Bt maize, to affect its prevalence. In contrast, the fungus *B. bassiana* attacks a wide range of insects and forms an endophytic relationship with maize plants (Bing and Lewis, 1993). As discussed by Elliot *et al.* (2000), plant endophytic–entomopathogen–host interactions are not well known, and assessment of transgenic alterations in plants which may influence these interactions warrants

Table 9.5. Survival (after 24 h) of *Helicoverpa zea* eggs placed in Bt cotton and non-Bt cotton fields during 2001 (250 eggs placed in each field on each date).

Location	Collection date	Treatment	% eggs surviving after 24 h in upper and middle third of plant	
			Upper third	Middle third
Marchant[a]	28 July	Bt	18.4	16.0
		Non-Bt	5.6	27.2
	25 August	Bt	86.4	86.4
		Non-Bt	76.0	87.2
Chula[b]	28 June	Bt	4.0	10.7
		Non-Bt	6.4	14.4
	2 August	Bt	53.3	58.7
		Non-Bt	47.3	25.3
Lindsey[c]	29 June	Bt	34.4	34.4
		Non-Bt	28.8	36.0
	15 August	Bt	87.2	80.8
		Non-Bt	76.8	72.8

[a]Non-Bt field treated with tralomethrin on 21 August; no insectide applied to Bt field.
[b]Non-Bt field treated with zeta-cypermethrin on 17 July and with tralomethrin on 16 August; Bt field treated with tralomethrin on 16 August.
[c]Non-Bt field treated with lambda-cyhalothrin on 3 August, and Bt field treated with dicrotophos on 3 August.

further investigation. A recent greenhouse and field study showed that *B. bassiana* establishes endophytic relationships with several transgenic Bt maize hybrids and does not cause any observable pathogenic effects (Lewis *et al.*, 2001). Thus, even though densities of *O. nubilalis* larvae are virtually eliminated in transgenic Bt maize fields, *B. bassiana* will probably persist in the maize agroecosystem. Phoofolo *et al.* (2001) reported that mortality levels caused by *B. bassiana* reached a maximum of 21% of *O. nubilalis* larvae during 1 year; however, during the other 2 years of the study, mortality caused by *B. bassiana* was relatively low.

B. bassiana appears to have its greatest impact on the overwintering stage (fifth instars) of *O. nubilalis*; infection levels can reach 84% (Bing and Lewis, 1993). The bioengineering of transgenic herbicide-tolerant Bt maize may have detrimental effects on the abundance of *B. bassiana*. A recent study showed that formulations of glyphosate used on herbicide-tolerant soybeans have fungicidal properties against four species of entomopathogenic fungi, including *B. bassiana* and *Neozygites floridana*, which infects spider mites, *T. urticae* (Morjan, 2001). Field applications of these formulations may have a negative effect on fungal epizootics of *N. floridana*; thus spider mite outbreaks may be more likely to occur in herbicide-tolerant soybean fields (Morjan, 2001). In transgenic maize agroecosystems,

Table 9.6. Survival (after 24 h) of *Helicoverpa zea* eggs placed in Bt cotton and non-Bt cotton fields during 2002 (250 eggs placed in each field on each date).

Location	Collection date	Treatment	% eggs surviving after 24 h in upper and middle third of plant	
			Present	Missing/collapsed
Ty Ty[a]	6 August	Bt	58.8	41.2
		Non-Bt	60.4	39.6
	13 August	Bt	72	28
		Non-Bt	84.8	15.2
Marchant[a,b]	6 August	Bt	28.8	71.2
		Non-Bt	43.2	56.8
Chula[a,b]	25 July	Bt	18	82
		Non-Bt	43.2	56.8
	8 August	Bt	40.8	59.2
		Non-Bt	62	38

[a]Non-Bt field treated with spinosad on 8 July (Ty Ty) and 9 July (Chula and Marchant).
[b]Non-Bt field treated with a pyrethroid in late July (prior to samples).

glyphosate formulations may reduce the levels of *B. basssiana* in the soil and on maize residues.

Risks and Benefits of Transgenic Insecticidal Crops for Natural Enemies

Transgenic insecticidal crops represent an extremely powerful new technology for integrated pest management. These crops have been bioengineered to produce highly selective oral toxins and more general digestive inhibitors, but, unfortunately, we have seen no new strategies for the use of this biotechnology. Currently, these transgenic crops are simply being used as a new method to deliver insecticides (van Emden, 1999). In many agroecosystems, this technology will reduce the use of broad-spectrum insecticides, and produce short-term benefits in pest suppression. However, it is likely that, due to the manner in which these biotechnologies are implemented, this technology will cause some degree of long-term disruptions of agroecosystems, although the overall magnitude of disruption may be less relative to the application of broad-spectrum insecticides. The use of transgenic insecticidal crops in a high-dosage strategy will simplify agroecosystems and may lead to breakdowns in the functioning of these ecosystems. Given the current high-dose strategy, the planting of these crops may create difficult problems for growers, involving reductions in abundance of biotic mortality factors which interact with emergent resistant pest populations (Johnson *et al.*, 1997), and unpredictable multiple effects on food webs (Puterka *et al.*, 2002, Turlings *et al.*, 2002). There are unexpected effects of transgenic modifications, e.g. increased lignin levels in transgenic Bt maize lines (Saxena and Stotzky, 2001) and production of Bt toxins by root tissues, resulting in within-season accumulation of Bt toxins in the soil (Stotzky, 2002). Understanding the interactions between biological control and biotechnology will greatly facilitate the integration of these two important pest management tactics and increase the probability of avoiding the extensive ecological disruptions associated with the rapid adoption and widespread use of insecticides during the 1950s and 1960s.

The use of transgenic insecticidal crops in pest management has focused narrowly on a high-dosage strategy of insecticidal toxicity, which is not based upon ecological principles of pest management (Hilbeck, 2001). However, this strategy may be appropriate for managing the development of resistance in target pest populations when high dosages can be consistently delivered to eliminate resistant heterozygotes (Comins, 1984). The underlying strategies and management of insect pests (Stern *et al.*, 1959; Lewis *et al.*, 1997) require a more comprehensive approach that considers the ecological complexity of agroecosystems. Examples of transgenic crops expressing moderate levels of insecticidal toxins that interact favourably with natural enemies provide guidance for a more balanced pest management approach (Chilcutt and Tabashnik, 1999; Bell *et al.*, 2001). Parasitoid activity may be preserved or enhanced, due to low levels of expression or transgenic resistance factors that cause chronic effects and not immediate mortality (Johnson and Gould, 1992; Johnson, 1997). However, there may be inherent risks in reduced-dosage strategies, e.g. spatial or temporal variation in expression that may make the crop more susceptible to attack by insect pests, necessitating further use of synthetic insecticides for management of insect pests. Transgenic modifications to enhance semiochemicals produced by plants, to increase attractiveness to beneficial species, may enhance pest suppression (Turlings *et al.*, 2002). Although such trait enhancement may be difficult, due to potentially complex genetic interactions, traits that enhance natural enemy efficacy would be invaluable and possibly critical to the large-scale success of a strategy involving a moderate level of expression.

Conclusions

1. The current high-dose strategy will reduce densities of natural enemies that tend to be more host specific, perhaps to a greater extent than the use of insecticides. The elimination of target species may cause local extinctions of host-specific natural enemies, creating metapopulation dynamics. As discussed previously (Bernal *et al.*, 2002), the widespread use of transgenic Bt crops will probably have greater negative effects on parasitoid species that tend to be more host specific, than on predatory species. However, to understand the full extent of these effects, several key questions must be addressed. What is the role of specialist natural enemies in population suppression of insect herbivores in annual cropping systems? How will elimination of specialized natural enemies affect levels of biological control? Will the long-term use of transgenic insecticidal crops simplify agroecosystems, reducing the connections within existing food webs?

2. In contrast to effects on specialist natural enemies, agroecosystems where transgenic crops reduce insecticide use will probably experience higher levels of generalist predators compared with systems receiving insecticides. However, significant reductions in the density of prey targeted by the transgenic crop may cause increased competition and intraguild predation. The net result for pest suppression will need to be quantified.

3. Owing to reduced insecticide usage and/or differential effects on non-target species and their natural enemies, densities of secondary pest populations may increase. These increased densities may benefit natural enemies, or alternatively, may require additional use of insecticides.

4. Levels of toxins in non-target herbivores, which will be influenced by the type of feeding behaviours and mode of action of the transgenic toxin, e.g. acute versus chronic effects, will influence natural enemy populations. Altered behaviours of insect herbivores exposed to transgenic toxins will affect natural enemies (Winterer and Bergelson, 2001; Gore *et al.*, 2002).

5. Long-term monitoring of population dynamics of natural enemy and host species in transgenic fields is required. For example, the absence of lepidopteran larvae from transgenic Bt rice fields may cause predator dispersal or switching behaviours, which may alter levels of biological control of the brown leafhopper, *N. lugens* (Bernal *et al.*, 2002).

6. Transgenic insecticidal crops provide a valuable experimental tool for determining the relative importance of individual pest species (Baute *et al.* 2002), and can provide a basis for a better understanding of multitrophic-level interactions in agroecosystems. Large-scale studies of the interactions between transgenic insecticidal crops, herbivores and natural enemies are needed to provide an understanding of the population dynamics of these interactions (Schuler *et al.*, 2001).

7. The high selection pressure resulting from the high-dosage strategy may enhance the development of tolerant pest populations, particularly for species with limited mobility and discrete generations, and may even lead to speciation, based upon developmental asynchrony between resistant and susceptible populations (see the discussion in Cerda and Wright, 2002).

References

Alomar, O. and Wiedenmann, R.N. (eds) (1996) *Zoophytophagous Heteroptera: Implications for Life History and Integrated Pest Management.* Thomas Say Publications in Entomology: Proceedings. Entomological Society of America, Lanham, Maryland.

Armer, C.A., Berry, R.E. and Kogan, M. (2000) Longevity of phytophagous heteropteran predators feeding on transgenic Btt-potato plants. *Entomologia Experimentalis et Applicata* 95, 329–333.

Ashouri, A., Overne, S., Michaud, D. and Cloutier, C. (1998) Fitness and feeding are affected in the two-spotted stinkbug, *Perillus bioculatus*, by the cysteine proteinase inhibitor, oryzacystatin I. *Archives of Insect Biochemistry and Physiology* 38, 74–83.

Ashouri, A., Michaud, D. and Cloutier, C. (2001a) Recombinant and classically selected factors of potato plant resistance to the Colorado potato beetle, *Leptinotarsa decemlineata*, variously affect the potato aphid parasitoid *Aphidius nigripes*. *BioControl* 46, 401–418.

Ashouri, A., Michaud, D. and Cloutier, C. (2001b) Unexpected effects of different potato resistance factors to the Colorado potato beetle (Coleoptera: Chrysomelidae) on the potato aphid (Homoptera: Aphididae). *Environmental Entomology* 30, 524–532.

Atwood, D.W., Young, S.Y. and Kring, T.J. (1997) Development of *Cotesia marginiventris* (Hymenoptera: Braconidae) in tobacco budworm (Lepidoptera: Noctuidae) larvae treated with *Bacillus thuringiensis* and thiodicarb. *Journal of Economic Entomology* 90, 751–756.

Atwood, D.W., Kring, T.J. and Young, S.Y. (1999) *Microplitis croceipes* (Hymenoptera: Braconidae) development in tobacco budworm (Lepidoptera: Noctuidae) larvae treated with *Bacillus thuringiensis* and thiodicarb. *Journal of Entomological Science* 34, 249–259.

Baute, T.S., Sears, M.K. and Schaafsma, A.W. (2002) Use of transgenic *Bacillus thuringiensis* Berliner corn hybrids to determine the direct economic impact of the European corn borer (Lepidoptera: Crambidae) on field corn in eastern Canada. *Journal of Economic Entomology* 95, 57–64.

Bell, H.A., Fitches, E.C., Down, R.E., Marris, G.C., Edwards, J.P., Gatehouse, J.A. and Gatehouse, A.M.R. (1999) The effect of snowdrop lectin (GNA) delivered via artificial diet and transgenic plants on *Eulophus pennicornis* (Hymenoptera: Eulophidae), a parasitoid of the tomato moth *Lacanobia oleracea* (Lepidoptera: Noctuidae). *Journal of Insect Physiology* 45, 983–991.

Bell, H.A., Fitches, E.C., Marris, G.C., Bell, J., Edwards, J.P., Gatehouse, J.A. and Gatehouse, A.M.R. (2001) Transgenic GNA expressing potato plants augment the beneficial biocontrol of *Lacanobia oleracea* (Lepidoptera: Noctuidae) by the parasitoid *Eulophus pennicornis* (Hymenoptera: Eulophidae). *Transgenic Research* 10, 35–42.

Beringer, J.E. (2000) Releasing genetically modified organisms: Will any harm outweigh any advantage? *Journal of Applied Ecology* 37, 207–214.

Bernal, C.C., Aguda, R.M. and Cohen, M.B. (2002) Effect of rice lines transformed with *Bacillus thuringiensis* toxin genes on the brown planthopper and its predator *Cyrtorhinus lividipennis*. *Entomologia Experimentalis et Applicata* 102, 21–28.

Bernal, J., Griset, J. and Gillogly, P.O. (2002) Impacts of developing on Bt maize-intoxicated hosts on fitness parameters of a stem borer parasitoid. *Journal of Entomological Science* 37, 27–40.

Bing, L.A. and Lewis, L.C. (1993) Occurrence of the entomopathogen *Beauveria bassiana* (Balsamo) Vuillemin in different tillage regimes and in *Zea mays* L., and virulence towards *Ostrinia nubilalis* (Hubner). *Agriculture, Ecosystems and Environment* 45, 147–156.

Birch, A.N.E., Geoghegan, I.E., Majerus, M.E.N., McNicol, J.W., Hackett, C., Gatehouse, A.M.R. and Gatehouse, J.A. (1999) Tri-trophic interactions involving pest aphids, predatory 2-spot ladybirds and transgenic potatoes expressing snowdrop lectin for aphid resistance. *Molecular Breeding* 5, 75–83.

Blumberg, D., Navon, A., Keren, S., Goldenberg, S. and Ferkovich, S.M. (1997) Interactions among *Helicoverpa armigera* (Lepidoptera: Noctuidae), its larval endopar-

asitoid *Microplitis croceipes* (Hymenoptera: Braconidae) and *Bacillus thuringiensis*. *Journal of Economic Entomology* 90, 1181–1186.
Brooks, W. (1993) Host–parasitoid–pathogen interactions. In: Beckage, N.E., Thompson, S.N. and Federici, B.A. (eds) *Parasites and Pathogens of Insects*, Vol. 2, *Pathogens*. Academic Press, New York, pp. 231–272.
Cannon, R.J.C. (2000) Bt transgenic crops: Risks and benefits. *Integrated Pest Management Reviews* 5, 151–173.
Cerda, H. and Wright, D.J. (2002) Could resistance to transgenic plants produce a new species of insect pest? *Agriculture, Ecosystems and Environment* 91, 1–3.
Chilcutt, C.F. and Tabashnik, B.E. (1999) Simulation of integration of *Bacillus thuringiensis* and the parasitoid *Cotesia plutellae* (Hymenoptera: Braconidae) for control of susceptible and resistant diamond back moth (Lepidoptera: Plutellidae). *Environmental Entomology* 28, 505–512.
Clark, T.L., Foster, J.E., Witkowski, J.F., Siegfried, B.D. and Spencer, T.A. (1997) Parasitoids recovered from European corn borer, *Ostrinia nubilalis* Hubner, (Lepidoptera: Pyralidae) larvae in Nebraska. *Journal of the Kansas Entomological Society* 70, 365–367.
Comins, H.N. (1984) The mathematical evaluation of options for managing pesticide resistance. In: Conway, G.R. (ed.) *Pest and Pathogen Control: Strategic, Tactical and Policy Models*. John Wiley & Sons, Chichester, pp. 454–469.
Couty, A. and Poppy, G.M. (2001) Does host-feeding on GNA-intoxicated aphids by *Aphelinus abdominalis* affect their longevity and/or fecundity? *Entomologia Experimentalis et Applicata* 100, 331–337.
Couty, A., de la Viña, G., Clarke, S.J., Kaiserd, L., Pham-Delegued, M.-H. and Poppy, G.M. (2001) Direct and indirect sublethal effects of *Galanthus nivalis* agglutinin (GNA) on the development of a potato-aphid parasitoid, *Aphelinus abdominalis* (Hymenoptera: Aphelinidae). *Journal of Insect Physiology* 47, 553–561.
Croft, B.A. (1990) *Arthropod Biological Control Agents and Pesticides*. John Wiley & Sons, New York.
Dogan, E.B., Berry, R.E., Reed, G.L. and Rossignol, P.A. (1996) Biological parameters of convergent lady beetle (Coleoptera: Coccinellidae) feeding on aphids (Homoptera: Aphididae) on transgenic plants. *Journal of Economic Entomology* 89, 1105–1108.
Down, R.E., Ford, L., Woodhourse, S.D., Raemaekers, R.J.M., Leitch, B., Gatehouse, J.A. and Gatehouse, A.M.R. (2000) Snowdrop lectin (GNA) has no acute toxic effects on a beneficial insect predator, the 2-spot ladybird (*Adalia bipunctata* L.). *Journal of Insect Physiology* 46, 379– 391.
Dutton, A., Klein, H., Romeis, J. and Bigler, F. (2002) Uptake of Bt-toxin by herbivores feeding on transgenic maize and consequences for the predator *Chrysoperla carnea*. *Ecological Entomology* 27, 441–447.
Elliot, S.L., Sabelis, M.W., Janssen, A., van der Geest, L.P.S., Beerling, E.A.M. and Fransen, J. (2000) Can plants use entomopathogens as bodyguards? *Ecology Letters* 3, 228–235.
Fitt, G.P. (2000) An Australian approach to IPM in cotton: integrating new technologies to minimise insecticide dependence. *Crop Protection* 19, 793–800.
Flexner, J.L., Lighthart, B. and Croft, B.A. (1986) The effects of microbial pesticides on non-target, beneficial arthropods. *Agriculture, Ecosysterms and Environment* 16, 203–254.
Flint, H.M., Henneberry, T.J., Wilson, F.D., Holguin, E., Parks, N. and Buehler, R.E. (1995) The effects of transgenic cotton, *Gossypium hirsutum* L., containing *Bacillus*

thuringiensis toxin genes for the control of pink bollworm *Pectinophora gossypiella* and other arthropods. *Southwestern Entomologist* 20, 281–292.

Gatehouse, A.M.R., Down, R.E., Powell, K.S., Newell, C.A., Hamilton, W.D.O. and Gatehouse, J.A. (1996) Transgenic potato plants with enhanced resistance to the peach-potato aphid *Myzus persicae. Entomologia Experimentalis et Applicata* 79, 47–52.

Giroux, S., Coderre, D., Vincent, C. and Cote, J.C. (1994) Effects of *Bacillus thuringiensis* var. *san diego* on predation effectiveness, development and mortality of *Coleomegilla maculata lengi* (Col.: Coccinellidae) larvae. *Entomophaga* 39, 61–69.

Glare, T.R. and O'Callaghan, M. (2000) Bacillus thuringiensis*: Biology, Ecology and Safety.* John Wiley & Sons, New York.

Gore, J., Leonard, B.R., Church, G.E. and Cook, D.R. (2002) Behavior of bollworm (Lepidoptera: Noctuidae) larvae on genetically engineered cotton. *Journal of Economic Entomology* 95, 763– 769.

Gould, F. (1998) Sustainability of transgenic insecticidal cultivars: Integrating pest genetics and ecology. *Annual Review of Entomology* 43, 701–726.

Gould, F., Kennedy, G.G. and Johnson, M.T. (1991) Effects of natural enemies on the rate of herbivore adaptation to resistant host plants. *Entomologia Experimentalis et Applicata* 58, 1–14.

Groot, A.T. and Dicke, M. (2002) Insect-resistant transgenic plants in a multi-trophic context. *The Plant Journal* 31, 387–406.

Hails, R.S. (2000) Genetically modified plants – the debate continues. *Trends in Ecology and Evolution* 15, 14–18.

Hare, J.D. (2002) Plant genetic variation in tritrophic interactions. In: Tscharntke, T. and Hawkins, B.A. (eds) *Multitrophic Level Interactions.* Cambridge University Press, Cambridge, UK, pp. 8–43.

Head, G., Brown, C.R., Groth, M.E. and Duan, J.J. (2001) Cry1Ab protein levels in phytophagous insects feeding on transgenic corn: implications for secondary exposure risk assessment. *Entomologia Experimentalis et Applicata* 99, 37–45.

Hilbeck, A. (2001) Implications of transgenic, insecticidal plants for insect and plant diversity. *Perspectives in Plant Ecology, Evolution, and Systematics* 4, 43–61.

Hilbeck, A. and Bigler, F. (2001) Effects of *Bacillus thuringiensis* via ingestion of transgenic corn-fed prey and purified proteins. In: McEwen, P.K., New, T.R. and Whittington, A.E. (eds) *Lacewings in the Crop Environment.* Cambridge University Press, Cambridge, UK, pp. 369–377.

Hilbeck, A., Baumgartner, M., Fried, P.M. and Bigler, F. (1998a) Effects of transgenic *Bacillus thuringiensis* corn-fed prey on mortality and development time of immature *Chrysoperla carnea* (Neuroptera: Chrysopidae). *Environmental Entomology* 27, 480–487.

Hilbeck, A., Moar, W.J., Pusztai-Carey, M., Filippini, A. and Bigler, F. (1998b) Toxicity of *Bacillus thuringiensis* Cry1Ab toxin to the predator *Chrysoperla carnea* (Neuroptera: Chrysopidae). *Environmental Entomology* 27, 1255–1263.

Hilbeck, A., Moar, W.J., Pusztai-Carey, M., Filippini, A. and Bigler, F. (1999) Prey-mediated effects of Cry1Ab toxin and protoxin and Cry2A protoxin on the predator *Chrysoperla carnea. Entomologia Experimentalis et Applicata* 91, 305–316

Hoy, C.W., Feldman, J., Gould, F., Kennedy, G.G., Reed, G. and Wyman, J.A. (1998) Naturally occurring biological controls in genetically engineered crops. In: Barbosa, P. (ed.) *Conservation Biological Control.* Academic Press, New York, pp. 185–205.

Johnson, M.T. (1997) Interaction of resistant plants and wasp parasitoids of tobacco budworm (Lepidoptera: Noctuidae). *Environmental Entomology* 26, 207–214.

Johnson, M.T. and Gould, F. (1992) Interaction of genetically engineered host-plant resistance and natural enemies of *Heliothis virescens* (Lepidoptera: Noctuidae) in tobacco. *Environmental Entomology* 21, 585–597.

Johnson, M.T., Gould, F. and Kennedy, G.G. (1997) Effects of natural enemies on relative fitness of *Heliothis virescens* genotypes adapted and not adapted to resistant host plants. *Entomologia Experimentalis et Applicata* 82, 219–230.

Jorgensen, H.B. and Lovei, G.L. (1999) Tri-trophic effect on predator feeding: consumption by the carabid *Harpalus affinis* of *Heliothis armigera* caterpillars fed on proteinase inhibitor-containing diet. *Entomologia Experimentalis et Applicata* 93, 113–116.

Kuhar, T.P., Wright, M.G., Hoffmann, M.P. and Chenus, S.A. (2002) Life table studies of European corn borer (Lepidoptera: Crambidae) with and without inoculative releases of *Trichogramma ostriniae* (Hymenoptera: Trichogrammatidae). *Environmental Entomology* 31, 482–489.

Laird, M., Lacey, L.A. and Davidson, E.W. (eds) (1990) *Safety of Microbial Insecticides*. CRC Press, Boca Raton, Florida.

Letourneau, D.K. and Burrows, B.E. (eds) (2002) *Genetically Engineered Organisms: Assessing Environmental and Human Health Effects*. CRC Press, Boca Raton, Florida.

Lewis, L.C. and Cossentine, J.E. (1986) Season long intraplant epizootics of entomopathogens, *Beauveria bassiana* and *Nosema pyrausta*, in a corn agroecosystem. *Entomophaga* 31, 363–369.

Lewis, L.C., Bruck, D.J., Gunnarson, R.D. and Bidne, K.G. (2001) Assessment of plant pathogenicity of endophytic *Beauveria bassiana* in Bt transgenic and non-transgenic corn. *Crop Science* 41, 1395–1400.

Lewis, W.J., van Lenteren, J.C., Phatak, S.C. and Tumlinson, J.H. (1997) A total system approach to sustainable pest management. *Proceedings of the National Academy of Sciences USA* 94, 12243–12248.

Lozzia, G.C. (1999) Biodiversity and structure of ground beetle assemblages (Coleoptera: Carabidae) in Bt corn and its effects on non target insects. *Bollettino di Zoologia Agraria e di Bachicoltura* 31, 37–58.

Lozzia, G.C., Furlanis, C., Manachini, B. and Rigamonti, I.E. (1998) Effects of Bt corn on *Rhopalosiphum padi* L. (Rhynchota Aphididae) and on its predator *Chrysoperla carnea* (Neuroptera Chrysopidae). *Bollettino di Zoologia Agraria e di Bachicoltura* 30, 153–164.

Malone, L.A. and Burgess, E.P.J. (2000) Interference of protease inhibitors on non-target organisms. In: Michaud, D. (ed.) *Recombinant Protease Inhibitors in Plants*. Eurekah.com, Georgetown, Texas, pp. 89–106.

Manachini, B. (2000) Ground beetle assemblages (Coleoptera: Carabidae) and plant dwelling non-target arthropods in isogenic and transgenic corn crops. *Bollettino di Zoologia Agraria e di Bachicoltura* 32, 181–198.

Manachini, B. and Lozzia, G.C. (2002) First investigations into the effects of Bt corn crop on nematofauna. *Bollettino di Zoologia Agraria e di Bachicoltura* 34, 85–96.

Marvier, M. (2001) Ecology of transgenic crops. *American Scientist* 89, 160–167.

Michaud, D. (ed.) (2000) *Recombinant Protease Inhibitors in Plants*. Eurekah.com. Georgetown, Texas.

Morjan, W.E. (2001) Effects of glyphosate and glyphosate-resistant soybeans on twospotted spider mite and green cloverworm. PhD thesis, Iowa State University, Ames, Iowa.

Obrycki, J.J., Losey, J.E., Taylor, O. and Hansen, L.C. (2001) Transgenic insecticidal corn: beyond insecticidal toxicity to ecological complexity. *BioScience* 51, 353–361.

Onstad, D.W., Siegel, J.P. and Maddox, J.V. (1991) Distribution of parasitism by

Macrocentrus grandii (Hymenoptera: Braconidae) in maize infested by *Ostrinia nubilalis* (Lepidoptera: Pyralidae). *Environmental Entomology* 20, 156–159.

Orr, D.B. and Landis, D.A. (1997) Oviposition of European corn borer (Lepidoptera: Pyralidae) and impact of natural enemy populations in transgenic versus isogenic corn. *Journal of Economic Entomology* 90, 905–909.

d'Oultremont, T. and Gutierrez, A.P. (2002) A multitrophic model of a rice–fish agroecosystem: II. Linking the flooded rice–fishpond systems. *Ecological Modelling* 155, 159–176.

Perlak, F.J., Oppenhuizen, M., Gustafson, K., Voth, R., Sivasupramaniam, S., Heering, D., Carey, B., Ihring, R.A. and Roberts, J.K. (2001) Development and commercial use of Bollgard® cotton in the USA – early promises versus today's reality. *The Plant Journal* 27, 489–501.

Persley, G.J. (1996) *Biotechnology and Integrated Pest Management.* CAB International, Wallingford, UK.

Phoofolo, M.W. (1997) Evaluation of natural enemies of the European corn borer, *Ostrinia nubilalis* (Lepidoptera: Pyralidae). PhD thesis, Iowa State University, Ames, Iowa.

Phoofolo, M., Obrycki, J.J. and Lewis, L.C. (2001) Quantitative assessment of biotic mortality factors of the European corn borer, *Ostrinia nubilalis* (Lepidoptera: Crambidae) in field corn. *Journal of Economic Entomology* 94, 617–622.

Pilcher, C.D. (1999) Phenological, physiological, and ecological influences of transgenic Bt corn on European corn borer management. PhD thesis, Iowa State University, Ames, Iowa.

Pilcher, C.D., Obrycki, J.J., Rice, M.E. and Lewis, L.C. (1997) Preimaginal development, survival, and field abundance of insect predators on transgenic *Bacillus thuringiensis* corn. *Environmental Entomology* 26, 446–454.

Poppy, G. (2000) GM crops: environmental risks and non-target effects. *Trends in Plant Science* 5, 4–6.

Pray, C.E., Huang, J., Hu, R. and Rozelle, S. (2002) Five years of Bt cotton in China – the benefits continue. *The Plant Journal* 31, 423–430.

Puterka, G.J., Bocchetti, C., Dang, P., Bell, R.L. and Scorza, R. (2002) Pear transformed with a lytic peptide gene for disease control affects nontarget organism, pear psylla (Homoptera: Psyllidae). *Journal of Economic Entomology* 95, 797–802.

Raps, A., Kehr, J., Gugerli, P., Moar, W.J., Bigler, F. and Hilbeck, A. (2001) Immunological analysis of phloem sap of *Bacillus thuringiensis* corn and of the nontarget herbivore *Rhopalosiphum padi* (Homoptera: Aphididae) for the presence of Cry1Ab. *Molecular Ecology* 10, 525–533.

Reed, G.L., Jensen, A.S., Riebe, J., Head, G. and Duan, J.J. (2001) Transgenic Bt potato and conventional insecticides for Colorado potato beetle management: comparative efficacy and non-target impacts. *Entomologia Experimentalis et Applicata* 100, 89–100.

Riddick, E.W. and Barbosa, P. (1998) Impact of Cry3A-intoxicated *Leptinotarsa decemlineata* (Coleoptera: Chrysomelidae) and pollen on consumption, development, and fecundity of *Coleomegilla maculata* (Coleoptera: Coccinellidae). *Annals of the Entomological Society of America* 91, 303–307.

Riddick, E.W., Dively, G. and Barbosa, P. (1998) Effect of a seedmix deployment of Cry3Aa-transgenic and nontransgenic potato on the abundance of *Lebia grandis* (Coleoptera: Carabidae) and *Coleomegilla maculata* (Coleoptera: Cocccinellidae). *Annals of the Entomological Society of America* 91, 647–653.

Rissler, J. and Mellon, M. (1996) *The Ecological Risks of Engineered Crops.* MIT Press, Cambridge, Massachusetts.

Romeis, J., Babendreier, D. and Wackers, F.L. (2003) Consumption of snowdrop lectin (*Galanthus nivalis*) agglutinin causes direct effects on adult parasitic wasps. *Oecologia* 134, 528–536.

Salama, H.S., Zaki, F.N. and Sharaby, A.F. (1982) Effect of *Bacillus thuringiensis* Berl. on parasites and predators of *Spodoptera littoralis* (Boisd.). *Zeitschrift für angewandte Entomologie* 94, 498– 504

Saxena, D. and Stotzky, G. (2001) Bt corn has a higher lignin content than non-Bt corn. *American Journal of Botany* 88, 1704–1706.

Schmidt, J.E.U. and Hilbeck, A. (2001) Ecology of transgenic crops expressing insecticidal Bt ∂-endotoxins – effects on trophic interactions and biodiversity of insect pollinators, non-target herbivores and natural enemies. *Bulletin of the Geobotanical Institute ETH* 67, 79–87.

Schuler, T.H., Poppy, G.M., Kerry, B.R. and Denholm, I. (1999a) Potential side effects of insect resistant transgenic plants on arthropod natural enemies. *Trends in Biotechnology* 17, 210–216.

Schuler, T.H., Potting, R.P.J., Denholm, I. and Poppy, G.M. (1999b) Parasitoid behaviour and Bt plants. *Nature* 400, 825–826.

Schuler, T.H., Denholm, I., Jouanin, L., Clark, S.J., Clark, A.J. and Poppy, G.M. (2001) Population-scale laboratory studies of the effect of transgenic plants on nontarget insects. *Molecular Ecology* 10, 1845–1853.

Setamou, M., Bernal, J.S., Legaspi, J.C., Mirkov, T.E. and Legaspi, B.C. Jr (2002a) Evaluation of lectin-expressing transgenic sugarcane against stalkborers (Lepidoptera: Pyralidae): Effects on life history parameters. *Journal of Economic Entomology* 95, 469–477.

Setamou, M., Bernal, J.S., Legaspi, J.C. and Mirkov, T.E. (2002b) Effects of snowdrop lectin (*Galanthus nivalis*) agglutinin expressed in transgenic sugarcane on fitness of *Cotesia flavipes* (Hymenoptera: Braconidae), a parasitoid of the nontarget pest *Diatraea saccharalis* (Lepidoptera: Crambidae). *Annals of the Entomological Society of America* 95, 75–83.

Shelton, A.M., Zhao, J.-Z. and Roush, R.T. (2002) Economic, ecological, food safety, and social consequences of the deployment of Bt transgenic crops. *Annual Review of Entomology* 47, 845– 881.

Siegel, J.P., Maddox, J.V. and Ruesink, W.G. (1987a) Survivorship of the European corn borer, *Ostrinia nubilalis* (Hübner) (Lepidoptera: Pyralidae) in Central Illinois. *Environmental Entomology* 16, 1071–1075.

Siegel, J.P., Maddox, J.V. and Ruesink, W.G. (1987b) Seasonal progress of *Nosema pyrausta* in the European corn borer, *Ostrinia nubilalis*. *Journal of Invertebrate Pathology* 52, 13–136.

Siegfried, B.D., Zoerb, A.C. and Spencer, T. (2001) Development of European corn borer larvae on Event 176 Bt corn: influence on survival and fitness. *Entomologia Experimentalis et Applicata* 100, 15–20.

Sims, S.T. (1995) *Bacillus thuringiensis* var *kurstaki* [CryIA(c)] protein expressed in transgenic cotton: effects on beneficial and other non-target insects. *Southwestern Entomologist* 20, 493–500.

Snow, A.A. and Palma, P.M. (1997) Commercialization of transgenic plants: potential ecological risks. *BioScience* 47, 86–96.

Stern, V.M., Smith, R.F., van den Bosch, R. and Hagen, K.S. (1959) The integrated control concept. *Hilgardia* 29, 81–101.

Stotzky, G. (2002) Release, persistence, and biological activity in soil of insecticidal proteins from *Bacillus thuringiensis*. In: Letourneau, D.K. and Burrows, B.E. (eds)

Genetically Engineered Organisms: Assessing Environmental and Human Health Effects. CRC Press, Boca Raton, Florida, pp. 187–222.

Thompson, S.N. and Hagen, K.S. (1999) Nutrition of entomophagous insects and other arthropods. In: Bellows, T.S. and Fisher, T.W. (eds) *Handbook of Biological Control.* Academic Press, New York, pp. 594–652.

Turlings, C.J., Gouinguene, S., Degen, T. and Fritzsche-Hoballah, M.E. (2002) The chemical ecology of plant–caterpillar–parasitoid interactions. In: Tscharntke, T. and Hawkins, B.A. (eds) *Multitrophic Level Interactions.* Cambridge University Press, Cambridge, UK, pp. 148–173.

van Emden, H.F. (1999) Transgenic host plant resistance to insects – Some reservations. *Annals of the Entomological Society of America* 92, 788–797.

Waage, J. (1997) What does biotechnology bring to integrated pest management? *Biotechnology and Development Monitor* 32, 19–21.

Watkinson, A.R., Freckleton, R.P., Robinson, R.A. and Sutherland, W.J. (2000) Predictions of biodiversity response to genetically modified herbicide-tolerant crops. *Science* 289, 1554–1557.

Way, M.J. and van Emden, H.F. (2000) Integrated pest management in practice – pathways toward successful application. *Crop Protection* 19, 81–103.

Wilson, F.D., Flint, H.M., Deaton, W.R., Fischoff, D.A., Perlak, F.J., Armstrong, T.A., Fuchs, R.L., Berberich, S.A., Parks, N.J. and Stapp, B.R. (1992) Cotton lines containing a *Bacillus thuringiensis* toxin to pink bollworm (Lepidoptera: Gelechiidae) and other insects. *Journal of Economic Entomology* 85, 1516–1521.

Winterer, J. and Bergelson, J. (2001) Diamondback moth compensatory consumption of protease inhibitor-transformed plants. *Molecular Ecology* 10, 1069–1074.

Wold, S.J., Burkness, E.C., Hutchison, W.D. and Venette, R.C. (2001) In-field monitoring of beneficial insect populations in transgenic corn expressing *Bacillus thuringiensis* toxin. *Journal of Entomological Science* 36, 177–187.

Wolfenbarger, L.L. and Phifer, P.R. (2000) The ecological risks and benefits of genetically engineered plants. *Science* 290, 2088–2093.

Zwahlen, C., Nentwig, W., Bigler, F. and Hilbeck, A. (2000) Tritrophic interactions of transgenic *Bacillus thuringiensis* corn, *Anaphothrips obscurus* (Thysanoptera: Thripidae), and the predator *Orius majusculus* (Heteroptera: Anthocoridae). *Environmental Entomology* 29, 846–850.

10

The GMO Guidelines Project: Development of International Scientific Environmental Biosafety Testing Guidelines for Transgenic Plants

A. Hilbeck and the Steering Committee of the GMO Guidelines Project*

Geobotanical Institute, Swiss Federal Institute of Technology (ETH), Zurichbergstrasse 38, CH-8044 Zurich, Switzerland

Introduction

The Cartagena Protocol on Biosafety of Living Modified Organisms (Biosafety Protocol) under the Convention on Biodiversity (CBD) and many other international forums identify a clear need in both developing and developed countries for comprehensive, transparent, scientific guidelines for meaningful pre-release testing and post-release monitoring of transgenic plants, to ensure their environmental safety and sustainable use. The lack of such guidelines globally, and the need for such guidelines in developing countries, has been repeatedly expressed by both the private and the public sector (CBD, 2000). For example, Chapter 16 of Agenda 21 recognizes that the maximum benefits of genetically modified crops can be achieved only if appropriate guidelines for their biosafety are in place and the relevant capacities to implement the guidelines are acquired (UN-DSD, 1999).

*The Steering Committee consists of the following members: David Andow, University of Minnesota, St Paul, USA; Nick Birch, Scottish Crop Research Institute, Dundee, UK; B.B. Bong, Vietnamese Ministry of Agriculture and Rural Development, Hanoi, Vietnam; Deise Capalbo, EMBRAPA Environment, Jaguariuna, Brazil; Gary Fitt, CSIRO Cotton Research Center, Narrabri, Australia; Eliana Fontes, EMBRAPA Genetic Resources and Biotechnology, Brazil; K.L. Heong, IRRI, Philippines; Jill Johnston, Ohio State University, Columbus, USA; Kristen Nelson, University of Minnesota, St Paul, USA; Ellie Osir, ICIPE, Nairobi, Kenya; Allison Snow, Ohio State University, Columbus, USA; Josephine Songa, KARI, Machakos, Kenya; Evelyn Underwood, ETH, Zurich, Switzerland; Fang Hao Wan, CAAS, Biological Control Institute, Beijing, China.

In addition, there is wide recognition that the regulatory and scientific capacity for conducting risk assessments needs to be strengthened in most countries. Most importantly, the needs of developing countries for capacity building and policy development must be addressed. Article 22 of the Biosafety Protocol requires that parties shall cooperate in the development and/or strengthening of human resources and institutional capacities in biosafety (UN-DSD, 1999). It is also recognized that this capacity-building activity will require significant investments, as many countries do not have the capability to make independent risk assessments or to evaluate independently submitted risk assessments on biosafety (CBD, 2000).

The GMO Guidelines Project 'Development of International Scientific Biosafety Testing Guidelines for Transgenic Plants' aims to develop international, scientific, conclusive and acceptable guidelines for assessing the environmental risks posed by a genetically modified organism (GMO). The project was launched by scientists of the Global Working Group on 'Transgenic Organisms in Integrated Pest Management and Biological Control' under the patronage of the International Organization for Biological Control (IOBC). We focus our efforts in capacity building on facilitating scientist-to-scientist exchange, because a sound, consistent science base may be easier to transfer among countries than a regulatory system. Consequently, the project focuses its capacity-building efforts on a few countries with reasonably developed scientific infrastructures. By strengthening the scientific capacities for risk assessment in these countries, we expect the expertise to diffuse more readily to neighbouring countries, and a good regional representation in the regional groups and on the advisory board of the project will assist this process.

The guidelines can be envisaged as a set of interlinked modules, consisting of scientific questions related to risk assessment and corresponding scientific methodologies to answer those questions. Table 10.1 illustrates what the guidelines are and what they are not. The guidelines will have no regulatory legitimacy themselves, but regulatory authorities can choose to implement parts, or all, of the guidelines as they desire or need, with confidence in the scientific soundness behind the information gathered using our identified methodologies. They are designed for use before approval is given for the GM plant. The compiled guidelines strive to address all issues and questions pertinent to the scientific sections, as comprehensively as possible. Not all of the questions must be addressed for every GMO. Based on the specific GMO and target region for release, for most GMOs, a case-specific subset of questions will need to be addressed for an assessment of their ecological impact. The guidelines will not be in the form of a decision tree, but they can easily be incorporated into one, which would have to be region, country or case specific, according to the regulatory situation and case-study requirements. The guidelines are not a guide to cost (or risk)–benefit analysis of transgenic organisms, as this presupposes a valuation system for comparing effects, which is outside the domain of scientific methodology, but they will guide the evaluation of any environmental effect, which can then be analysed in the appropriate political context (Table 10.1, section 1).

Table 10.1. What the GMO Guidelines are and what they are not.

The guidelines are:	The guidelines are not:
1 Comprehensive scientific questions of which a case-specific subset can be extracted for ecological impact assessment, depending on GMO and target region for release	Regulatory guidelines (e.g. decision tree or decision guide)
2 Relevant to the environmental and agricultural impacts of GMOs	For evaluation of human health impacts or ethical implications
3 All or some questions can be considered (many may prove to be irrelevant in a particular country's context)	All questions must always be answered in all cases
4 Protocols that will lead to scientifically defensible results that can be used in risk assessment	Prescriptive protocols that must always be used to answer a particular scientific question
5 Serve as a scientific standard for the data that support risk assessment	Validation of poor scientific methods for generating data for risk assessment
6 Guidelines that provide the necessary scientific information to address a scientific question related to risk assessment	Guidelines that require the generation of data that are not necessary for risk assessment

They will cover the environmental and agricultural impacts of GMOs, and they will not evaluate human health impacts or ethical implications. The guidelines should be applicable to most GM plants, but are at present focused on those GM plants modified to produce a novel gene product currently used in pest control. There is more information available on this class of GMO than any other class, and the expertise of the group is concentrated in the disciplines related to pest control (Table 10.1, section 2).

The guidelines will be designed for use on a case-by-case basis, as specified in the Biosafety Protocol and the EU Directive on release of GM plants. A case-by-case approach is necessary because there is insufficient experience available to allow aggregate analysis and assessment. Each GM plant and ecosystem must be looked at separately. The relevant questions will therefore differ on a case-by-case and country-by-country basis, but with any of the selected questions and associated protocols, the user of the guidelines can have confidence that the protocols will produce scientifically sound data (Table 10.1, sections 3 and 4).

Existing data can be evaluated against the guidelines, so that the strengths and weaknesses of the existing data can be clarified. They will therefore also provide guidance for judging other possible protocols. By setting a clear scientific standard, the data for risk assessment is likely to converge and stabilize around that standard. The protocols will indicate the information that is essential for answering the question, and will provide additional guidance on collecting data that is not essential but, none the less, useful (see Table 10.1, sections 5 and 6).

The scientific scope of the guidelines is divided into five scientific sections: needs assessment, plant characterization, non-target and biodiversity impacts, gene flow and its consequences, and resistance management. In the following part of the chapter, we summarize the areas covered by each section. This summary represents the current status of the project, and will be subject to change, as the project is an ongoing process.

Problem Formulation and Options Assessment

The problem formulation and options assessment (PFOA) section provides a framework for evaluating the need for the transgenic plant in specific crop production contexts, and comparing it against other potential solutions to the problem. A science-driven assessment must be a deliberative process designed to provide for social reflection and discussion about transgenic organisms (Forester, 1999; Susskind *et al.*, 2000). This includes providing an approach for evaluating projected changes in crop production practices that result from the implementation of the GMO or alternative solution(s). This section sets the context for the environmental analyses that follow in the non-target, gene flow and resistance sections. It defines the target agroecosystems in which the use of the GMO or alternative solution is proposed, including the crop system, farming system and ecological and structural context, and the people who will be affected by the use of the GMO or alternative solution. It establishes the need for the GMO to perform a particular function in the target agroecosystem, and addresses how potential alternatives compare with the GMO proposal. The assessment is based on an application of the precautionary approach or principle.

For many political systems in the world, the legitimating authority exists to incorporate PFOA in a legislative or regulatory context. For some legislative or regulatory situations (e.g. US regulations), an assessment can be incorporated into the public consultative process prior to regulation, or it may be added as an alternative process that informs the debate in traditional decision-making bodies. Stakeholder participation will also be addressed.

The first part of the assessment addresses the question of problem identification, and addresses questions such as: What is the agricultural problem the GMO is designed to address? How extensive and severe is the problem? Identification and characterization of affected people, farming systems and agroecosystems, perception of affected people: do they regard the agricultural problem as a core need? What factors do they see as important impacts on their agricultural success? What needs do they have that might conflict with proposed solutions?

The assessment also focuses on identifying potential problem solutions. The GMO is one potential alternative, but other potential solutions may include changes in farm management practices, changes in local community practices, changes in government support and structures, and/or changes in the structure of agricultural production. The section will also provide a systematic approach for evaluating the potential efficacy of the alternative solutions and how they meet the needs identified previously. Potential solutions can be compared on the basis of effectiveness, efficiency, efficacy and sustainability.

Finally, the PFOA addresses the potential socio-economic effects of the identified solutions: What are the potential consequences from implementing the solution? Possible impacts on internal to the farming system include changes in crop management practices. Possible impacts external to the farming system include impact on other nearby cropping systems, structure of agricultural production, market prices and availability. Which impacts are potentially adverse and are they reversible? Can adverse impacts be mitigated by the imposition of restrictions on the user? Do structures exist that can monitor compliance with any management strategies or guidelines and report adverse impacts?

The assessment will result in a range of feasible alternative solutions, of which the GMO is one, and an evaluation of their efficacy or suitability. It will also provide background information on the target cropping systems, non-target cropping systems, target farmers and current practices, etc., to focus the analysis in the subsequent sections.

Transgene Expression and Locus Structure

This section will specify methodologies to determine the stability of the genotype (structural stability), the phenotype (stability of expression) and stability during inheritance. Genotypic stability may be related to the nature of the insertion, the insertion location or the nature of the surrounding DNA. It can be evaluated by determining the number of transgene insertions and sequencing the transgene(s) and regions flanking the insert(s).

Furthermore, the section will specify the phenotypic effects of the transgene in the plant, including position, pleiotrophic and epistatic effects, genotype by environment interaction, transcription products of both marker genes and target genes, and how, what plant parts and when product concentrations should be measured in transgenic plants to facilitate risk assessment and management of environmental effects. Some of these characteristics should be measured in the field, and others can be measured in the lab. There is a need to develop uniform reporting standards for gene product concentrations in plant tissues, so that they can be used in any regulatory oversight system in the world. The plant phenological stages and the tissues that should be sampled during those stages have been variously reported (Agbios, 2003; compare reports of Bt expression in various transgenic Bt events and crop plants). Additional work is needed to clarify what is essential for evaluating non-target effects and resistance management.

Finally, the section will determine whether the phenotype of the transgene is inherited stably over multiple generations. A transgene can be stably integrated into the genome but its phenotype can be altered because the genetic background changes or because the gene is silenced or enhanced. However, nutritional characterization of the harvested GMO product is outside the scope of this project, unless it results in some environmental effect.

Non-target and Biodiversity Effects

The central problem for non-target and biodiversity impact assessments is how to focus assessment procedures appropriately. In the Guidelines Project, this is done through two main tasks. First, the section will specify procedures to determine which non-target species, structural characteristics of the biota and/or function/processes (e.g. ecosystem functions and processes) should be tested (= selection procedures). Secondly, the section will specify scientific procedures for testing these species/structures/functions/processes (= testing procedures). Presently, these testing procedures concentrate on the methodologies for tier 1 testing, i.e. short-term, acute ecotoxicology tests for hazard identification of pesticides. The procedures also consider how to evaluate cumulative effects on successive generations. Results will include an estimation of maximum potential hazard for key non-target species (the maximum mortality, growth suppression and sub-lethal effects the transgenic plant can cause), and estimations of the potential impact of the transgenic crop on key non-target species and ecosystem functions.

Identified categories for non-target testing include the following: non-target pests or potential pests, biological control agents of pests (natural enemies), crop pollination by animals, soil ecosystem functions, species of conservation concern and species of cultural significance. This list may be subject to change during the project, since a number of issues are under discussion. For example, the definition of target and non-target pests is not as trivial as it seems. All target pests should be 'controlled' (i.e. negatively affected) by the transgenic crop, while non-target pests may or may not be 'controlled' in this way, but the impacts of the non-target pests may be exacerbated by the effects of the transgenic crop, creating additional control problems.

A challenging question is what is the role of other 'neutral' or 'value unknown' non-target species? The vast majority of species found in an agricultural field are 'neutral' or 'value unknown' species. Criteria need to be developed to determine when some of these 'neutral' or 'value unknown' species need to be evaluated, such as if the species is important in another crop or in natural areas, or if the species is an important alternative food source for polyphagous natural enemies.

Five criteria have been identified for selecting non-target species to be tested: co-occurrence (What is the species assemblage of the target agroecosystem?), abundance (What species or suite of species are most abundant in the agroecosystem?), association (What species are more constantly associated with the target crop?), trophic linkage (What species have a strong link with the crop? For example, herbivores or saprovores that feed on GM plant material, pollen or exudates, or their natural enemies, or species that feed on excreta or exudates of herbivores or saprovores) and significance to humans (known non-target pests, particularly those that reach levels where crop damage occurs or are candidate secondary pests, biological control organisms and pollinators). In what follows, the biological control (through arthropod natural enemies) and soil functions categories will be described in more detail to illustrate the approach used.

Arthropod natural enemies (NE) can be affected by the novel compounds in GM plants via multiple pathways (Fig. 10.1). These pathways involve a variety of trophic connections, and the potential effects of metabolites of the novel GM compound. In addition, these compounds and metabolites can interact with existing primary and secondary metabolites of the plant. For many natural enemies the primary exposure pathway will be mediated via the herbivore, as prey or host (Wolfenbarger and Phifer, 2000; Hilbeck, 2001). However, many natural enemy species also feed on plants or plant parts, consuming pollen, nectar, sap, aphid honeydew and/or intracellular fluids. The plant can also influence the herbivore–natural enemy interaction, which can complicate procedures to test for the effects of the herbivore on natural enemies. The potential interactions within plants and prey/hosts can alter plant or prey/host physiology. Alterations in physiology can act indirectly on natural enemies, and often exert themselves by changing the quality of the food resource for the natural enemy (Hilbeck, 2002). These indirect effects of altered food quality (including interaction effects with natural secondary compounds in plants and their herbivorous insects) are not developed in detail in Fig. 10.1, but are subsumed within the effect of the whole plant or the herbivore. The transgene products or metabolites might also interact with other secondary plant compounds, magnifying or neutralizing the effects of the transgene products on the herbivore and/or its natural enemies. All of the major potential pathways by which a transgene could affect a natural enemy are illustrated in Fig. 10.1. This can then serve as a basis for further exposure analysis.

In recommending testing procedures, we need to consider which of the natural enemies identified can be exposed to the toxin. Of those that can be exposed, the guiding principle is to expose them to the toxin in an ecologically relevant way at higher than typical concentrations. This requires characterizing the material to which the natural enemy would be exposed and characterizing the effects of the toxins on target and non-target pest species over time. In addition, it may mean that risk is assessed in a tri-trophic system, a bi-trophic system, or both. Data on ecological effects caused by reduced prey quality, quantity and availability also need to be included in biosafety testing guidelines. Higher than typical concentrations are desired to enable characterization of maximum hazard to the non-target species in tier 1 tests.

Another part of the non-target impacts section will consider soil issues. Developing testing guidelines for impact of transgenic plants on soil organisms and soil ecosystem processes is a great challenge. This is primarily because we know only little about soil-inhabiting microorganisms and their functions (*c.* 1–5% that can be cultured). We have only a slightly better working knowledge and understanding of the soil macro-biota; we do know that both macro- and micro-biota have a crucial role in soil ecosystem functioning.

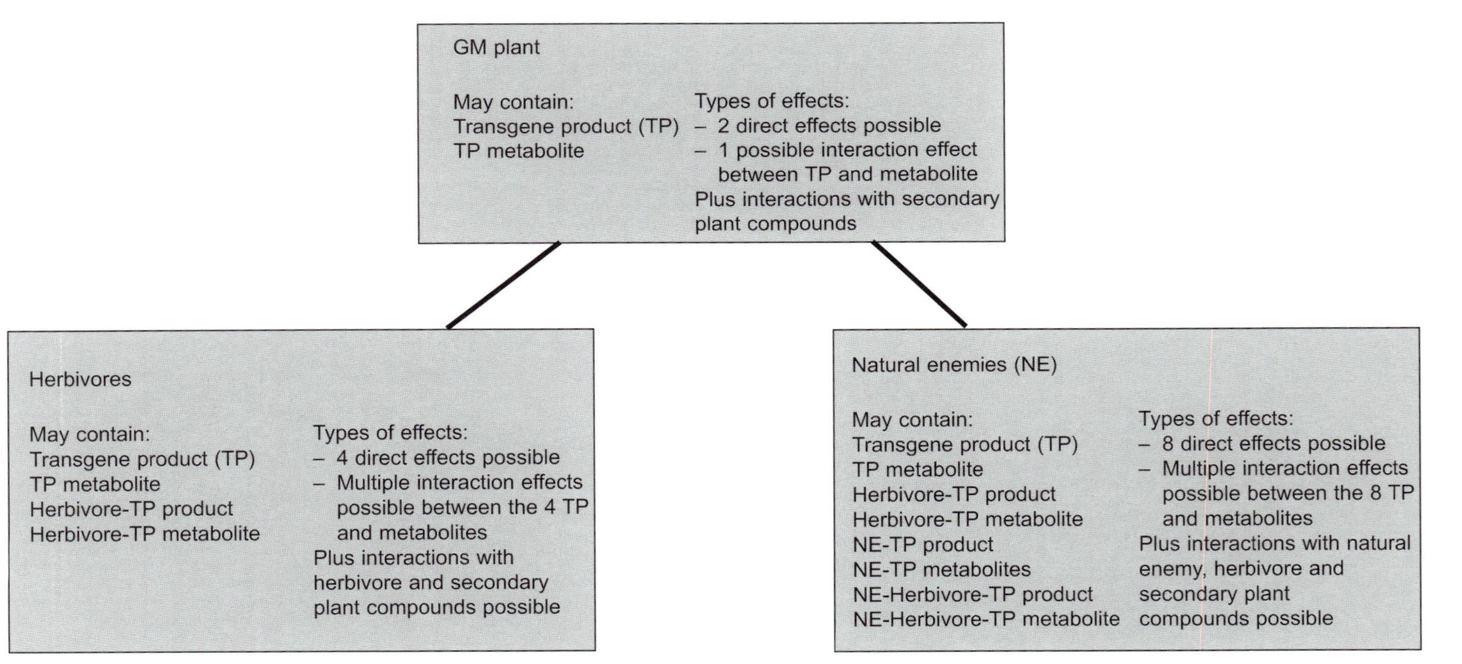

Fig. 10.1. The various bi- and tri-trophic routes, types of compounds and effects that can possibly impact natural enemies. GM, genetically modified; NE, natural enemies; TP, transgene product (e.g. novel protein encoded by transgene); TP metabolite, a plant-produced metabolite of the transgene product; Herbivore-TP product, a herbivore-produced metabolite of the plant-produced transgene product (e.g. the novel plant protein); Herbivore-TP metabolite, a herbivore-produced metabolite of the plant-produced transgene metabolite; NE-TP product, a NE-produced metabolite of the plant-produced transgene product; NE-TP metabolite, a NE-produced metabolite of the plant-produced transgene metabolite; NE-Herbivore-TP product, a NE-produced metabolite of the herbivore-produced plant transgene product; NE-Herbivore-TP metabolite, a NE-produced metabolite of the herbivore-produced plant transgene metabolite.

Because of these scientific knowledge gaps, the guidelines pursue two routes simultaneously for safety testing of transgenic plants in soil: a taxon-specific route, which will probably work for known macro-biota, and an ecosystem function route. The taxon-specific route will proceed via: (i) defining selection criteria for testing species; (ii) conducting exposure analyses; and (iii) developing testing procedures for the identified species. Soil ecosystem functions can include respiration assays to measure the level of activity in soil, which is related to soil microbial biomass, basal respiration, substrate-induced respiration, nitrogen mineralization including ammonification, which is the crucial first step in nitrogen cycling, mineralization or decomposition of plant material (transgenic and non-transgenic), and phosphorus immobilization. Impacts on functional microbial diversity can be measured by changes in microbial genetic diversity community composition. Methods are being developed to determine species genetic diversity of microorganisms by DNA fingerprinting and/or other molecular techniques.

The diversity of soil macroorganisms must be considered. The selection criteria discussed for non-target testing above can be applied, including abundance and functional significance (for example, direct consumers of plant residue (e.g. earthworms, beetles, mites) that degrade large pieces of residue into smaller pieces, thereby facilitating/enhancing microbial degradation; organisms that are important for soil physico-chemical structure (e.g. creating soil macro- and micro-pores influencing water drainage, leaching, soil aeration, etc.), and soil natural enemies).

Exposure analysis of soil macroorganisms is necessary, but conducting a meaningful exposure analysis in soil ecosystems is complicated because the organisms are exposed to dead or decaying plant material and, to a lesser extent, to living plant material, including plant secretions such as root exudates. Furthermore, once released from the plant tissue, the novel transgene products/proteins will interact physically and chemically with the soil and its constituent components, such as humic acids, clay minerals and colloids (Stotzky, 2002). Far less is known about the fate of complex organic molecules, such as proteins, in soils, compared with smaller organic or inorganic chemicals, such as pesticides or industrial pollutants. Therefore, as the basis for an exposure analysis, the following parameters would need to be determined. First, the various routes of movement and transport of transgenic plant material and its novel transgene products need to be identified (e.g. root exudates or leakages, movement associated with plant residues, including roots, release of proteins from the plant residues, etc.). This requires good working knowledge of protein expression, the soil–plant interface, and residue movement and management. Secondly, the fate of such plant material, and of the novel protein(s) released from it, needs to be understood and quantified (e.g. adsorption to clay minerals, humic acids, etc., resulting potentially in accumulation of the protein, immobilization or leaching of transgenic proteins, etc.). Testing procedures should then be developed using a process similar to that suggested for other non-target organisms, i.e. determining endpoints of an exposure analysis and developing protocols that simulate, as closely as possible, an ecologically relevant delivery system.

Gene Flow and its Consequences

The gene flow section is determining protocols for establishing: (i) the likelihood of intra- and interspecific gene flow; (ii) the possibility of subsequent geographic spread of transgenes; and (iii) the potential ecological effects resulting from gene flow, such as invasiveness, weediness, effects on non-target species, agricultural production and biodiversity, reduction of genetic diversity and impact on conservation issues.

Gene flow between genetically modified (GM) crops and their wild relatives has been cited as one of the main ecological risks associated with the application of biotechnology to crop production (Rissler and Mellon, 1996; Ellstrand *et al.*, 1999). Here, we propose a framework for identifying risk issues related to gene flow. Several types of recipient populations are considered, such as wild species that interbreed with the crop (including weedy relatives and rare relatives), and non-GM crop varieties. The impetus for considering risks associated with gene flow is that once transgenes have moved into new populations the process cannot be reversed. With appropriate DNA markers, it is possible to detect gene flow from transgenic crops, but is difficult to predict the ecological effects of transgenes that are integrated into different genetic backgrounds or expressed in different ecological contexts. Plants that acquire transgenes will continue to evolve, subject to natural and artificial selection pressures in the agricultural setting and beyond. The hazards or consequences of transgene escape may vary widely, depending on the type of trait, the type of population into which it becomes integrated and the ecological context. The first steps in assessing the possible risks of gene flow are determining whether pollen-mediated gene flow can occur among both cultivated and wild relatives in a given region, and whether transgenes could also be dispersed via seeds and/or vegetative propagules. If this is the case, we then ask whether transgenes will increase in frequency due to natural and/or artificial selection, and what ecological consequences of this process should be considered, including what the recipient ecosystem will be, and what organisms in that recipient ecosystem could be affected.

Resistance Management

The resistance management section will determine the resistance risk (the risk that a pest will evolve resistance to the transgenic crop or cropping practices associated with the crop), and management responses needed to reduce this risk. The resistance section is developing methodologies to specify, for each of the target and other affected pests to be considered. This section has developed a stepwise approach to the data needs for risk analysis. The first step involves data collection that should occur prior to field release. The second step specifies the data that needs to be collected during the field-testing stage of development and before commercial release. The final step occurs after commercial release.

Data needs prior to field testing:

- Potential resistance risk: species ranking based on exposure and future exposure, selective intensity (potential maximum hazard from non-target section), and history of resistance problems.
- Potential consequences of resistance: factors for prioritizing species (key pests, pests that are difficult to control or require management practices with significant human health or environmental risks).
- Operational definition of resistance.
- Possibility for resistance management.
- Possibility for monitoring and response plan.

Data needs during field testing, but before commercial release:

- Resistance risk: improve understanding of exposure, hazard, resistance frequency and potential dominance in the field.
- Refine operational definition of resistance.
- Develop an implementable resistance management plan.
- Specify monitoring methods, response plan and roles of stakeholders.

Data needs after commercialization:

- Quality control of monitoring effort.
- Using monitoring information – processing and reporting.
- Implementation of the response plan.

Conclusion

While the task is large and complex, the project is also critically important, as outlined in the introduction. We welcome involvement of all public-sector scientists and encourage interested researchers to contact us at the given addresses. The more people become involved and engage in developing the product, the better the guidelines we will be able to produce. Simultaneously, the guidelines will be more widely spread and recognized. Interested public-sector scientists can enrol in the project at http://www.gmo-guidelines.info, but can also contact us directly at the e-mail and postal addresses provided on the above website.

References

Agbios (2003) *Essential Biosafety*, 2nd edn. Available on CD or online at: http://www.essentialbiosafety.info

CBD (Secretariat of the Convention on Biological Diversity) (2000) *Cartagena Protocol on Biosafety to the Convention on Biological Diversity: Text and Annexes.* Secretariat of the Convention on Biological Diversity, Montreal. www.biodiv.org/doc/legal/cartagena-protocol-en-pdf

Ellstrand, N.C., Prentice, H.C. and Hancock, J.F. (1999) Gene flow and introgression from domesticated plants into their wild relatives. *Annual Review of Ecology and Systematics* 30, 539–563.

Forester, J. (1999) *The Deliberative Practitioner: Encouraging Participatory Planning Processes.* MIT Press, Cambridge, Massachusetts.

Hilbeck, A. (2001) Implications of transgenic, insecticidal plants for insect and plant biodiversity. *Perspectives in Plant Ecology, Evolution and Systematics* 4, 43–61.

Hilbeck, A. (2002) Transgenic host plant resistance and non-target effects. In: Letourneau, D.K. and Burrows, B.E. (eds) *Genetically Engineered Organisms – Assessing Environmental and Human Health Effects.* CRC Press, Boca Raton, Florida, pp. 167–185.

Rissler, J. and Mellon, M. (1996) *The Ecological Risks of Engineered Crops.* MIT Press, Cambridge, Massachusetts.

Stotzky, G. (2002) Release, persistence, and biological activity in soil of insecticidal proteins from *Bacillus thuringiensis.* In: Letourneau, D.K. and Burrows, B.E. (eds) *Genetically Engineered Organisms – Assessing Environmental and Human Health Effects.* CRC Press, Boca Raton, Florida, pp. 187–222.

Susskind, L., Levy, P. and Thomas-Larmer, J. (2000) *Negotiating Environmental Agreements: How to Avoid Escalating Confrontation, Needless Costs, and Unnecessary Litigation.* Island Press, Washington, DC.

UN-DSD (Division for Sustainable Development) (1999) *Agenda 21. Chapter 16: Environmentally Sound Management of Biotechnology.* http://www.un.org/esa/sustdev/agenda21chapter16.htm

Wolfenbarger, L.L. and Phifer, P.R. (2000) The ecological risks and benefits of genetically engineered plants. *Science* 290, 2088–2093.

11

Genetic Manipulation of Natural Enemies: Can We Improve Biological Control by Manipulating the Parasitoid and/or the Plant?

G.M. Poppy[1] and W. Powell[2]

[1]*Biodiversity and Ecology Division, School of Biological Sciences, University of Southampton, Bassett Crescent East, Southampton SO16 7PX, UK;* [2]*Plant and Invertebrate Ecology Division, Rothamsted Research, Harpenden, Herts AL5 2JQ, UK*

Introduction

The use of natural enemies for controlling herbivorous pest insects has been practised since ancient times in China (DeBach, 1974) and is still one of the major approaches in biological control (DeBach and Rosen, 1991). During the 20th century, there were widespread attempts to control insect pests by releasing natural enemies into new geographic regions (so-called classical biological control), and, more recently, strategies based on inundative, augmentative releases and manipulation of wild populations have become increasingly practised. The past few years have seen a significant revival in research and development related to biological control, due to financial and political pressures for sustainable agriculture. Thus, there are new opportunities to demonstrate the practicality and usefulness of biological control in integrated pest management (IPM) programmes, and we need to explore novel ways to improve the efficacy and/or reliability of biological control using insect natural enemies.

It is widely recognized that a number of both biotic and abiotic factors are critical to the success of any natural enemy. We will focus on parasitoids, as they are important biological control agents of a wide range of insect pests, and are the most frequently used natural enemy group in biological control programmes (van Driesche and Bellows, 1996). Major challenges facing biological control are

 Genetics, Evolution and Biological Control (eds L.E. Ehler, R. Sforza and T. Mateille)

how to improve the low success rate, and advance the robustness and reliability of control. The use of genetics to enhance the efficiency of natural enemies has attracted a lot of discussion (Roush, 1990; Hopper *et al.*, 1993; Wajnberg, Chapter 2, this volume), although, so far, it has delivered very little. During the 21st century, which is predicted to be the century for genetics, post-genomic knowledge will allow exciting developments in health care, environmental protection and agricultural practice.

The genetic manipulation of parasitoids has the potential to significantly improve biological control. Considering the tri-trophic nature of interactions between plant, herbivorous host and parasitoid (Price *et al.*, 1980; Poppy, 1997), there are two obvious ways to genetically manipulate the parasitoid. The most obvious method is to directly manipulate the genetics of the parasitoid itself, but a less obvious, and possibly easier, method is to exploit our growing knowledge of the influence of the plant on parasitoid foraging behaviour and genetically manipulate the plant to improve parasitoid efficiency.

Parasitoids show a remarkable phenotypic plasticity due to associative learning (Turlings *et al.*, 1993), and the interaction between innate, conditioned and learnt behavioural responses (Poppy *et al.*, 1997) is critical in determining parasitoid effectiveness, and needs to be taken into account whenever considering their genetic manipulation. The genetic control of learning and the ability to select parasitoids for learning abilities is a very exciting prospect, which may soon be a possibility due to current research on the honeybee genome. It is therefore important to consider the 'gene × environment' interactions before deciding which traits should be selected and how to ensure they have maximum impact on biological control.

Genetic Manipulation

It is beyond the scope of this chapter to describe in detail genetic modification techniques, but it is useful to highlight some of the options currently available. Humans have been genetically modifying plants and animals for many thousands of years, but the methods for manipulating the genetics are ever changing (Halford, 2000). During the past century, methods have become increasingly sophisticated and have been instrumental to the success of the green revolution (Reeves and Cassaday, 2002).

As we enter the post-genomic era, the ability to genetically manipulate plants and parasitoids to express traits previously unattainable via selective breeding becomes a reality. The use of recombinant DNA technology allows plants to be engineered with genes controlling traits originating from a range of species, not just those from closely related plant species (Halford and Shewry, 2000; Sharma *et al.*, 2002). The discovery of transposable elements for *Drosophila* has opened the doors once again on the possibilities for genetically engineering insects (Atkinson *et al.*, 2001). In spite of much excitement about the potential to engineer beneficial insects, such as the production of predatory mites resistant to insecticides

(Hoy, 1976, 1985, 1992), there have been few such real conclusive demonstrations of this technology. Recent advances in genetic engineering have allowed insects from a range of orders to be engineered with a range of traits, and thus genetic engineering of parasitoids and predators is becoming a real possibility again (Atkinson *et al.*, 2001). However, it must be borne in mind that the public are sceptical about modification of plants, so the manipulation of virulence or other fitness traits in biological control agents such as parasitoids may well be a step too far for the public to accept. Therefore, the first generation of genetically engineered parasitoids is likely to be resistant to insecticides rather than super biocontrol agents, even though the benefits of the latter may outweigh any risks.

Most of the examples discussed in this chapter do not involve the use of recombinant DNA technology, but in plants this technology is currently having, and is likely to have, a tremendous impact on agricultural practice. We are already seeing the role of parasitoids and predators as important biosafety indicators for this new technology (Schuler *et al.*, 1999a; Poppy, 2000). As well as considering the negative side of genetically modified (GM) plants, it is important to consider how such plants may in fact aid biological control by parasitoids (Poppy and Sutherland, 2003).

Genetic Manipulation of Parasitoids

The importance of intraspecific, genetic variability in influencing the performance of parasitoids released in 'classical' biological control programmes, and hence in contributing to the success or failure of such attempts, has often been highlighted (Hoy, 1985; Roush, 1990; Hopper *et al.*, 1993; Wajnberg, Chapter 2, this volume). However, genetic factors also need to be considered in the development and implementation of biological control and integrated pest management strategies based on short-term, augmentative releases of parasitoids and in the enhancement of wild populations by means of conservation biological control. Augmentation, involving either inundative or inoculative releases of parasitoids and predators, has been widely used against pests on protected horticultural crops, whereas conservation biological control, based on habitat management to conserve populations of natural enemies, linked with behavioural manipulation to enhance their impact on target pests, is gaining popularity for combating pests on field crops (van Driesche and Bellows, 1996).

Augmentative releases involve the mass-rearing of parasitoids, which often necessitates careful design of rearing protocols to minimize the potential detrimental effects of genetic factors such as founder effects, inbreeding, genetic bottlenecks and genetic drift (Hopper *et al.*, 1993). On the other hand, opportunities exist for genetic manipulation or selective breeding to optimize biological and behavioural traits that enhance performance after release. Because augmentative releases are intended to function in the short term, potential problems of releasing selected strains, such as reduced fitness for long-term survival in the wild or loss of selected traits through interbreeding with wild populations, are unlikely to arise. In the case of the conservation and manipulation of wild parasitoid populations, it is important to determine the within-species and within-population genetic variability of behavioural traits that affect the response of individuals to environmental foraging cues, particularly semiochemical stimuli, and to understand how such variability could influence local movement and gene-flow patterns that determine the genetic structure of local populations, in relation to sub-populations in different habitats or on different host species. Opportunities also exist for the genetic manipulation of the biological sources of semiochemical foraging cues, especially those arising from plants, in order to increase their effectiveness at concentrating parasitoid activity in target areas.

For many parasitoids, semiochemical information plays a key role in their long-range and short-range foraging behaviour and in host recognition (Vinson, 1976; Tumlinson *et al.*, 1992; Vet and Dicke, 1992; Stowe *et al.*, 1995). This information can be provided by the host itself, either directly or via host products such as excreted frass, or by the food plant of the host, or can result from the interaction between the herbivore host and its food plant (Vet and Dicke, 1992). The intraspecific behavioural responses of parasitoids to semiochemical stimuli are variable and, as well as genetical control, are influenced by phenotypic plasticity, physiological state and environmental and experiential factors (Vet and Dicke, 1992; Poppy *et al.*, 1997). Vet and Dicke (1992) hypothesize that responses to host-derived stimuli are more likely to be congenitally fixed in parasitoids, whereas responses to plant-derived stimuli are more likely to be influenced by experiential factors, such as developmental conditioning or associative learning. Very few studies have yet been done to investigate the genetic basis of variability in parasitoid responses to semiochemical foraging cues, partly because of the difficulty of separating genetic and experiential influences (Poppy *et al.*, 1997; Wajnberg, Chapter 2, this volume). However, the few studies that have been conducted also indicate the existence of genetic variability (Prévost and Lewis, 1990; Potting *et al.*, 1997; Perez-Maluf *et al.*, 1998; Gu and Dorn, 2000).

Evidence that parasitoid responses to host-derived cues can be innately determined as opposed to learnt was provided by studies of aphidiine parasitoid responses to aphid sex pheromones (Poppy *et al.*, 1997). Naïve females (i.e. having had no contact with either aphids or plants since emergence) of both specialist and generalist species responded strongly to aphid sex pheromone components in wind tunnel bioassays (Glinwood *et al.*, 1999a,b). Because these parasitoids had been reared in asexual aphid morphs, they could not have been exposed to sex pheromone chemicals during development or during emergence from the host, thus removing the possibility of developmental conditioning or post-emergence experience inducing the responses. Identification of genes that determine behavioural responses to specific chemical cues could advance future possibilities for the genetic manipulation of parasitoids designed to increase their attraction to, or retention in, target areas.

Strong evidence for genetic determination of the variation in behavioural traits, including responses to semiochemical foraging cues, can be provided by crossing experiments. Crossing is one of the most reliable methods for detecting genetic contributions to phenotypic variability and for providing information on the genetic structure of a trait (Hopper *et al.*, 1993). Crosses between the closely related species *Aphidius ervi* and *Aphidius microlophii* (at the time thought to be two host-associated races of *A. ervi*) indicated that parasitoid performance on different hosts, measured as mummy production, was strongly determined by inherited factors (Powell and Wright, 1988). *A. ervi* reared on the pea aphid, *Acyrthosiphon pisum*, produced very few mummies when confined with the nettle aphid, *Microlophium carnosum* (the only known host of *A. microlophii*). However, the offspring of crosses between *A. ervi* females and *A. microlophii* males regularly performed as well on nettle aphids as they did on pea aphids, even though they were reared on the latter. Further experiments revealed that *A. ervi* made very few oviposition attacks on nettle aphids, whereas the hybrid females attacked this host as frequently as did *A. microlophii* females (Fig. 11.1). Pennacchio *et al.* (1994) also demonstrated that, although *A. microlophii* females readily attacked pea aphids, they seldom released eggs in this host, whereas *A. ervi* × *A. microlophii* hybrids did. The results of these crossing experiments indicate that there is a strong genetic basis in the variation in both host recognition (determining attack responses) and host acceptance during an attack (determining egg release). Identifying genes that determine the ability to detect and respond to specific host recognition and acceptance cues could provide opportunities for manipulating parasitoid host

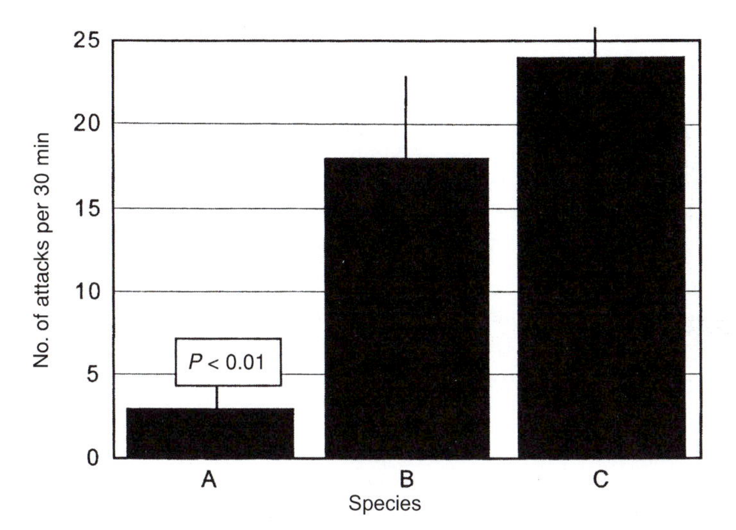

Fig. 11.1. Host attack rate for the parasitoids *Aphidius ervi* and *Aphidius microlophii* against the nettle aphid *Microlophium carnosum* in Petri dish arenas (number of attacks in 30 min when isolated females are provided with 20 aphids). A, *A. ervi* reared on the pea aphid *Acyrthosiphon pisum*. B, *A. microlophii* reared on *M. carnosum*. C, *A. ervi* (mother) × *A. microlophii* (father) hybrids reared on *A. pisum* ($n = 20$). *A. ervi* attacked nettle aphids significantly less ($P < 0.01$; *t*-test) than either *A. microlophii* or the hybrids.

preferences, and so increase the effectiveness of mass-reared parasitoids released against specific hosts.

The level of response to host recognition cues is also variable amongst individuals of the same species, and this can be quantified in standardized attack rate trials (Powell and Wright, 1992). One method of demonstrating that there is a heritable genetic component to the observed phenotypic variability in such quantitative traits is to test for parent–offspring regressions (Wajnberg, Chapter 2, this volume). At Rothamsted, we combined the methods of parent–offspring regression and isofemale lines. We established isofemale lines of the aphid parasitoid *Praon volucre* using females that varied in their attack rates against the English grain aphid *Sitobion avenae*, on which they were reared (Rehman, 1999; Rehman and Powell, unpublished data). These lines were then reared through two generations on several different aphid hosts. Mother–daughter, daughter–granddaughter and mother–granddaughter regressions were tested for attack rates on *S. avenae, A. pisum, Myzus persicae* and *Aulocorthum solani*. Statistically significant regressions were obtained in 11 of the 12 tests, 7 of which were highly significant ($P < 0.001$) (Table 11.1). Data for the attack rates on *M. persicae* are presented in Fig. 11.2 as an example. Attack rate performance was obviously influenced by an inherited genetic factor that, in theory, could be selected for in order to produce more efficient parasitoid strains. Interestingly, both semiochemical and visual cues are known to play a role in host recognition for aphid parasitoids (Michaud and Mackauer, 1994; Battaglia *et al.*, 1995, 2000; Powell *et al.*, 1998) and their variations are almost certainly influenced by independent genetic determinism.

Table 11.1. Statistical significance of parent–offspring regressions of attack rates against different hosts for isofemale lines of the aphid parasitoid *Praon volucre*. All mothers were reared on *Sitobion avenae* whereas daughters and granddaughters were reared on the relevant test species.

Host	Numbers tested M–D–GD[a]	Mother–daughter	Daughter–granddaughter	Mother–granddaughter
Sitobion avenae	10–91–243	$P < 0.05$	$P < 0.001$	$P < 0.05$
Acyrthosiphon pisum	10–50–250	$P < 0.01$	$P < 0.001$	$P > 0.05$
Myzus persicae	10–47–222	$P < 0.001$	$P < 0.001$	$P < 0.001$
Aulocorthum solani	10–50–175	$P < 0.001$	$P < 0.001$	$P < 0.05$

[a]Mothers–daughters–granddaughters.

Chemical information relating to the host food plant is obtained by the female parasitoid at the time of emergence from its mummified host, and this often determines initial host plant preferences during post-emergence foraging (van Emden *et al.*, 1996; Storeck *et al.*, 2000). This raises the possibility of treat-

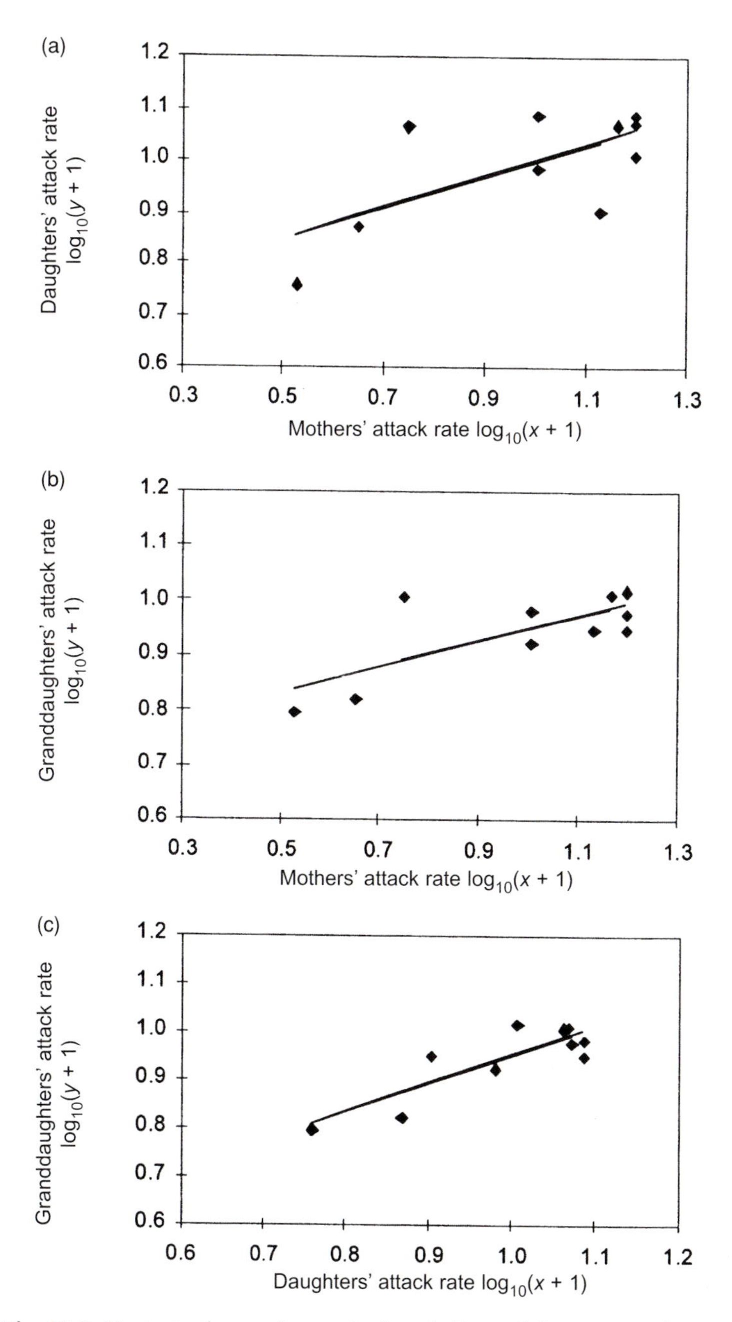

Fig. 11.2. Host attack rates for ten isofemale lines of the parasitoid *Praon volucre* against the aphid *Myzus persicae* in Petri dish arenas (number of attacks in 30 min when isolated females are provided with 20 aphids). (a) Mother–daughter regression ($P < 0.001$); (b) mother–granddaughter regression ($P < 0.001$); (c) daughter–granddaughter regression ($P < 0.001$).

ing mass-reared mummies with specific, plant-derived semiochemicals to tailor the foraging preferences of the emerging parasitoids for specific target crops. It would be useful therefore to investigate potential intraspecific genetic variability in behavioural responses to such plant-derived semiochemicals. However, aphid parasitoids readily change their plant preferences as a result of successful foraging experiences, providing phenotypic plasticity that allows them to cope with temporal and spatial changes in resource availability within ecosystems (Guerrieri *et al.*, 1997; Powell *et al.*, 1998; Storeck *et al.*, 2000). There is no information on the potential role of genetics in determining variation in the learning abilities of individual parasitoids, and this is also an area that merits future investigation.

Parasitoid Manipulation via the Plant

Parasitoids live and/or search for herbivores on plants – thus plant characteristics affect the ecology, behaviour and physiology of parasitoids. Peter Price and his colleagues were the first to outline how plant characteristics could influence top-down regulation in food chains via multitrophic interactions (Price *et al.*, 1980). Plants can regulate herbivore numbers both directly, using internal plant defences (bottom-up), and indirectly, by interacting with parasitoids and predators via external, volatile, chemical signals (top-down). Since this seminal review, there has been considerable research into tri-trophic interactions, but most investigations have focused on studying only a few model interactions involving crop plants. The research has also tended to focus on laboratory studies to understand the proximate mechanisms regulating these interactions (Chadwick and Goode, 1999). Recently, there has been an attempt to look at more adaptive questions, and the ecology and evolution of these plant–parasitoid interactions is beginning to be studied in more detail (Turlings *et al.*, 2002). Such ecological and evolutionary understanding is very important, since we now have the ability to genetically manipulate plants to express novel traits. It is possible to manipulate them to have increased direct and/or indirect defences, but we need to understand the consequences of such actions because the traits will have the possibility to move around in the environment via both inter- and intraspecific gene flow.

Although it is possible to speculate about which traits could have an influence on parasitoids, it is more useful to highlight good examples illustrating how plant characteristics are known to affect the behaviour and/or effectiveness of parasitoids.

The morphology of the plant surface can be critical in tri-trophic interactions. For example, the effects of quantity and quality of leaf domatia and/or trichomes on natural enemy behaviour have been well documented, and plant breeders have selected for plant cultivars which affect herbivores both directly, and indirectly via parasitoids (Obrycki, 1986; van Lenteren and de Ponti, 1990). Efficacy of the whitefly parasitoid *Encarsia formosa* was improved on half-haired varieties of cucumbers, due to a reduction in their walking speed, compared with the speed of parasitoids on hairless varieties, and a lack of interference in their foraging behaviour, which did occur on hairy varieties (Hulspas-Jordan and van Lenteren, 1978).

Plant domatia have also been shown to play an important role in tri-trophic interactions, for example affecting predator–prey interactions in cotton (Agrawal and Karban, 1997). When domatia were artificially added to cotton plants, the number of phytophagous mites declined as the number of predaceous bugs increased, resulting in an increase in boll productivity. Thus, manipulating the domatia on plants can result in improved productivity, due to an increase in mite natural enemies (Agrawal and Karban, 1997).

Plants can also provide nutritional benefits to predators and parasitoids, and so manipulation of pollen, nectar (Wackers *et al.*, 2001) or other plant secretions, such as the lipid-rich secretions from acacia plants (Janzen, 1973), can influence tri-trophic interactions. Herbivory can affect foliar nectar production, both spatially and temporally, and such nectar induction influences predator and parasitoid behaviour (Wackers *et al.*, 2001). The use of nectaried cultivars of cotton has clearly been shown to increase predation/parasitism of herbivores in field studies (Schuster and Calderon, 1986). Therefore, genetically manipulating the quantity and quality of nectar produced by plants could be a powerful way to attract and retain parasitoids and predators and thus to enhance biological control of herbivores.

The effects of constitutive plant chemicals on parasitoids and predators are well known, and there are several classic studies that clearly demonstrate their effectiveness. If a plant is bred to directly defend itself against a herbivore, using toxins or digestibility reducers, this can affect the natural enemy in a number of ways, which can be hard to predict. If these plants cause an increase in the development time of the herbivorous host, this would provide a larger 'window of opportunity' for the parasitoid to attack its host. However, parasitoids may not always benefit from an increased host development time, because the leaves that reduce herbivory may not be conducive to parasitoid behaviour (Damman, 1987; Leather and Walsh, 1993).

Plants which constitutively produce chemicals that attract parasitoids may stimulate increased parasitism. Van Emden demonstrated that susceptible cabbage cultivars were attacked less by aphids in the presence of the parasitoid *Diaeretiella rapae*, due to this cultivar producing more of the allyl isothiocyanate, which attracted the parasitoid (van Emden, 1986). One of the difficulties of breeding plants to produce chemicals of benefit to parasitoids is the complexity of breeding for such polygenic traits or for traits that are hard to assess during breeding programmes. Information gained from genomic studies, coupled with recombinant DNA technology, will give us more control over how we can manipulate plants and initiate the possibilities of engineering novel pathways into plants. Plants have already been engineered to produce compounds derived from pathways not previously found in that plant. A good example of this is the production of *Arabidopsis* plants that make a cyanogenic glucosinolate because they contain genes from sorghum (Tattershall *et al.*, 2001). These plants have been shown to affect the flea beetle, *Phyllotreta nemorum*, and we are currently assessing the tri-trophic interactions on this plant, using the diamondback moth, *Plutella xylostella*, and its parasitoid, *Cotesia plutellae* (Barker *et al.*, 2001).

The past decade has seen a number of investigations of the role of herbivore-induced chemical signals in tri-trophic interactions. Unfortunately, most of these have been principally laboratory-based studies that have used model crop plants, although this has increased our proximate understanding of herbivore-induced plant signals (Poppy, 1999; Pickett and Poppy, 2001). In order to best exploit these plant signals in biocontrol, we need to increase our adaptive understanding of the ecology and evolution of these signals, and there needs to be more field studies conducted at the appropriate spatial–temporal scale, as recently conducted in tomato fields (for a review see Thaler, 1999) and maize plants (Bernasconi Ockroy *et al.*, 2001).

In maize and cotton, the volatiles induced by lepidopteran larvae have been compared between wild relatives and modern elite cultivars. In cotton, seven times more volatiles were induced by herbivory of a 'wild' naturalized cotton compared with modern varieties (Loughrin *et al.*, 1995). Chemical analysis of *Spodoptera littoralis*-induced maize plants and behavioural bioassays with its parasitoid, *Cotesia marginiventris*, have shown that there are quantitative and qualitative differences in the volatile profiles of eight different Mexican varieties of maize, and that this variation may be useful in biological control programmes (Fritzsche-Hoballah *et al.*, 2002). Further research is required to quantify this variation under field conditions and to determine whether these traits can be bred into agronomically important elite varieties via traditional means, or by genetically engineering into the plants.

The use of *Arabidopsis* as a model plant is becoming increasingly common, due to the genetic tractability of this plant, but it has not been exploited in many insect–plant studies (Poppy, 1999). However, recent genomic-type studies have demonstrated that it is a useful tool for studying tri-trophic interactions in crucifers (van Poecke *et al.*, 2001; van Poecke and Dicke, 2002), and recent work in our laboratory has shown that it is a useful model for looking at tri-trophic interactions involving the aphid *M. persicae* and its parasitoid, *D. rapae* (R. Girling, unpublished results), and *P. xylostella* and its parasitoid, *C. plutellae* (Barker *et al.*, 2001). Although of limited use in ecological studies, *Arabidopsis* may allow us to establish genetically the function of traits influencing tri-trophic interactions. The challenge then is to translate the information from *Arabidopsis* to other economically or ecologically more important crucifers.

Conclusions

There is no doubt that we now have the technology and knowledge to genetically manipulate parasitoids and/or plants to improve biological control. However, our knowledge of the proximate mechanisms far exceeds our understanding of the adaptive mechanisms of tri-trophic interactions. The use of recombinant DNA technology allows us to be more precise in our manipulation and to alter traits previously not possible by conventional breeding. Currently, we are assessing the impacts of GM crops on beneficial organisms, such as parasitoids and predators, as part of the risk assessment of GM crops (Schuler *et al.*, 1999b, 2001; Poppy, 2000; Couty *et al.*, 2001). Although this reactive approach is sensible for the current generation of plants expressing proteins such as Bt and protease inhibitors (for a review, see Schuler *et al.*, 1998), we are overlooking the opportunity of engineering plants that act synergistically with parasitoids. How far we can exploit GM technology to benefit parasitoids, and become an integral part of IPM, remains unclear, but there is scientific and political pressure for such developments (Poppy and Sutherland, 2003). However, we will need to have a more complete understanding of the complexities expressed at all trophic levels if we are to really exploit the opportunities while minimizing the risks.

If we are to genetically manipulate parasitoids, another critical issue to consider is the possible effects on genetic and organismic biodiversity. By manipulating the genetics of parasitoids, either directly or via the plant, the dynamics of tri-trophic interactions will be affected, and we need to understand how this will affect the stability of interactions and at which spatial scale such effects may operate. For example, if we genetically alter a generalist parasitoid to increase its impact on a particular herbivorous host–plant combination, do we alter the ability of this parasitoid to be a generalist and attack other herbivorous host–plant combinations? Modelling can help us to address some of these questions, but there is a lack of empirical data for developing and validating appropriate models, and few people are collecting such data. We need to ensure that laboratory and field studies become more integrated and that proximate and adaptive mechanims are explored together. Only then will we be able to genetically manipulate parasitoids with the confidence that we are maximizing the benefits and minimizing the risks.

References

Agrawal, A.A. and Karban, R. (1997) Domatia mediate plant–arthropod mutualism. *Nature* 387, 562–563.

Atkinson, P.W., Pinkerton, A.C. and O'Brochta, D.A. (2001) Genetic transformation systems in insects. *Annual Review of Entomology* 46, 317–346.

Barker, J., Poppy, G. and Payne, C. (2001) Development of *Plutella xylostella* on *Arabidopsis thaliana*. *Proceedings of the 4th international workshop on the management of diamondback moth and other crucifer pests*. Melbourne, Australia.

Battaglia, D., Pennacchio, F., Romano, A. and Tranfaglia, A. (1995) The role of physical cues in the regulation of host recognition and acceptance behavior of *Aphidius ervi* Haliday (Hymenoptera: Braconidae). *Journal of Insect Behavior* 8, 739–750.

Battaglia, D., Poppy, G., Powell, W., Romano, A., Tranfaglia, A. and Pennacchio, F. (2000) Physical and chemical cues influencing the oviposition behaviour of *Aphidius ervi* (Hymenoptera: Braconidae). *Entomologia Experimentalis et Applicata* 94, 219–227.

Bernasconi Ockroy, M.L., Turlings, T.C.J., Edwards, P.J., Fritzche-Hoballah, M.E., Ambrosetti, L., Bassetti, P. and Dorn, S. (2001) Responses of natural populations of predators and parasitoids to artificially induced volatile emissions in maize plants (*Zea mays* L.). *Agricultural and Forest Entomology* 3, 1–10.

Chadwick, D. and Goode, J. (1999) *Insect–Plant Interactions and Induced Plant Defence*. Novartis Foundation Symposium No. 223. John Wiley & Sons, Chichester.

Couty, A., de la Viña, G.L., Clark, S.J., Kaiser, L., Pham-Delègue, M.-H. and Poppy, G.M. (2001) Direct and indirect sublethal effects of *Galanthus nivalis* (GNA) on the development of a potato-aphid parasitoid *Aphelinus abdominalis*. *Journal of Insect Physiology* 47, 553–561.

Damman, H. (1987) Leaf quality and enemy avoidance by the larvae of a pyralid moth. *Ecology* 68, 88–97.

DeBach, P. (1974) *Biological Control by Natural Enemies*. Cambridge University Press, London.

DeBach, P. and Rosen, D. (1991) *Biological Control by Natural Enemies*, 2nd edn. Cambridge University Press, Cambridge.

Fritzsche-Hoballah, M.E., Tamo, C. and Turlings, T.C.J. (2002) Differential attractiveness of induced odors emitted by eight maize varieties for the parasitoid *Cotesia marginiventris*: is quality or quantity important? *Journal of Chemical Ecology* 28, 951–968.

Glinwood, R.T., Du, Y.-J., Smiley, D.W.M. and Powell, W. (1999a) Comparative responses of parasitoids to synthetic and plant-extracted nepetalactone component of aphid sex pheromones. *Journal of Chemical Ecology* 25, 1481–1488.

Glinwood, R.T., Du, Y.-J. and Powell, W. (1999b) Responses to aphid sex pheromones by the pea aphid parasitoids *Aphidius ervi* and *Aphidius eadyi*. *Entomologia Experimentalis et Applicata* 92, 227–232.

Gu, H. and Dorn, S. (2000) Genetic variation in behavioral response to herbivore-infested plants in the parasitic wasp, *Cotesia glomerata* (L.) (Hymenoptera: Braconidae). *Journal of Insect Behavior* 13, 141–156.

Guerrieri, E., Pennacchio, F. and Tremblay, E. (1997) Effect of adult experience on in-flight orientation to plant and plant–host complex volatiles in *Aphidius ervi* Haliday (Hymenoptera: Braconidae). *Biological Control* 10, 159–165.

Halford, N.G. (2000) Genetically modified plants. *Biological Sciences Review* 12, 2–7.

Halford, N.G. and Shewry, P.R. (2000) Genetically modified crops: methodology, benefits, regulation amd public concerns. *British Medical Bulletin* 56, 62–73.

Hopper, K.R., Roush, R.T. and Powell, W. (1993) Management of genetics of biological control introductions. *Annual Review of Entomology* 38, 27–51.

Hoy, M.A. (1976) Genetic improvement of insects: fact or fantasy. *Environmental Entomology* 5, 833–839.

Hoy, M.A. (1985) Improving establishment of arthropod natural enemies. In: Hoy, M.A. and Herzog, D.C. (eds) *Biological Control in Agricultural IPM Systems*. Academic Press, New York, pp. 151–166.

Hoy, M.A. (1992) Biological control of arthropods: genetic engineering and environmental risk. *Biological Control* 2, 166–170.

Hulspas-Jordan, P.M. and van Lenteren, J.C. (1978) The relationship between host-plant leaf structure and parasitization efficiency of the parasitic wasp *Encarsia formosa* Gahan (Hymenoptera: Aphelinidae). *Mededelingent Faculteit Landbouwwetenfchappen Rijksuniversiteit (Gent)* 43, 431–440.

Janzen, D.H. (1973) Interaction of the bull's horn acacia (*Acacia cornigera* L.) with an ant inhabitant (*Pseudomyrmex ferruginea* F. Smith) in eastern Mexico. *University of Kansas Science Bulletin* 47, 315–358.

Leather, S.R. and Walsh, P.J. (1993) Sub-lethal plant defences: the paradox remains. *Oecologia* 93, 153–155.

Loughrin, J.H., Manukian, A., Heath, R.R. and Tumlinson, J.H. (1995) Volatiles emitted by different cotton varieties damaged by feeding beet armyworm larvae. *Journal of Chemical Ecology* 21, 1217–1227.

Michaud, J.P. and Mackauer, M. (1994) The use of visual cues in host evaluation by aphidiid wasps. I. Comparison between three *Aphidius* parasitoids of the pea aphid. *Entomologia Experimentalis et Applicata* 70, 273–283.

Obrycki, J.J. (1986) The influence of foliar pubescence on entomophagous species. In: Boethal, D.J. and Eikenbary, R.D. (eds) *Interactions of Plant Resistance and Parasitoids and Predators of Insects.* Ellis Harwood, Chichester, UK, pp. 61–83.

Pennacchio, F., Digilio, M.C., Tremblay, E. and Tranfaglia, A. (1994) Host recognition and acceptance behaviour in two aphid parasitoid species: *Aphidius ervi* and *Aphidius microlophii* (Hymenoptera: Braconidae). *Bulletin of Entomological Research* 84, 57–64.

Perez-Maluf, R., Kaiser, L., Wajnberg, E., Carton, Y. and Pham-Delégue, M.-H. (1998) Genetic variability of conditioned probing to a fruit odor in *Leptopilina boulardi* (Hymenoptera: Eucoilidae), a *Drosophila* parasitoid. *Behavioral Genetics* 28, 67–73.

Pickett, J.A. and Poppy, G.M. (2001) Switching on plant genes by external chemical signals. *Trends in Plant Science* 6, 137–139.

Poppy, G.M. (1997) Tritrophic interactions: improving ecological understanding and biological control? *Endeavour* 21, 61–65.

Poppy, G.M. (1999) A clearer understanding of the raison d'être of secondary plant substances. *Trends in Plant Science* 4, 82–83.

Poppy, G.M. (2000) GM crops: environmental risks and non-target effects. *Trends in Plant Science* 5, 4–6.

Poppy, G.M. and Sutherland, J.P. (2003) Can biological control benefit from GM? Tritrophic interactions on insect-resistant transgenic plants? In: *Insects and Disease, Proceedings of the Royal Entomological Society's 21st Symposium.* CAB International, Wallingford, UK.

Poppy, G.M., Powell, W. and Pennacchio, F. (1997) Aphid parasitoid responses to semiochemicals – genetic, conditioned or learnt? *Entomophaga* 42, 193–199.

Potting, R.P.J., Vet, L.E.M. and Overholt, W.A. (1997) Geographic variation in host selection behaviour and reproductive success in the stemborer parasitoid *Cotesia flavipes* (Hymenoptera: Braconidae). *Bulletin of Entomological Research* 87, 515–524.

Powell, W. and Wright, A.F. (1988) The abilities of the aphid parasitoids *Aphidius ervi* Haliday and *A. rhopalosiphi* De Stefani Perez (Hymenoptera: Braconidae) to transfer between different known host species and the implications for the use of alternative hosts in pest control strategies. *Bulletin of Entomological Research* 78, 683–693.

Powell, W. and Wright, A.F. (1992) The influence of host food plants on host recognition by four aphidiine parasitoids (Hymenoptera: Braconidae). *Bulletin of Entomological Research* 81, 449–453.

Powell, W., Pennacchio, F., Poppy, G.M. and Tremblay, E. (1998) Strategies involved in the location of hosts by the parasitoid *Aphidius ervi* Haliday (Hymenoptera: Braconidae: Aphidiinae). *Biological Control* 11, 104–112.

Prévost, G. and Lewis, W.J. (1990) Heritability differences in the response of the braconid wasp *Microplitis croceipes* to volatile allelochemicals. *Journal of Insect Behavior* 3, 277–287.

Price, P.W., Bouton, C.E., Gross, P., McPheron, B.A., Thompson, J.N. and Weis, A.E. (1980) Interactions among three trophic levels: influence of plant on interactions between insect herbivores and natural enemies. *Annual Review of Ecology and Systematics* 11, 41–65.

Reeves, T.G. and Cassaday, K. (2002) History and past achievements of plant breeding. *Australian Journal of Agricultural Research* 53, 851–863.

Rehman, A. (1999) The host relationship of aphid parasitoids of the genus *Praon* (Hymenoptera: Aphidiinae) in agroecosystems. PhD thesis, University of Reading, UK.

Roush, R.T. (1990) Genetic variation in natural enemies: critical issues for colonization in biological control. In: Mackauer, M., Ehler, L.E. and Roland, J. (eds) *Critical Issues in Biological Control.* Intercept Ltd, Andover, UK, pp. 263–288.

Schuler, T.H., Poppy, G.M., Kerry, B.R. and Denholm, I. (1998) Insect resistant transgenic plants. *Trends in Biotechnology* 16, 168–175.

Schuler, T.H., Poppy, G.M., Kerry, B.R. and Denholm, I. (1999a) Potential side effects of insect-resistant transgenic plants on arthropod natural enemies. *Trends in Biotechnology* 17, 210–216.

Schuler, T.H., Potting, R.P.J., Denholm, I. and Poppy, G.M. (1999b) Parasitoid behaviour and Bt plants. *Nature* 400, 825–826.

Schuler, T.H., Denholm, I., Jouanin, L., Clark, S.J., Clark, A. and Poppy, G.M. (2001) Population-scale laboratory studies of the effect of transgenic plants on nontarget insects. *Molecular Ecology* 10, 1845–1853.

Schuster, M.F. and Calderon, M. (1986) Interactions of host plant resistant genotypes and beneficial insects in cotton ecosystems. In: Boethel, D.J. and Eikenbary, R.D. (eds) *Interactions of Plant Resistance and Parasitoids and Predators of Insects.* Ellis Harwood, Chichester, UK, pp. 84–97.

Sharma, H.C., Crouch, J.H., Shrama, K.K., Seetharama, N. and Hash, C.T. (2002) Applications of biotechnology for crop improvement: prospects and constraints. *Plant Science* 163, 381–395.

Storeck, A., Poppy, G.M., van Emden, H.F. and Powell, W. (2000) The role of plant chemical cues in determining host preference in the generalist aphid parasitoid *Aphidius colemani. Entomologia Experimentalis et Applicata* 97, 41–46.

Stowe, M.K., Turlings, T.C.J., Loughrin, J.H., Lewis, W.J. and Tumlinson, J.H. (1995) The chemistry of eavesdropping, alarm, and deceit. *Proceedings of the National Academy of Sciences USA* 92, 23–28.

Tattershall, D.B., Bak, S., Jones, P.R., Olsen, C.E., Nielsen, J.K., Hansen, M.L., Høj, P.B. and Møller, B.L. (2001) Resistance to an herbivore through engineered cyanogenic glucoside synthesis. *Science* 293, 1826–1828.

Thaler, J.S. (1999) Jasmonic acid mediated interactions between plants, herbivores, parasitoids, and pathogens: a review of field experiments in tomato. In: Agrawal, A.A., Tuzun, S. and Bent, E. (eds) *Induced Plant Defenses Against Pathogens and Herbivores. Biochemistry, Ecology, and Agriculture.* APS Press, St Paul, Minnesota, pp. 319–334.

Tumlinson, J.H., Turlings, T.C.J. and Lewis, W.J. (1992) The semiochemical complexes that mediate insect parasitoid foraging. *Agricultural Zoology Reviews* 5, 221–252.

Turlings, T.C.J., Wäckers, F., Vet, L.E.M., Lewis, W.J. and Tumlinson, J.H. (1993) Learning of host-finding cues by hymenopterous parasitoids. In: Papaj, D.R. and Lewis, A. (eds) *Insect Learning: Ecological and Evolutionary Perspectives.* Chapman & Hall, New York, pp. 51–78.

Turlings, T.C.J., Gouinguené, S., Degen, T. and Fritzsche-Hoballah, M.E. (2002). The chemical ecology of plant–caterpillar–parasitoid interactions. In: Tscharntke, T. and Hawkins, B. (eds) *Multitrophic Level Interactions.* Cambridge University Press, Cambridge, pp. 148–173.

van Driesche, R.G. and Bellows, T.S. (1996) *Biological Control.* Chapman & Hall, New York.

van Emden, H.F. (1986) The interaction of plant resistance and natural enemies: effects on populations of sucking insects. In: Boethel, D.J. and Eikenbary, R.D. (eds) *Interactions of Plant Resistance and Parasitoids and Predators of Insects.* Ellis Harwood, Chichester, UK, pp. 138–150.

van Emden, H.F., Sponagl, B., Wagner, E., Baker, T., Ganguly, S. and Douloumpaka, S. (1996) Hopkins 'host selection principle', another nail in its coffin. *Physiological Entomology* 21, 325–328.

van Lenteren, J.C. and de Ponti, O.M.B. (1990) Plant-leaf morphology, host-plant resistance and biological control. *Symposia Biologica Hungarica* 39, 365–386.

van Poecke, R.M.P. and Dicke, M. (2002) Induced parasitoid attraction by *Arabidopsis thaliana*: involvement of the octadeconoid and salicylic acid pathway. *Journal of Experimental Botany* 375, 173–179.

van Poecke, R.M.P., Posthumus, M.A. and Dicke, M. (2001) Herbivore induced volatile production by *Arabidopsis thaliana* leads to attraction of the parasitoid *Cotesia rubecula*: chemical, behavioural and gene expression analysis. *Journal of Chemical Ecology* 27, 1911–1928.

Vet, L.E.M. and Dicke, M. (1992) Ecology of infochemical use by natural enemies in a tritrophic context. *Annual Review of Entomology* 37, 141–172.

Vinson, S.B. (1976) Host selection by insect parasitoids. *Annual Review of Entomology* 21, 109–133.

Wackers, F.L., Zuber, D., Wunderlin, R. and Keller, F. (2001) The effect of herbivory on temporal and spatial dynamics of foliar nectar production in cotton and castor. *Annals of Botany* 87, 365–370.

12

Sex-ratio Distorters and Other Selfish Genetic Elements: Implications for Biological Control

R. Stouthamer

Department of Entomology, University of California, Riverside, CA 92521, USA

Introduction

Selfish genetic elements are a class of heritable elements that manipulate their host's reproduction to enhance their own transmission. This manipulation can be either neutral or detrimental to the host (Werren *et al.*, 1988). Examples of such elements are transposable elements, heritable microorganisms and microsporidia. The most obvious selfish genetic elements are sex-ratio-distorting factors. These factors are preferentially transmitted through only one sex and they manipulate their host to produce either more or fitter individuals of this sex. The most common factors are female-biasing factors, such as *Wolbachia* bacteria, which are inherited from mother to the offspring exclusively through egg cytoplasm (Werren, 1997). The complete parthenogenetic (thelytokous, unisexual, uniparental) reproduction, well known from parasitoid wasps used in biological control, is often induced by bacterial infections (Stouthamer, 1997; Zchori-Fein *et al.*, 2001). Over the past 10 years it has become clear that factors such as heritable bacterial infections are extremely common in insects, and may have profound effects on their host's reproduction (O'Neill *et al.*, 1997). It has been estimated that *Wolbachia* bacteria, which cause a series of different reproductive anomalies, may occur in as many as three-quarters of all insects (Jeyaprakash and Hoy, 2000). Other bacteria that manipulate the reproduction of insects are also known (Hurst and Jiggins, 2000; Zchori-Fein *et al.*, 2001). This leads to the conclusion that these reproductive parasites may be present in an extremely high proportion of all insects.

Some of the better-known sex-ratio distorters with potential application in biological control are heritable bacteria that induce thelytokous parthenogenesis (i.e. virgin females produce daughters). The advantage of parthenogenesis for biological control was already realized by Clausen in 1924 (Timberlake and Clausen, 1924), although the involvement of heritable bacteria in thelytoky was not known at that time:

> It is evident, therefore, that frequent or general reproduction in this way among those species which are thelyotokous would result in a much greater rate of increase than if portion of the progeny were males. This may be illustrated by calculating the possible rate of increase of *Aschrysopophagus modestus* Timberlake for one year, allowing five generations of fifty each. One female reproducing parthenogenetically for this period of time would give rise to 312,500,000 individuals, providing there were no mortality. If, however, a female were fertilized and gave rise to progeny of both sexes in the ratio of one to three, the total number of females at the end of the year would be 74,2000,000. This of course is purely theoretical, yet it has a practical bearing upon the practices and possibilities of parasite introduction and colonization. In case of the introduction of a species which is thelyotokous into a new locality, it would be decidedly advantageous to eliminate the males, as illustrated above. No evidence has been produced to show that the absence of the male contributes in any way to a weakening of the race. In certain cases the advantage is offset by a reduction in the average number of progeny because of a lack of fertilization. This has been demonstrated in several cases but is not known to be a general condition. As a matter of fact, in those species which are able to produce females without fertilization the male is very seldom present under natural conditions.

This chapter will give an overview of reproductive parasites that may be important to biological control. Can we improve natural enemies by introducing these reproductive parasites, or would it be better to 'cure' infected natural enemies before they are introduced? Would it be a good idea to introduce these reproductive parasites in pest populations that we want to control? What should we know about potential effects of reproductive parasites when we introduce different collections of the same species for biocontrol?

Overview of Sex-ratio Distorters and Their Potential Application

Female-biasing sex-ratio distorters

Several heritable bacteria cause female-biased sex ratios; they accomplish this by killing the male offspring (male-killers) (Hurst, 1991), rendering genetic males into functional females (feminizing) (Rigaud, 1997) or by inducing parthenogenetic development (parthenogenesis induction) (Stouthamer, 1997). In addition, one unknown factor in the wasp *Nasonia vitripennis* causes female-biased sex ratios by somehow influencing the mothers' fertilization behaviour (Skinner, 1982). These wasps will fertilize practically all of their eggs and therefore produce almost exclusively female offspring when mated.

Male killing

The phenotype male killing appears to be adopted by many phylogenetically unrelated bacteria. Male-killers are known from the alpha proteobacteria (*Wolbachia, Rickettsia*) (Werren *et al.*, 1994; Hurst *et al.*, 1999), gamma proteobacteria (*Arsenophonus*) (Werren *et al.*, 1986), mollicutes (*Spiroplasma*) (Hackett *et al.*, 1986) and the Cytophaga–Flavobacterium–Bacteroides group (Hurst *et al.*, 1997a). In all cases, males die in their early development. The male-killing behaviour might benefit the bacterium because: (i) it reduces the competition between females that can transmit the infection and males that do not pass it on; (ii) it enhances the survival of females, by allowing them to feed on their dead brothers upon emergence; and (iii) it could lead to an avoidance of inbreeding depression, by making infected females mate with unrelated males (Hurst *et al.*, 1997b). In many ladybirds, male-killing bacteria of various kind are found. For instance, in the Moscow population of *Adalia bipunctata*, at least four different male-killing bacteria coexist (von der Schulenburg *et al.*, 2002). One of the obvious risks of a spreading male-killer infection is that there will be mostly infected females and very few males to fertilize the females. In most ladybird populations known so far, the infection frequency has not reached dramatic frequencies, and sufficient males appear to be present to assure the mating of most females. However, in some populations of the butterfly *Acraea encedana* the frequency of *Wolbachia* infection has reached very high proportions, and many females in the population remain virgin because of the lack of males (Jiggins *et al.*, 2000, 2001). In these populations, interesting changes have evolved in the mating behaviour of males and females. In heavily infected populations, females form leks where males choose females to mate with (Jiggins *et al.*, 2000). These high infection frequencies may be maintained if males in these biased populations mate more frequently with uninfected females than with infected females (Randerson *et al.*, 2000). The fitness of the male is served much better if it mates with uninfected females, because they are able to produce male offspring that again will have plenty of mating possibilities. If the infection frequency reaches extreme frequencies, the growth rate of the population is much reduced compared with the uninfected population, and populations may run the risk of extinction. Very effective male-killers may have led to the extinction of their host populations. Such effective male-killers may exist or have existed, but they will be difficult to find because their host population will go extinct after a short time.

From the standpoint of biological control, would it be a good idea to introduce male-killing infections into the pest population? First of all, it appears not yet possible to introduce these bacteria at will in species that do not carry them. In addition, even if introduced into a new species, the phenotype that the bacterium will induce in these new hosts is not always the same as it induced in its original host (Fujii *et al.*, 2001). Finally, we are not able to predict to what levels of infection these bacteria will spread in a new species; if the transmission of the bacterium from mother to her offspring is not perfect, we may increase the population growth rate of the pest population. Furthermore, we also expect high infection frequencies to put very strong selective pressures on the rest of the host's genome to evolve traits that will overcome the effects of the bacteria. Hurst and Jiggins (2000) suggest that introducing male-killers in conjunction with sterile male releases could improve control of pest populations.

Can these bacteria be used for the improvement of natural enemies? There does not seem to be an easily envisioned scenario of how male-killing infections can improve the efficiency of natural enemies. If anything, they may impair the effectiveness of natural enemies. The high-prevalence male-killers in several insect groups, among others ladybirds, should be kept in mind by biocontrol practitioners when they import these insects for releases. Curing them of their infection before release may improve their biological control efficiency.

Feminization

Several heritable microorganisms are known to cause feminization, in which infected genetic male individuals develop into functional females (Rigaud, 1997). The advantage of this trans-sexuality for the bacterium is clear, only individuals that lay eggs transmit the bacterium. Similar to a spread of the male-killing bacteria, the feminizing bacteria cause populations to become extremely female biased. Very successful cases of feminization must have led to the extinction of their host populations, because of the lack of males. In the extant populations carrying feminizing bacteria, some factor must keep the bacteria from going to fixation. These suppressing factors can act through a lowering of the transmission efficiency of the bacterium, through killing the bacterium, or through suppressing the phenotype the bacterium induces in its host.

Wolbachia causing feminization have been found in isopods (Rigaud, 1997) and in Lepidoptera (Fujii *et al.*, 2001; Kageyama *et al.*, 2002). The mechanisms of this manipulation are, in general, not understood. The best-studied cases of feminization are in the isopods, where the *Wolbachia* infects the androgenic gland. This gland produces a hormone that allows embryos to develop into males – with this gland not functioning, the default developmental pathway is initiated, which leads to the development of females (Rigaud, 1997). Compared with the method of male killing, this sex-ratio distorter can, in principle, spread much more rapidly, because a larger number of individuals will become infected. Recently a bacterium of the Cytophaga–Flavobacterium–Bacteroides group has been discovered in mites, which causes feminization in a phytophagous mite species (Weeks and Breeuwer, 2001). So far no cases of feminization have been discovered in insects used for biological control. The use of this phenotype for controlling insect pest populations is inadvisable, for the same reasons as applying a male-killer for this purpose (see above). However, it may be useful for the mass-rearing of insects used in inundative biological control.

Thelytokous parthenogenesis induction

Thelytokous parthenogenesis, in which unfertilized eggs develop into females, is well known amongst practitioners of biological control. As described in the excerpt from Clausen's paper in the introduction, the advantages of thelytoky appear obvious. In many cases parthenogenesis in Hymenoptera is caused by bacterial infection. The best-known cause of parthenogenesis is *Wolbachia* (Stouthamer, 1997), but recently Zchori-Fein *et al.* (2001) discovered a new group of Cytophaga–Flavobacterium–Bacteroides (CFB) bacteria that also induces thelytokous parthenogenesis in some wasps. The cytology of the parthenogenesis induction is known from a number of *Wolbachia*-infected wasp species. In general, the bacteria cause a modification of the normal mitotic divisions, leading to the fusion of two identical sets of chromosomes (Stille and Davring, 1980; Stouthamer and Kazmer, 1994; Gottlieb *et al.*, 2002). The resulting individuals are completely homozygous. The induction of parthenogenesis by *Wolbachia* is not limited to this mechanism of gamete duplication – in the mite species *Bryobia* a different mechanism allows unfertilized eggs to become diploid (Weeks and Breeuwer, 2001). No information is as yet available on the cytology induced by the CFB bacterium. Interestingly, a CFB bacterium is also known to cause feminization in a haploid mite species, where haploid individuals grow out to be females. In this case, feminization and parthenogenesis induction have become the same (Weeks *et al.*, 2001).

Populations of wasps infected with thelytokous parthenogenesis-inducing bacteria consist of either only infected females or a mixture of infected and uninfected individuals. The former case appears to be the most common, but it has been argued that this is a sampling artefact (Stouthamer, 1997). In many cases, populations are declared to be sexual only because males are present. To determine if populations consist of both infected and uninfected females, isofemale lines need to be initiated with field-collected females and their mode of reproduction needs to be established. This was done in collections of *Trichogramma* species, and there a substantial number of populations appear to consist of both infected and uninfected individuals (Pinto, 1998). In addition, the biocontrol literature mentions some cases where apparently sexual populations 'turn' into thelytokous populations, either during mass-rearing or following release (Day and Hedlund, 1988). These latter cases are most probably also mixtures of arrhenotokous and thelytokous individuals, where, during the conditions subsequent to the release or during the mass-rearing, the thelytokous population outcompetes the arrhenotokous population. The selective advantages of one mode of reproduction can be quite substantial under the artificial conditions of laboratory rearing.

One example illustrating the confusion that originated from mixtures of thelytokous and arrhenotokous individuals in collections of parasitoid wasps used for biocontrol is the case of *Encarsia perniciosi.* These wasps somehow arrived in North America long after their host, San José scale (*Quadrapidiotus perniciosum*), had become a pest. Flanders (1944) imported material from the east coast of the USA to California, and found that males were present in the field-collected material. During subsequent laboratory rearing the culture became thelytokous. Flanders blamed this either on the plant material on which the host was reared or on the constant temperatures under which the cultures were kept in the laboratory (Flanders, 1945). Subsequently, when these wasps were imported from various places in the USA to Europe several authors found that they would get all-female cultures, which they interpreted as being an indication of them exhibiting thelytokous parthenogenesis (Chumakova and Goryunova, 1963; Babushkina, 1978). However, after the first all-female generation the wasps reproduced poorly, and the cultures eventually died out. Subsequently, the parthenogenetic form of *E. perniciosi* was thought to be an inferior parasitoid because their cultures did so poorly in the laboratory (Neuffer, 1964). Much effort was then put into collecting the sexual form and keeping it sexual during laboratory rearing. To avoid the cultures from becoming parthenogenetic, they were reared at fluctuating temperatures (Neuffer, 1964), and indeed sexual populations could be maintained, and upon release they successfully controlled the San José scale in Germany. Later, Stouthamer and Luck (1991), in an effort to duplicate the transition from sexual to thelytokous reproduction in this species, failed to find any effect of constant temperatures, and they hypothesized that, while in the case of Flanders' original observations the culture must indeed have consisted of a mixture of sexual and thelytokous individuals, in the later importations to Europe the all-female generation that was observed was caused by the asynchrony of production of male and female offspring in sexual *E. perniciosi*. In this species the daughters are produced as primary parasitoids on scale insects, while sons can only be produced as hyperparasitoids on developing female larvae of their own species (heteronomous hyperparasitism) (Walter, 1983). Consequently, when mated females are exposed to a batch of scale insects, they can initially produce only female offspring; male offspring can only be produced once the female larvae have developed to the stage where they are suitable for parasitization. This can be several weeks later. This example illustrates how thelytoky can occur in mixed populations, and that the expectation of thelytoky can sometimes lead to much confusion and failures in establishment of laboratory cultures.

There has been some debate about the use of thelytokous forms in classical biological control and, particularly, what form to use when both arrhenotokous and thelytokous forms exists of a particular species. Aeschlimann (1990) suggested releasing the thelytokous form first, and subsequently a sexual form. He argued that the thelytokous form would be better able to establish itself and reduce the pest density; subsequently, the release of the sexual form would increase the genetic variation within the population and would result in better-adapted forms. Stouthamer (1993) argued that the reverse order would be better in classical biological control. The arrhenotokous form would be able to adapt better to the circumstances in the release area, and subsequently the release of the thelytokous form should result in mating between thelytokous females and arrhenotokous males. This should result in the presence of well-adapted thelytokous forms in the field, having the advantage of both thelytokous reproduction and adaptation to the local circumstances. So far, this argument has been purely theoretical in the case of classical biological control, and it is questionable whether we could ever test these hypotheses, because mating between arrhenotokous males and thelytokous females is possible in only a minority of the PI-*Wolbachia* infections (Stouthamer, 1997).

For inundative biological control, thelytoky has the advantage that all the wasps produced in mass-rearing will be females. Because only female parasitoids are effective biocontrol agents, the cost of producing a female should be lower for thelytokous than for arrhenotokous females. Only a single test was done in greenhouses to determine how arrhenotokous and thelytokous forms of the same species compared. These experiments showed that the thelytokous and arrhenotokous forms of *Trichogramma deion* and *Trichogramma cordubensis* were equally capable of finding host patches in the experimental set-up (Silva *et al.*, 2000). However, the arrhenotokous forms parasitized significantly more host eggs within a patch. If the cost of producing wasps was included in the calculations – in which the assumption was that per parasitized host from the mass-rearing a single wasp would emerge – the cost of using thelytokous wasps was substantially lower than that of arrhenotokous wasps. The authors concluded that the use of thelytokous forms would be more economical. Since that time it has become clear that in some thelytokous forms of *Trichogramma* the mortality of unfertilized infected eggs can be substantial and that therefore the assumption of a single wasp emerging per host egg may not hold (Tagami *et al.*, 2001). Consequently, the experiments will need to be repeated to determine the average number of wasps emerging per host egg.

It may be possible to make use of thelytoky to select the best genetic lines for biocontrol in the field (Fig. 12.1). This selection requires that both thelytokous and arrhenotokous forms of a species exist and that thelytokous females are still capable of mating. Under these circumstances it is possible to create a large number of genetically distinct lines. This is done by mating an infected female with an unrelated male (P). The offspring this female produces from fertilized eggs will also be infected and will be heterozygous (F_1). If we allow the heterozygous-infected virgin females to produce offspring, their offspring (F_2) will each be a different combination of the genome of their grandparents' genome. In addition, each of these F_2 females will be 100% homozygous (Stouthamer and Kazmer, 1994) because of the gamete duplication induced by *Wolbachia*. If we initiate isofemale lines from these F_2 females and mass-rear these lines, the relative performance of these different genotypes can be tested by releasing several of the lines simultaneously in the field and by collecting hosts from the release area to determine which lines have successfully parasitized hosts. These tests can be done in detail if we have the genetic tools (f.i.microsatellite markers) to distinguish the different isofemale lines. A lower-tech version of the same experiment can be done by simply recollecting the wasps from the field and setting them up as isofemale lines again, and subsequently releasing them again in equal numbers. After a number of iterations of this process, those genotypes that are the best at finding hosts in the field will be selected for. These can then be masseared for application in biocontrol.

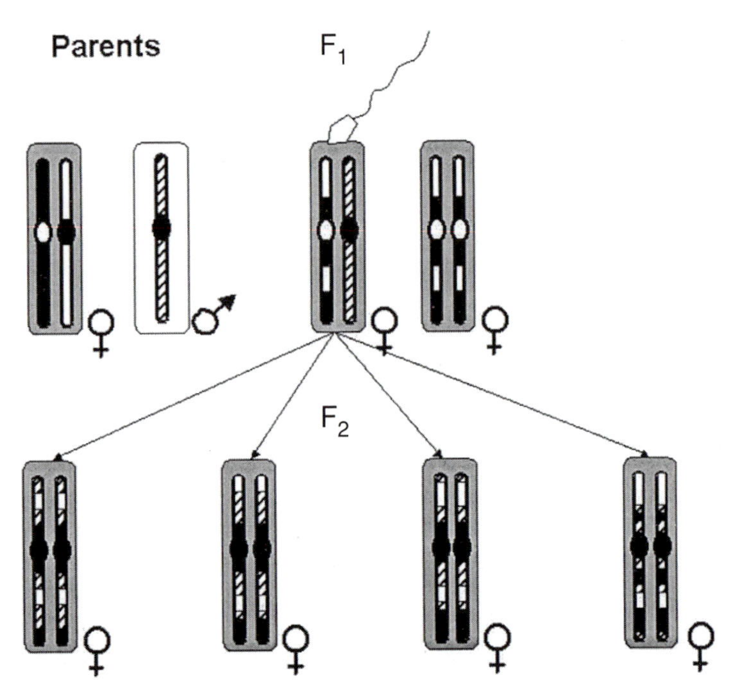

Fig. 12.1. Mating scheme to select the optimal genotype for field release. Infected mother (indicated by dark background) mates with male. Their F_1 offspring will consist of daughters: the fertilized eggs will contain a set of chromosomes from both parents; the unfertilized eggs, only the mother's chromosomes. These unfertilized eggs will become females through *Wolbachia*-induced gamete duplication. If a virgin F_1 heterozygous female is allowed to reproduce, she will produce F_2 offspring that are each a different combination of her two sets of chromosomes and are completely homozygous.

For those species where no thelytokous forms are available, it appears to be difficult to infect them with the parthenogenesis-inducing bacteria. Transmission of *Wolbachia* and other parthenogenesis bacteria from infected to uninfected wasps has been tried only on a limited scale. Grenier *et al.* (1998) showed that the species *Trichogramma dendrolimi* could be infected through microinjection of the *Wolbachia* from an infected *Trichogramma pretiosum*. The transmission and expression of thelytoky in the new host was low. So although it did result in a low level of expression of parthenogenesis, it illustrates that, most probably, there is a specificity of the relationship of the *Wolbachia* to its host. Huigens *et al.* (2000) showed that within the species *Trichogramma kaykai* it is possible to transmit the infection horizontally when both infected and uninfected larvae share the same host. Under these circumstances approximately 40% of the offspring of uninfected females became infected. Again, this horizontal transmission does not seem to be a general phenomenon, because when similar experiments were tried intraspecifically with other species, or interspecifically, no lasting transmission was recorded (de Almeida and Huigens, Wageningen, 2002, personal communication). These studies were all done with *Wolbachia* from *Trichogramma*, in which the infection is generally found in populations where both infected and uninfected individuals coexist. In species where the infection has gone to fixation, we expect the coevolution between the parthenogenesis-inducing bacterium and its host to be much more advanced, and therefore it may be much more difficult to transmit such bacteria successfully. Only a single case (van Meer and Stouthamer, 1999) has been published where transmission of a parthenogenesis inducing *Wolbachia* from the species *Muscidifurax uniraptor* to another insect, i.e the fly, *Drosophila simulans*, was attempted. This was done through microinjection of the *Wolbachia* into fly eggs, and resulted in the establishment of the infection; however, its transmission was very low and after seven generations the infection was lost. No obvious phenotype appeared to be induced by this infection. At present we have no evidence that routine transmissions of parthenogenesis-inducing bacteria will be feasible in the near future; however, with the rapid development of our knowledge of the *Wolbachia* genome this may change.

In summary, the advantage of thelytoky over arrhenotoky for biological control has not yet been tested stringently. Although a number of our most successful biological control agents are thelytokous (for instance the whitefly parasitoid *Encarsia formosa*), no general statements can be made regarding the advantage or disadvantage of thelytoky over arrhenotoky for biological control.

Male-biasing sex-ratio distorters

Male-biasing sex-ratio distorters cause a higher proportion of males in the offspring of females. Such factors are not suitable for improving the natural enemies used in biological control. However, knowledge of their presence is important to avoid importing such a factor in the wasps used for release in biocontrol projects, and these factors may be useful for introduction into pest populations to reduce their growth rate. Male-biasing factors are rarer than female-biasing factors. Two classes of male-biasing factors are recognized: sex chromosome meiotic drive elements and B-chromosomes. The meiotic drive elements cause an over-representation of chromosomes carrying the driving element, either in sperm or in the egg. They have only been found in different dipteran species; suspected cases of female meiotic drive outside Diptera appear to be cases of infection with male-killing *Wolbachia* (Jiggins *et al.*, 1999). Male meiotic drive has been found in several mosquito species (Wood and Newton, 1991). Consequently these factors appear to have little potential for application outside the suppression of dipteran species.

The B-chromosome male-biasing factor has been found in two species of parasitoid wasps: *N. vitripennis* (Werren, 1991) and *T. kaykai* (Stouthamer *et al.*, 2001). This factor is a small B-chromosome that is transmitted through sperm from males to their offspring. In eggs fertilized with sperm carrying the B-chromosome, the B-chromosome somehow causes the destruction of the paternal set of chromosomes, and the resulting egg grows out to become a haploid male, again carrying the B-chromosome (Reed, 1993). The effect of this is that the offspring of females that have mated with a male carrying this B-chromosome all become males; the fertilized eggs become males carrying the B-chromosome; and the unfertilized eggs become normal haploid males (Fig. 12.2). Consequently this factor has been named paternal sex ratio (PSR). Because of its mode of action, this chromosome has been called one of the most selfish genetic elements known so far (Werren *et al.*, 1988). The mode of action of PSR also allows this factor to cross species boundaries without too many problems. In interspecific fertilizations, the PSR often appears to work in a similar way in its new hosts as in its original host. Such crosses result in males carrying the PSR chromosome of the donor species and the A chromosome of the new host. Because these individuals have a complete set of chromosomes of the new host, they can develop normally into males that are again carriers of the PSR. Such interspecific transmission has been shown among the different species of *Nasonia* (Dobson and Tanouye,

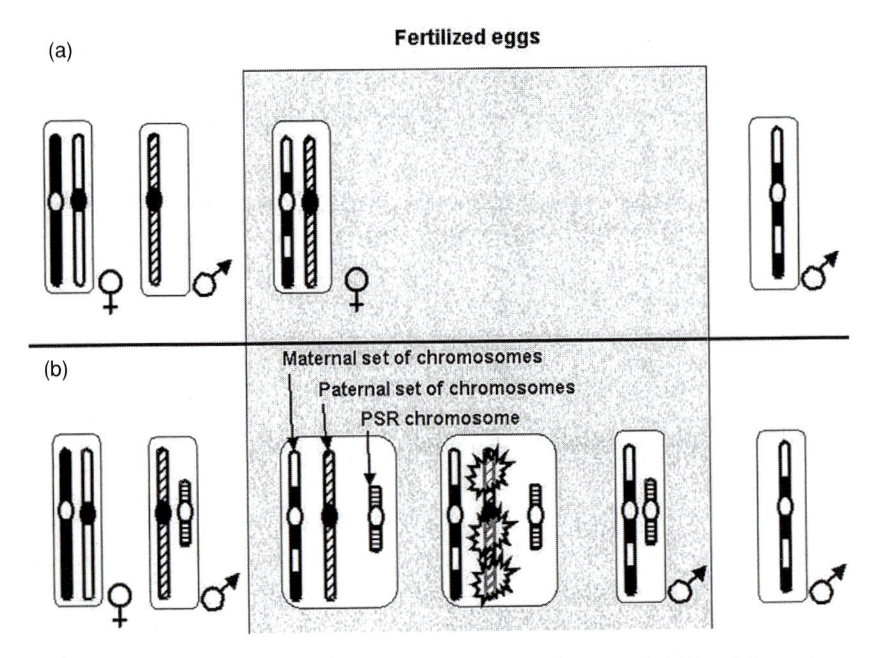

Fig. 12.2. Haplo-diploid sex determination. (a) Females are diploid, while males are haploid. Fertilized eggs grow out to be diploid females, while unfertilized eggs become haploid males. (b) When a female mates with a paternal sex ratio (PSR) chromosome-carrying male her fertilized eggs grow out to be PSR males, because the presence of the PSR chromosome in the sperm causes the destruction of the normal paternal set of chromosomes. The resulting PSR male carries the maternal set of chromosomes plus the PSR chromosome. Unfertilized eggs again give rise to haploid males.

1998; Beukeboom and Werren, 2000), as well as in *Trichogramma* (Stouthamer and van Vugt, unpublished). In addition to its ability to cross species boundaries relatively easily, the PSR may also drive itself through populations to reach equilibrium frequencies, depending on the mating structure of the population, the fertilization proportion of the eggs and the presence of female-biasing sex-ratio distorters (Werren and Beukeboom, 1993; Stouthamer *et al.*, 2001). For a long time PSR factors were thought to be rare and limited to *Nasonia*; however, with the discovery of the PSR in *T. kaykai*, it is likely that these factors may be much more common than previously thought.

Application of PSR factors in biocontrol

PSR factors can only be used to reduce the population growth rate of insects with a haplo-diploid sex-determination system. These insects include Hymenoptera, thrips, whiteflies and some mite species. One of the obvious applications of a PSR factor for insect control would be to introduce these factors into pest species with a haplo-diploid sex-determination system. In particular, exotic species may be targeted for such control. In systems where biological control by primary parasitoids is limited by the presence of hyperparasites, the introduction of a PSR chromosome into the hyperparasitoid population may reduce its growth rate.

Other Heritable Bacteria of Importance to Biological Control

Cytoplasmic incompatibility-inducing *Wolbachia*

Another very common heritable bacterium is the cytoplasmic incompatibility (CI)-inducing (CI) *Wolbachia* (Stouthamer *et al.*, 1999). This bacterium causes incompatibility in crosses between individuals of different infection status (Table 12.1). In its simplest form, the sperm from an infected male is incompatible with uninfected eggs and causes these eggs die. More complex situations can exist when two lines are infected with different *Wolbachia*; in these cases any cross between these lines leads to no offspring (Table 12.1c). Finally, multiple infections within individuals are also common; in these cases sperm from males that are double infected is incompatible with single-infected eggs. In haplo-diploid organisms the incompatibility can lead to the death of the egg or result in the development of a male (Vavre *et al.*, 1999a). Much of the early biocontrol work involved crossing studies between different collections of closely related natural enemies. These crosses were done to establish the species status of the different collections. In several cases compatibility patterns were found in these crossing studies that are consistent with CI-*Wolbachia* infection of conspecific lines. For instance, the work done on determining the status of different *Aphytis* collections resulted in a large number of one-way incompatible crosses that could readily be

Table 12.1. Different types of cytoplasmic incompatibility induced by *Wolbachia* in insects with a diplo-diploid or haplo-diploid sex-determination system.

Diplo-diploid	Parents		Offspring	
	Mother	Father	Daughters	Sons
(a) Unidirectional incompatibility (diplo-diploids): single *Wolbachia*-infected individual, I, and uninfected individual, U	I	I	I	I
	I	U	I	I
	U	U	U	U
	U	I	x	x
(b) Unidirectional incompatibility (diplo-diploids): single *Wolbachia*-infected individuals, I_1, and double-infected individuals, $I_{1,2}$	$I_{1,2}$	$I_{1,2}$	$I_{1,2}$	$I_{1,2}$
	$I_{1,2}$	I_1	$I_{1,2}$	$I_{1,2}$
	I_1	I_1	I_1	I_1
	I_1	$I_{1,2}$	x	x
(c) Bidirectional incompatibility (diplo-diploids): individuals infected with *Wolbachia* 1(I_1), and individuals infected with *Wolbachia* 2 (I_2)	I_1	I_1	I_1	I_1
	I_1	I_2	x	x
	I_2	I_2	I_2	I_2
	I_2	I_1	x	x

Haplo-diploid	Parents		Offspring	
	Mother	Father	Fertilized eggs	Unfertilized eggs
(d) Unidirectional incompatibility (haplo-diploids): female mortality cytoplasmic incompatibility	I	I	Daughter I	Son I
	I	U	Daughter I	Son I
	U	U	Daughter U	Son U
	U	I	x	Son U
(e) Unidirectional incompatibility (haplo-diploids): male development cytoplasmic incompatibility in haplo-diploids	I	I	Daughter I	Son I
	I	U	Daughter I	Son I
	U	U	Daughter U	Son U
	U	I	Son U	Son U

explained by different infection status of the collections (Rao and DeBach, 1969; Khasimuddin and DeBach, 1976). However, not all cases of one-way incompatibility or complete incompatibility between morphologically similar species are caused by infection with heritable bacteria (Pinto *et al.*, 1991; Stouthamer *et al.*, 1996, 2000).

The presence of CI-*Wolbachia* may also impact the rearing of natural enemies. Addition of certain antibiotics to the diet, exposure to elevated rearing temperatures (Johanowicz and Hoy, 1998; Opijnen and Breeuwer, 1999) or diapause (Perrot Minnot *et al.*, 1996) can result in partial curing of the insects. This again can cause incompatibility between the infected males and the cured females, with the resulting reduction in offspring production by these females, or the production of more males in the case of haplo-diploid insects. It is therefore important to check the imported natural enemies for the presence of these bacteria. Simple PCR-based protocols are available to determine if infection with *Wolbachia* or some of the other heritable bacteria is present (Stouthamer *et al.*, 2002).

When wasps from different collections are mixed, one should also be aware of the possible impact of infection with CI-*Wolbachia*. Finally, when wasps are released for augmentative biological control, the infection status of the resident population of the natural enemy should match that of the natural enemies that will be released. Field mixing of infected and uninfected populations can lead to a prolonged depression in the growth rate of the natural enemy, the length of which depends on the mating system, egg fertilization frequency and the relative size of the released population to the resident population (Mochiah *et al.*, 2002).

Detection of Sex-ratio Distorters and CI-*Wolbachia* in Natural Enemies

Several methods can be used to detect sex-ratio distorters and other heritable bacteria in natural enemies. The simplest method is to determine the sex ratio of the offspring of individual females, for male-killing and feminizing bacteria, and, for the presence of PSR-like chromosomes, we expect a polymorphism in the population and therefore differences between different families should be indicative of the presence of these factors. For infections with CI-causing *Wolbachia*, no variation in sex ratios are expected between different families. The second approach to determine whether heritable bacteria are present is to screen individuals for infection with different known sex-ratio-distorting bacteria, using PCR reactions. If only specific primers are used in this screening for finding these factors, there is a risk that new, unknown reproductive parasites are not recognized. However, in many cases this will be the easiest way to detect infections from groups that are known reproductive parasites. The third approach is to feed the natural enemies antibiotics and to determine whether the offspring sex ratios from infected and uninfected lines differ (potentially indicative of sex-ratio distorters), or if, in crosses between infected and uninfected individuals, patterns emerge that are consistent with cytoplasmic incompatibility. For a more detailed description of how to find sex-ratio-distorting factors in populations, see Stouthamer *et al.* (2002).

Discussion

Sex-ratio-distorting factors that induce such phenotypes as feminization and male-killing in insects released for biological control purposes can reduce their effectiveness. Similarly, the presence of CI-inducing bacteria may cause a reduction in the biological control potential of natural enemies when they are mixed, either in mass-rearing or when released into populations of the same species with a different infection status. In general, it would be a good idea for practitioners of biological control to be aware of the infection status of the natural enemies that are imported. While, in the past, detecting these bacteria may have involved elaborate crossing experiments, at present detecting certain bacteria is very easy, using PCR with specific primers (Stouthamer *et al.*, 2002). However, determining the exact phenotype induced by these bacteria will still require detailed experiments. Curing all imported natural enemies of their heritable bacteria as a standard procedure should not be done without at least checking for the phenotype they induce. In a number of cases the infection with heritable bacteria can enhance offspring production (Vavre *et al.*, 1999b), or the natural enemies may be unable to survive without them (Dedeine *et al.*, 2001).

Heritable bacteria inducing parthenogenesis in natural enemies seem very promising for biological control, because for their potential for increased population growth rate. However, in practical terms we are only able to manipulate these bacteria in very few species. In most species the infection appears to have gone to fixation. In species with fixed infections, arrhenotokous lines cannot be established, even though males can be induced through antibiotic treatment. Males and females of such species simply seem not capable of successful matings, and/or the females do not fertilize their eggs. In those species where both forms are present, the advantage of thelytokous parthenogenesis may not be as straightforward as Clausen (Timberlake and Clausen, 1924) assumed, because some cost may be involved in carrying the symbionts (Stouthamer and Luck, 1993; Tagami *et al.*, 2001).

Male-biasing factors seem unlikely as a source for improvement of natural enemies. They may have a role in suppressing pest populations that reproduce by haplo-diploidy. These male-biasing factors may be more common than thought so far, and for wasps used in biological control they should be avoided, because they can severely reduce the population growth rate.

At least three genome projects are now under way to sequence the complete genome of several *Wolbachia* strains. It is the expectation that in the near future techniques will be developed to modify these symbionts. How such modified bacteria may play a role in improved biological control remains to be seen.

References

Aeschlimann, J.P. (1990) Simultaneous occurrence of thelytoky and bisexuality in Hymenopteran species, and its implications for the biological control of pests. *Entomophaga* 35, 3–5.

Babushkina, N.G. (1978) On the effects of temperature on the sex ratios in the populations of far eastern *Prospaltella*, a parasite of San Jose scale. *Byulleten VSES. Nauchno-issled. inst. Zashch. Rast.* 44, 13–18.

Beukeboom, L.W. and Werren, J.H. (2000) The paternal-sex-ratio (PSR) chromosome in natural populations of *Nasonia* (Hymenoptera : Chalcidoidea). *Journal of Evolutionary Biology* 13, 967–975.

Chumakova, B.M. and Goryunova, Z.S. (1963) Development of males of *Prospaltella perniciosi* parasite of San Jose scale. *Entomological Review* 42, 178–181.

Day, W.H. and Hedlund, R.C. (1988) Biological comparisons between arrhenotokous and thelytokous biotypes of *Mesochorus nigripes*. *Entomophaga* 33, 201–210.

Dedeine, F., Vavre, F., Fleury, F., Loppin, B., Hochberg, M.E. and Bouletreau, M. (2001) Removing symbiotic *Wolbachia* bacteria specifically inhibits oogenesis in a parasitic wasp. *Proceedings of the National Academy of Sciences USA* 98, 6247–6252.

Dobson, S.L. and Tanouye, M.A. (1998) Interspecific movement of the paternal sex ratio chromosome. *Heredity* 81, 261–269.

Flanders, S.E. (1944) Observations on *Prospaltella perniciosi* and its mass production. *Journal of Economic Entomology* 37, 105.

Flanders, S.E. (1945) The bisexuality of uniparental Hymenoptera, a function of the environment. *American Naturalist* 79, 122–141.

Fujii, Y., Kageyama, D., Hoshizaki, S., Ishikawa, H. and Sasaki, T. (2001) Transfection of *Wolbachia* in Lepidoptera: the feminizer of the adzuki bean borer *Ostrinia scapulalis* causes male killing in the Mediterranean flour moth *Ephestia kuehniella*. *Proceedings of the Royal Society of London. Series B, Biological Sciences* 268, 855–859.

Gottlieb, Y., Zchori-Fein, E., Werren, J.H. and Karr, T.L. (2002) Diploidy restoration in *Wolbachia*-infected *Muscidifurax uniraptor. Journal of Invertebrate Pathology* 81, 166–174.

Grenier, S., Pintureau, B., Heddi, A., Lassabliere, F., Jager, C., Louis, C. and Khatchadourian, C. (1998) Successful horizontal transfer of *Wolbachia* symbionts between *Trichogramma* wasps. *Proceedings of The Royal Society of London. Series B, Biological Sciences* 265, 1441–1445.

Hackett, K.J., Lynn, D.E., Williamson, D.L., Ginsberg, A.S. and Whitcomb, R.F. (1986) Cultivation of the *Drosophila* sex-ratio spiroplasma. *Science* 232, 1253–1255.

Huigens, M.E., Luck, R.F., Klaassen, R.G.H., Maas, F.M.P.M., Timmermans, M.J.T.N. and Stouthamer, R. (2000) Infectious parthenogenesis. *Nature* 405, 178–179.

Hurst, G.D.D. and Jiggins, F.M. (2000) Male-killing bacteria in insects: mechanisms, incidence, and implications. *Emerging Infectious Diseases* 6, 329–336.

Hurst, G.D.D., Hammarton, T.C., Bandi, C., Majerus, T.M.O., Bertrand, D. and Majerus, M.E.N. (1997a) The diversity of inherited parasites of insects: the male-killing agent of the ladybird beetle *Coleomegilla maculata* is a member of the Flavobacteria. *Genetical Research* 70, 1–6.

Hurst, G.D.D., Hurst, L.D. and Majerus, M.E.N. (1997b) Cytoplasmic sex-ratio distorters. In: O'Neill, S.L., Hoffmann, A.A. and Werren, J.H. (eds) *Influential Passengers, Inherited Microorganisms and Arthropod Reproduction.* Oxford University Press, Oxford, pp. 125–154.

Hurst, G.D.D., von der Schulenburg, J.H.G., Majerus, T.M.O., Bertrand, D., Zakharov, I. A., Baungaard, J., Volkl, W., Stouthamer, R. and Majerus, M.E.N. (1999) Invasion of one insect species, *Adalia bipunctata*, by two different male-killing bacteria. *Insect Molecular Biology* 8, 133–139.

Hurst, L.D. (1991) The incidences and evolution of cytoplasmic male killers. *Proceedings of the Royal Society of London. Series B* 244, 91–99.

Jeyaprakash, A. and Hoy, M.A. (2000) Long PCR improves *Wolbachia* DNA amplification: wsp sequences found in 76% of sixty-three arthropod species. *Insect Molecular Biology* 9, 393–405.

Jiggins, F.M., Hurst, G.D.D. and Majerus, M.E.N. (1999) How common are meiotically driving sex chromosomes in insects? *American Naturalist* 154, 481–483.

Jiggins, F.M., Hurst, G.D.D. and Majerus, M.E.N. (2000) Sex-ratio-distorting *Wolbachia* causes sex-role reversal in its butterfly host. *Proceedings of the Royal Society of London. Series B, Biological Sciences* 266, 69–73.

Jiggins, F.M., Hurst, G.D.D., Schulenburg, J. and Majerus, M.E.N. (2001) Two male-killing *Wolbachia* strains coexist within a population of the butterfly *Acraea encedon*. *Heredity* 86, 161–166.

Johanowicz, D.L. and Hoy, M.A. (1998) Experimental induction and termination of non-reciprocal reproductive incompatibilities in a parahaploid mite. *Entomologia Experimentalis et Applicata* 87, 51–58.

Kageyama, D., Nishimura, G., Hoshizaki, S. and Ishikawa, Y. (2002) Feminizing *Wolbachia* in an insect, *Ostrinia furnacalis* (Lepidoptera: Crambidae). *Heredity* 88, 444–449.

Khasimuddin, S. and DeBach, P. (1976) Hybridization tests: a method for establishing biosystematic statuses of cryptic species of some parasitic Hymenoptera. *Annals of the Entomological Society of America* 69, 15–20.

Mochiah, M.B., Ngi-Song, A.J., Overholt, W.A. and Stouthamer, R. (2002) *Wolbachia* infections in some populations of *Cotesia sesamiae*: implications for biological control. *Biological Control* 25, 74–80.

Neuffer, G. (1964) Bemerkungen zur Parasitenfauna von *Quadraspidiotus perniciosus* und zur Zucht bisexueller *Prospaltella perniciosi* im Insektarium. *Zeitschrift fur Pflannzenkrankheiten und Pflanzenzucht* 71, 1–11.

O'Neill, S.L., Hoffman, A.A. and Werren, J.H. (1997) *Influential Passengers: Inherited Microorganisms and Arthropod Reproduction.* Oxford University Press, Oxford, UK.

Opijnen, T. v. and Breeuwer, J.A.J. (1999) High temperatures eliminate *Wolbachia*, a cytoplasmic incompatibility inducing endosymbiont, from the two-spotted spider mite. *Experimental and Applied Acarology* 23, 871–881.

Perrot Minnot, M.J., Guo, L., Werren, J.H. and Guo, L.R. (1996) Single and double infections with *Wolbachia* in the parasitic wasp *Nasonia vitripennis*: effects on compatibility. *Genetics* 143, 961–972.

Pinto, J.D. (1998) Systematics of the North American species of *Trichogramma*. *Memoirs of the Entomological Society of Washington* 22, 1–287.

Pinto, J.D., Stouthamer, R., Platner, G.R. and Oatman, E.R. (1991) Variation in reproductive compatibility in *Trichogramma* and its taxonomic significance. *Annals of the Entomological Society of America* 84, 37–46.

Randerson, J.P., Jiggins, F.M. and Hurst, L.D. (2000) Male killing can select for male mate choice: a novel solution to the paradox of the lek. *Proceedings of the Royal Society of London. Series B, Biological Sciences* 267, 867–874.

Rao, S.V. and DeBach, P. (1969) Experimental studies on hybridization and sexual isolation between some *Aphytis* species. *Hilgardia* 39, 515–556.

Reed, K. (1993) Cytogenetic analysis of the paternal sex ratio chromosome of *Nasonia vitripennis*. *Genome* 36, 157–161.

Rigaud, T. (1997) Inherited microorganisms and sex determination of arthropod hosts. In: O'Neill, S.L., Hoffman, A.A. and Werren, J.H. (eds) *Influential Passengers: Inherited Microorganisms and Arthropod Reproduction.* Oxford University Press, Oxford, UK, pp. 81–101.

Silva, I., van Meer, M.M.M., Roskam, M.M., Hoogenboom, A., Gort, G. and Stouthamer, R. (2000) Biological control potential of *Wolbachia*-infected versus uninfected wasps: laboratory and greenhouse evaluation of *Trichogramma cordubensis* and *T. deion* strains. *Biocontrol Science and Technology* 10, 223–238.

Skinner, S.W. (1982) Maternally inherited sex ratio in the parasitoid wasp *Nasonia vitripennis*. *Science* 215, 1133–1134.

Stille, B. and Davring, L. (1980) Meiosis and reproductive stategy in the parthenogenetic gall wasp *Diplolepis rosae*. *Heriditas* 92, 353–362.

Stouthamer, R. (1993) The use of sexual versus asexual wasps in biological control. *Entomophaga* 38, 3–6.

Stouthamer, R. (1997) *Wolbachia*-induced parthenogenesis. In: O'Neill, S.L., Hoffman, A. A. and Werren, J.H. (eds) *Influential Passengers: Inherited Microorganisms and Arthropod Reproduction*. Oxford University Press, Oxford, UK, pp. 102–124.

Stouthamer, R. and Kazmer, D.J. (1994) Cytogenetics of microbe-associated parthenogenesis and its consequences for gene flow in *Trichogramma* wasps. *Heredity* 73, 317–327.

Stouthamer, R. and Luck, R.F. (1991) Transition from bisexual to unisexual cultures in *Encarsia perniciosi* (Hymenoptera: Aphelinidae): new data and a reinterpretation. *Annals of the Entomological Society of America* 84, 150–157.

Stouthamer, R. and Luck, R.F. (1993) Influence of microbe-associated parthenogenesis on the fecundity of *Trichogramma deion* and *Trichogramma pretiosum*. *Entomologia Experimentalis et Applicata* 67, 183–192.

Stouthamer, R., Luck, R.F., Pinto, J.D., Platner, G.R. and Stephens, B. (1996) Non-reciprocal cross-incompatibility in *Trichogramma deion*. *Entomologia Experimentalis et Applicata* 80, 481–489.

Stouthamer, R., Breeuwer, J.A.J. and Hurst, G.D.D. (1999) *Wolbachia pipientis*: microbial manipulator of arthropod reproduction. *Annual Review of Microbiology* 53, 71–102.

Stouthamer, R., Jochemsen, P., Platner, G.R. and Pinto, J.D. (2000) Crossing incompatibility between *Trichogramma minutum* and *T. platneri* (Hymenoptera: Trichogrammatidae): implications for application in biological control. *Environmental Entomology* 29, 832–837.

Stouthamer, R., Tilborg, M. v., Jong, J.H. d., Nunney, L. and Luck, R.F. (2001) Selfish element maintains sex in natural populations of a parasitoid wasp. *Proceedings of the Royal Society of London. Series B, Biological Sciences* 268, 617–622.

Stouthamer, R., Hurst, G.D.D. and Breeuwer, J.A.J. (2002) Sex ratio distorters and their detection. In: Hardy, I.C.W. (ed.) *Sex Ratios: Concepts and Research Methods*. Cambridge University Press, Cambridge, UK, pp. 195–215.

Tagami, Y., Miura, K. and Stouthamer, R. (2001) How does infection with parthenogenesis-inducing *Wolbachia* reduce the fitness of *Trichogramma*? *Journal of Invertebrate Pathology* 78, 267–271.

Timberlake, P.H. and Clausen, C.P. (1924) The parasites of *Pseudococcus maritimus* (Ehrhorn) in California. *University of California Publications in Entomology* 3, 223–292.

van Meer, M.M.M. and Stouthamer, R. (1999) Cross-order transfer of *Wolbachia* from *Muscidifurax uniraptor* (Hymenoptera: Pteromalidae) to *Drosophila simulans* (Diptera: Drosophilidae). *Heredity* 2, 163–169.

Vavre, F., Allemand, R., Fleury, F., Fouillet, P. and Bouletreau, M. (1999a) Evidence of a new cytoplasmic type of incompatibility due to *Wolbachia* in haplodiploid insects. *Annales de la Société Entomologique de France* 35, 133–135.

Vavre, F., Girin, C. and Bouletreau, M. (1999b) Phylogenetic status of a fecundity-enhancing *Wolbachia* that does not induce thelytoky in *Trichogramma*. *Insect Molecular Biology* 8, 67–72.

von der Schulenburg, J.H.G., Hurst, G.D.D., Tetzlaff, D., Booth, G.E., Zakharov, I.A. and Majerus, M.E.N. (2002) History of infection with different male-killing bacteria in the two-spot ladybird beetle *Adalia bipunctata* revealed through mitochondrial DNA sequence analysis. *Genetics* 160, 1075–1086.

Walter, G.H. (1983) Differences in host relationships between male and female heteronomous parasitoids: a review of host location, oviposition and pre-imaginal physiology and morphology. *Journal of the Entomological Society of South Africa* 46, 261–282.

Weeks, A.R. and Breeuwer, J.A.J. (2001) *Wolbachia*-induced parthenogenesis in a genus of phytophagous mites. *Proceedings of the Royal Society of London. Series B, Biological Sciences* 268, 2245–2251.

Weeks, A.R., Marec, F. and Breeuwer, J.A.J. (2001) A mite species that consists entirely of haploid females. *Science* 292, 2479–2482.

Werren, J.H. (1991) The paternal-sex-ratio chromosome of *Nasonia*. *American Naturalist* 137, 392–402.

Werren, J.H. (1997) Biology of *Wolbachia*. *Annual Review of Entomology* 42, 587–609.

Werren, J.H. and Beukeboom, L.W. (1993) Population genetics of a parasitic chromosome: theoretical analysis of PSR in subdivided populations. *American Naturalist* 142, 224–241.

Werren, J.H., Skinner, S.W. and Huger, A.M. (1986) Male-killing bacteria in a parasitic wasp. *Science* 231, 990–992.

Werren, J.H., Nur, U. and Wu, C.I. (1988) Selfish genetic elements. *Trends in Evolution and Ecology* 3, 297–302.

Werren, J.H., Hurst, G.D.D., Zhang, W., Breeuwer, J.A.J., Stouthamer, R. and Majerus, M.E.N. (1994) Rickettsial relative associated with male killing in the ladybird beetle (*Adalia bipunctata*). *Journal of Bacteriology* 176, 388–394.

Wood, R.J. and Newton, M.E. (1991) Sex-ratio distortion caused by meiotic drive in mosquitoes. *American Naturalist* 137, 379–391.

Zchori-Fein, E., Gottlieb, Y., Kelly, S.E., Brown, J.K., Wilson, J.M., Karr, T.L. and Hunter, M.S. (2001) A newly discovered bacterium associated with parthenogenesis and a change in host selection behavior in parasitoid wasps. *Proceedings of the National Academy of Sciences USA* 98, 12555–12560.

Index

Page numbers in *italics* indicate tables.